AF698004

INVERSE PROBLEMS IN DIFFUSION PROCESSES

SIAM PROCEEDINGS SERIES LIST

Neustadt, L. W., Proceedings of the First International Congress on Programming and Control (1966)

Hull, T. E., Studies in Optimization (1970)

Day, R. H. and Robinson, S. M., Mathematical Topics in Economic Theory and Computation (1972)

Proschan, F. and Serfling, R. J., Reliability and Biometry: Statistical Analysis of Lifelength (1974)

Barlow, R. E., Reliability & Fault Tree Analysis: Theoretical & Applied Aspects of System Reliability & Safety Assessment (1975)

Fussell, J. B. and Burdick, G. R., Nuclear Systems Reliability Engineering and Risk Assessment (1977)

Erisman, A. M., Neves, K. W., and Dwarakanath, M. H., Electric Power Problems: The Mathematical Challenge (1981)

Bednar, J. B., Redner, R., Robinson, E., and Weglein, A., Conference on Inverse Scattering: Theory and Application (1983)

Santosa, Fadil, Symes, William W., Pao, Yih-Hsing, and Holland, Charles, Inverse Problems of Acoustic and Elastic Waves (1984)

Gross, Kenneth I., Mathematical Methods in Energy Research (1984)

Babuska, I., Chandra, J., and Flaherty, J., Adaptive Computational Methods for Partial Differential Equations (1984)

Boggs, Paul T., Byrd, Richard H., and Schnabel, Robert B., Numerical Optimization 1984 (1985)

Angrand, F., Dervieux, A., Desideri, J. A., and Glowinski, R., Numerical Methods for Euler Equations of Fluid Dynamics (1985)

Wouk, Arthur, New Computing Environments: Parallel, Vector and Systolic (1986)

Fitzgibbon, William E., Mathematical and Computational Methods in Seismic Exploration and Reservoir Modeling (1986)

Drew, Donald A. and Flaherty, Joseph E., Mathematics Applied to Fluid Mechanics and Stability: Proceedings of a Conference Dedicated to R.C. DiPrima (1986)

Heath, Michael T., Hypercube Multiprocessors 1986 (1986)

Papanicolaou, George, Advances in Multiphase Flow and Related Problems (1987)

Wouk, Arthur, New Computing Environments: Microcomputers in Large-Scale Computing (1987)

Chandra, Jagdish and Srivastav, Ram, Constitutive Models of Deformation (1987)

Heath, Michael T., Hypercube Multiprocessors 1987 (1987)

Glowinski, R., Golub, G. H., Meurant, G. A., and Periaux, J., First International Conference on Domain Decomposition Methods for Partial Differential Equations (1988)

Salam, Fathi M. A. and Levi, Mark L., Dynamical Systems Approaches to Nonlinear Problems in Systems and Circuits (1988)

Datta, B., Johnson, C., Kaashoek, M., Plemmons, R., and Sontag, E., Linear Algebra in Signals, Systems and Control (1988)

Ringeisen, Richard D. and Roberts, Fred S., Applications of Discrete Mathematics (1988)

McKenna, James and Temam, Roger, ICIAM 87: Proceedings of the First International Conference on Industrial and Applied Mathematics (1988)

Rodrigue, Garry, Parallel Processing for Scientific Computing (1989)

Caflish, Russel E., Mathematical Aspects of Vortex Dynamics (1989)

Wouk, Arthur, Parallel Processing and Medium-Scale Multiprocessors (1989)

Flaherty, Joseph E., Paslow, Pamela J., Shephard, Mark S., and Vasilakis, John D., Adaptive Methods for Partial Differential Equations (1989)

Kohn, Robert V. and Milton, Graeme W., Random Media and Composites (1989)

Mandel, Jan, McCormick, S. F., Dendy, J. E., Jr., Farhat, Charbel, Lonsdale, Guy, Parter, Seymour V., Ruge, John W., and Stüben, Klaus, Proceedings of the Fourth Copper Mountain Conference on Multigrid Methods (1989)

Colton, David, Ewing, Richard, and Rundell, William, Inverse Problems in Partial Differential Equations (1990)

Chan, Tony F., Glowinski, Roland, Periaux, Jacques, and Widlund, Olof B., Third International Symposium on Domain Decomposition Methods for Partial Differential Equations (1990)

Dongarra, Jack, Messina, Paul, Sorensen, Danny C., and Voigt, Robert G., Proceedings of the Fourth SIAM Conference on Parallel Processing for Scientific Computing (1990)

Glowinski, Roland and Lichnewsky, Alain, Computing Methods in Applied Sciences and Engineering (1990)

Coleman, Thomas F. and Li, Yuying, Large-Scale Numerical Optimization (1990)

Aggarwal, Alok, Borodin, Allan, Gabow, Harold, N., Galil, Zvi, Karp, Richard M., Kleitman, Daniel J., Odlyzko, Andrew M., Pulleyblank, William R., Tardos, Eva, and Vishkin, Uzi, Proceedings of the Second Annual ACM-SIAM Symposium on Discrete Algorithms (1990)

Cohen, Gary, Halpern, Laurence, and Joly, Patrick, Mathematical and Numerical Aspects of Wave Propagation Phenomena (1991)

Gómez, S., Hennart, J. P., and Tapia, R. A., Advances in Numerical Partial Differential Equations and Optimization: Proceedings of the Fifth Mexico-United States Workshop (1991)

Glowinski, Roland, Kuznetsov, Yuri A., Meurant, Gérard, Périaux, Jacques, and Widlund, Olof B., Fourth International Symposium on Domain Decomposition Methods for Partial Differential Equations (1991)

Alavi, Y., Chung, F. R. K., Graham, R. L., and Hsu, D. F., Graph Theory, Combinatorics, Algorithms, and Applications (1991)

Wu, Julian J., Ting, T. C. T., and Barnett, David M., Modern Theory of Anisotropic Elasticity and Applications (1991)

Shearer, Michael, Viscous Profiles and Numerical Methods for Shock Waves (1991)

Griewank, Andreas and Corliss, George F., Automatic Differentiation of Algorithms: Theory, Implementation, and Application (1991)

Frederickson, Greg, Graham, Ron, Hochbaum, Dorit S., Johnson, Ellis, Kosaraju, S. Rao, Luby, Michael, Megiddo, Nimrod, Schieber, Baruch, Vaidya, Pravin, and Yao, Frances, Proceedings of the Third Annual ACM-SIAM Symposium on Discrete Algorithms (1992)

Field, David A. and Komkov, Vadim, Theoretical Aspects of Industrial Design (1992)

Field, David A. and Komkov, Vadim, Geometric Aspects of Industrial Design (1992)

Bednar, J. Bee, Lines, L. R., Stolt, R. H., and Weglein, A. B., Geophysical Inversion (1992)

O'Malley, Robert E. Jr., ICIAM 91: Proceedings of the Second International Conference on Industrial and Applied Mathematics (1992)

Keyes, David E., Chan, Tony F., Meurant, Gérard, Scroggs, Jeffrey S., and Voigt, Robert G., Fifth International Symposium on Domain Decomposition Methods for Partial Differential Equations (1992)

Dongarra, Jack, Messina, Paul, Kennedy, Ken, Sorensen, Danny C., and Voigt, Robert G., Proceedings of the Fifth SIAM Conference on Parallel Processing for Scientific Computing (1992)

Corones, James P., Kristensson, Gerhard, Nelson, Paul, and Seth, Daniel L., Invariant Imbedding and Inverse Problems (1992)

Ramachandran, Vijaya, Bentley, Jon, Cole, Richard, Cunningham, William H., Guibas, Leo, King, Valerie, Lawler, Eugene, Lenstra, Arjen, Mulmuley, Ketan, Sleator, Daniel D., and Yannakakis, Mihalis, Proceedings of the Fourth Annual ACM-SIAM Symposium on Discrete Algorithms (1993)

Kleinman, Ralph, Angell, Thomas, Colton, David, Santosa, Fadil, and Stakgold, Ivar, Second International Conference on Mathematical and Numerical Aspects of Wave Propagation (1993)

Banks, H. T., Fabiano, R. H., and Ito, K., Identification and Control in Systems Governed by Partial Differential Equations (1993)

Sleator, Daniel D., Bern, Marshall W., Clarkson, Kenneth L., Cook, William J., Karlin, Anna, Klein, Philip N., Lagarias, Jeffrey C., Lawler, Eugene L., Maggs, Bruce, Milenkovic, Victor J., and Winkler, Peter, Proceedings of the Fifth Annual ACM-SIAM Symposium on Discrete Algorithms (1994)

Lewis, John G., Proceedings of the Fifth SIAM Conference on Applied Linear Algebra (1994)

Brown, J. David, Chu, Moody T., Ellison, Donald C., and Plemmons, Robert J., Proceedings of the Cornelius Lanczos International Centenary Conference (1994)

Dongarra, Jack J. and Tourancheau, B., Proceedings of the Second Workshop on Environments and Tools for Parallel Scientific Computing (1994)

Bailey, David H., Bjørstad, Petter E., Gilbert, John R., Mascagni, Michael V., Schreiber, Robert S., Simon, Horst D., Torczon, Virginia J., & Watson, Layne T., Proceedings of the Seventh SIAM Conference on Parallel Processing for Scientific Computing (1995)

Clarkson, Kenneth, Agarwal, Pankaj K., Atallah, Mikhail, Frieze, Alan, Goldberg, Andrew, Karloff, Howard, Manber, Udi, Munro, Ian, Raghavan, Prabhakar, Schmidt, Jeanette, and Young, Moti, Proceedings of the Sixth Annual ACM-SIAM Symposium on Discrete Algorithms (1995)

Cohen, Gary, Joly, Patrick, Bécache, Eliane, and Roberts, Elizabeth, Third International Conference on Mathematical and Numerical Aspects of Wave Propagation (1995)

Engl, Heinz W., and Rundell, W., Inverse Problems in Diffusion Processes, Proceedings of the GAMM-SIAM Symposium (1995)

INVERSE PROBLEMS IN DIFFUSION PROCESSES

PROCEEDINGS OF THE GAMM-SIAM SYMPOSIUM

Edited by Heinz W. Engl
Johannes-Kepler Universität
William Rundell
Texas A & M University

Society for Industrial and Applied Mathematics
Philadelphia

Gesellschaft für Angewandte Mathematik und Mechanik
Regensburg, Germany

INVERSE PROBLEMS IN DIFFUSION PROCESSES

PROCEEDINGS OF THE GAMM-SIAM SYMPOSIUM

Proceedings of the GAMM-SIAM Symposium held on Lake St. Wolfgang in the Austrian Alps, June 27-July 1, 1994.

This conference was conducted by Gesellschaft für Angewandte Mathematik und Mechanik (GAMM) with the cooperation of the Society for Industrial and Applied Mathematics (SIAM)

10 9 8 7 6 5 4 3 2 1

Library of Congress Cataloging-in-Publication Data

Society for Industrial and Applied Mathematics.
Inverse problems in diffusion processes : proceedings of the GAMM –SIAM Symposium / edited by Heinz W. Engl, William Rundell.
p. cm.
Symposium held on Lake St. Wolfgang in the Austrian Alps, June 27-July 1, 1994.
Includes bibliographical references (p.).
ISBN 0-89871-351-X (pbk.)
1. Diffusion--Mathematics--Congresses. 2. Inverse problems (Differential equations)--Congresses. I. Engl, Heinz W. II. Rundell, William. III. Gesellschaft für Angewandte Mathematik und Mechanik. IV. GAMM–SIAM Symposium (1994 : Lake St. Wolfgang)
QC185.S63 1995
530.4'75'0151535--dc20 95-19089

PREFACE

1 Preface

Inverse problems in mathematics are as old as the discipline itself; from earliest times questions have been posed that require the determination of an underlying structure from auxiliary information. Old chestnuts such as "find the next number in the sequence ..." or if the answer to a question was "...", then what was the question, are in this category. However, inverse problems in differential equations have a much more recent history. While the concept of differential equation dates from the time of Newton and the main analytic techniques for solving them are at least a hundred years old, the study of inverse problems in differential equations is, in reality, less than fifty years old with a real explosion only in the last ten or twenty years. It has been fashionable to lay some of the blame on this relatively late start for the discipline on Hadamard, who required that any problem involving a differential equation should be *well posed;* that is, there should exist a unique solution that depends continuously on the data. It is true that the goal of almost any work on inverse problems strives to find conditions that guarantee (unique) identifiability of the unknown, but it is almost never the case that this quantity will exhibit continuous dependence of the data, and for this reason the question of existence of a solution is often the wrong question to ask.

While the influence of Hadamard's opinions was undeniable, a more critical reason lies in the fact that the mathematics required to meaningfully attack these problems had not been available until the middle of the century. Many of the classical problems of mathematical physics are linear, but the vast majority of inverse problems are nonlinear. This required the development of nonlinear functional analysis, a research area which in itself is currently a source of considerable activity. One of the goals of nineteenth century applied mathematics was to obtain "closed form" representations of solutions. The issue of numerical solution schemes did not arise. However, the solutions of inverse problems in differential equations are almost never amenable to useful representation formulae and knowledge of modern, sophisticated numerical analysis is essential.

A standard paradigm in applied mathematics problems is to make assumptions based on the underlying physics to obtain, perhaps after some approximation, a differential equation that has to be solved for a state variable. Connected with this equation will be certain natural conditions such as the initial configuration of the system or the values of quantities on the boundary of the region. For example, we can recover the temperature in a domain Ω by using the (simplified) physical "law" that heat flux is proportional to the gradient of the temperature, the proportionality parameter being the diffusion coefficient, and the law of conservation of heat mass. This leads to the partial differential equation $\nabla.(a\nabla u) = f$, where $u(x)$ is the temperature at the point $x \in \Omega$, $a(x)$ is the diffusion coefficient. We may assume that the temperature is known on the boundary of Ω. The standard *direct* or *forward* problem assumes that the region Ω and its material parameters, including the diffusion coefficient a and the source f, are known, and seeks the value of the state variable u throughout Ω. This is a classical problem. An associated *inverse* problem might seek the value of the diffusion coefficient in addition to the temperature u. In order to make up for the loss of information, additional data must be imposed and this could take on a variety of forms, but is usually some functional of u. For example, in hydrological applications it is often possible to measure u at selected points within the domain Ω; in nondestructive evaluation situations such intrusion is prohibited and instead information about u on the boundary of the domain (such as the heat flux) may be measured. Of course even in a simple situation such as the one described there are several possible inverse problems. We may not know the exact shape of the region Ω or may be forced to consider a model where the flux had a more complicated relationship to the temperature than the simple postulate of linear dependence. In the latter case the inverse problem might be to discover the form of this relationship by means of additional data measurements of u.

If, following the above example, we denote the unknown quantity (coefficient, domain or other system parameter) by a, the measured data, some functional $\Phi(u)$ of the state variable, by g, and the map that takes a given parameter a to exact data by $F(a)$, then formally, the inverse problem seeks a solution of the equation $F(a) = g$. As noted above, the strategy in resolving the inverse problem would seek a functional Φ that guaranteed that this map had at most one solution. Of course, in practice the data are always finite and we may have to restrict our allowable solution set to a finite dimensional space in order to restore the possibility of unique determination. There is another difficulty, namely that the measured data g may not correspond to $F(a)$ for any admissible a. For problems in a differential equations setting it is frequently the case that $F(a)$ is a smooth function. Measured data, especially when they contain considerable error, are unlikely to be well approximated by a smooth function. Thus in a practical sense, existence of a solution of the equation $F(a) = g$ is an unlikely event. One resolution is to seek a function a from an admissible class $\mathcal{A}$ such that in a certain norm to be chosen, the quantity $\|F(a) - g\|_H$ is minimized. This leads to the very common technique of least squares, adapted, of course, to the fact that the mapping F is nonlinear. This nonlinearity adds considerable complications to the problem for there might be many local minima, and indeed this turns out to be the case for many inverse problems of practical interest.

There is yet a third difficulty to be overcome. One can imagine that, in the case of our heat flow example, a change in the value of the diffusion coefficient in a small region near the centre of the body would have negligible effect on any measurements we might make on the function u in a neighbourhood of the boundary. This is a statement of the fact that small changes in the measured data can lead to large changes in the value of the coefficient, or the problem is ill-posed in the (third) sense of Hadamard. To adequately resolve this requires that the solution obtained, either by solving $F(a) = g$ in some approximate sense or by minimizing some functional of this quantity should be stabilized. There are many ways to approach this; one of the earliest attempts, due to Tikhonov, was to add a penalty term to the minimization where one sought instead to minimize the quantity $\|F(a) - g\|_H^2 + \alpha\|a\|_V^2$ where V is some subspace of H. The constant α is called the regularization parameter. For example, if $H = L^2$ then one might choose $V = H^2$ to reflect the fact that one is seeking a smooth function and with this particular choice is in fact seeking one which has a "small" second derivative. More exactly, the choice of the minimizing functional guarantees that functions with a large second derivative will be penalized when attempting to decrease the value of $\|F(a) - g\|_H$. The magnitude of this penalty term is controlled by the parameter α. In practice, finding optimal choices of the spaces H and V and of the regularization parameter α is an extremely difficult task.

The list of important applications of inverse problems spans the entire range of applied science. Many of them, such as medical imaging (for example, CAT scans) and geophysical prospecting (for example, location of oil reservoirs) are known well beyond the scientific community. This volume is the first in a series that hopes to encompass some of this range. It is devoted to inverse problems that arise in diffusion processes. Even here the scope of applications, both potential and those already in practice, is extremely wide and includes groundwater flow, heat conduction and nondestructive evaluation. In these proceedings the reader will find a variety of applications and techniques. We will give a brief overview of the contents.

One of the now classical problems of the subject is that which seeks to determine the internal temperature of a body with prescribed values of the temperature on the surface from measurements of flux on part of the boundary surface. In a single space dimension where the region is the half line $x > 0$, it is the problem of determining the temperature u, and some material parameter such as thermal conductivity or diffusivity from an initial condition, and the value of the temperature and the heat flux at $x = 0$. This is sometimes referred to as the sideways heat problem. It is notoriously ill conditioned and any solution method must pay considerable attention to this facet. There are three papers that address the heat conduction problem. Beck presents the "function specification method" that also bears his name and a numerical study about its behaviour. The paper by Elden presents a numerical scheme based on the Tikhonov regularization method and provides clear insight into why the particular scheme is so successful. Murio's paper gives an account of his mollification method for the stabilization of the inverse problem and contains several illustrative numerical examples.

There are four papers that consider the more theoretical aspects of regularization schemes. That of Chavent offers a survey of some recent results on the regularization of nonlinear least squares problems, and in particular draws a comparison between those problems that have a unique stable global minimum and those that have parasitic local minima. The paper by Kunisch treats the inverse problem as a constrained minimization problem by the so-called augmented Lagrangian method. The ideas are illustrated by reference to the recovery of the diffusion coefficient from the steady state case. In the paper by Vainikko regularization methods are studied with reference to a diffusion problem in hydrology. Here the function a corresponds to the filtration coefficient and the state variable u is the piezometric head. Finally, Seidman discusses various techniques for determining approximations for a class of ill-posed problems, in particular, the interplay between optimal error bounds and *a priori* information about the problem.

The paper by Lorenzi is concerned with the behaviour of some dielectric materials related to rigid bodies involving dispersive phenomena. Quantities that must be identified include the dielectric coefficient, the magnetic permeability and the conductivity. This leads to a very complex inverse problem.

There are three papers that consider inverse diffusion problems from a more classical aspect. Isakov presents some recent results concerning the identification of parameters and unknown source terms in parabolic equations where additional measurements are made on the lateral boundary of the region or, possibly, at some later time. Yin gives several examples in the use of his method which seeks to recover parameters in semilinear parabolic equations by combining the differential equation and the additional conditions into a single functional equation. In his paper, Lowe considers diffusion problems where coefficients can be determined by making a series of flux measurements over part of the boundary. Each measurement corresponds to the input of a different source term. Analytical and numerical results based on a quasi-Newton scheme are presented.

The papers presented in this volume are based on some of the invited talks given at the GAMM – SIAM Conference on "Inverse Problems in Diffusion Processes" which took place June 27 – July 1, 1994, on Lake St.Wolfgang in the Austrian Alps, which was attended by 69 scientists (48 of whom gave talks) from 14 countries. The program committee of that conference consisted of the editors of this volume and of D. Colton and A. Louis. The Scientific Advisory Board of the planned conference series, of which this conference was the first event, includes, in addition, M. Bertero, G. Chavent, M. Cheney, R. Ewing, A.

Friedman, R. Kress, K. Kunisch, P. Sabatier, and W. Symes. The conference was organized locally by H. W. Engl and O. Scherzer (Linz) and financially supported by the Austrian Federal Ministry for Science and Research, the Upper Austrian State Government, the Linzer Hochschulfonds, and GAMM. In addition to these institutions, we thank the staff of the Bundesinstitut für Erwachsenenbildung for the kind atmosphere they provided for the conference.

CONTENTS

Function Specification Method for Solution of the Inverse Heat Conduction Problem*

James V. Beck[†]

Abstract

The function specification method is discussed for the solution of the inverse heat conduction problem. Insight through the use of sensitivity coefficients is given. Several specified functions are suggested and discussed. Some experimental data is analyzed for which the heat flux is known. The residual principle used with this data works very well even though the real measurement errors are quite different from ideal errors.

1 Introduction

The inverse heat conduction problem (IHCP) is the estimation of the transient surface heat flux, q(t), or surface temperature, T(0,t), from one or more measured interior temperature histories in a heat conducting solid. The determination of the surface heat flux is emphasized rather than T(0,t) because q(t) is the quantity usually of most practical interest and also because it is more difficult to obtain than the surface temperature. (The heat flux implies a spatial derivative of temperature.) The determination of the surface heat flux is also related to determining the surface heat transfer coefficient, which is needed from transient temperature measurements inside quenched solids [10]; quenching involves rapidly cooling metal parts to cause metallurgical changes.

The function specification method (FSM) for solving the inverse heat conduction problem is a powerful method and has many desirable features. Two such features are that the FSM lets problems which are parabolic remain "almost parabolic" and that linear problems remain linear. Not all methods have these features. Furthermore the method can be used for convolution integral equations and various inverse boundary or volumetric source problems in diffusion that are formulated using partial differential equations.

One of the important aspects of direct transient heat conduction problems is that the partial differential equation is parabolic; as a consequence, the temperature, T, at position x and time t, depends only on the previous times, or for $0 < t' < t$, where t' is an arbitrary time less than t. In the FSM for the solution of the IHCP, the surface heat flux at time t is found using a limited amount of "future" data, that is, $r - 1$ time steps beyond t; r is frequently a small integer, such as 4 to 10, (which can be determined using the

[†]Department of Mechanical Engineering, Michigan State University, East Lansing, MI 48824

*Partially supported under US Air Force contract no. FY1456-91-N0058 and State of Michigan Research Excellence Fund through Composites Materials and Structures Center at Michigan State University.

residual principle, [6]). Future information is used because the temperature at an interior location is damped and delayed compared to the heated surface. (Theoretically the speed of propagation of the heat conduction equation is infinite, but diffusive nature of the equation causes an effective "delay".) By the same token, estimating the heat flux, q, at time t using an interior sensor requires measurements after time t, which are termed "future values". By using a limited number of time steps, the basic parabolic character of the problem is not changed. It is changed when the whole time domain is considered in the solution. A consequence of using limited future information is that the FSM for the IHCP is sequential, which causes the number of computations to increase only linearly proportional to the number of time steps. Furthermore, the early-time q estimates are independent of the estimates of q at the final times which is not necessarily true for non-sequential methods.

There are many other methods of solution of the IHCP. See [8], [12] and [17] for surveys and comparisons. A brief list is now given. The simplest conceptual method is exact matching of measured and calculated values. However, this method is usually unsatisfactory, particularly as the time steps become small. Numerical space marching has been investigated [12], [17] and [20]. Whole domain Tikhonov regularization has been used by many and is discussed in [8] and [17]. Tikhonov regularization can also be cast into a sequential form, which produces nearly the same results as the whole domain form [8]. Murio [17] is a proponent of the mollification method which involves digitally filtering the measurements; many methods can benefit from some mollification, particularly for small dimensionless time steps. Alifanov [1], [2], Jarny, Ozisik and Bardon [13] and others have used forms of iterative regularization, which uses adjoint equations. The Kalman filter has been used [16]. Additional methods are listed in [12].

A brief survey of this paper is now given. In the next section a mathematical description of the IHCP is given. Section 3 gives some desirable features of solution methods for the IHCP. Section 4 gives some ideas coming from the study of sensitivity coefficients. Section 5 provides another derivation of the function specification method. Section 6 shows some results using the FSM with some actual experimental data. The paper finishes with a summary and conclusions section.

2 Mathematical Description of the IHCP

A schematic of an actual experimental geometry with three contacting heat-conducting plates (and each of a different material) is shown in Fig. 1. One-half of a mica heater, a composite specimen and ceramic insulation are shown. The thermal properties of each material are known from previous parameter estimation of them. (By thermal properties is meant thermal conductivity, k, and volumetric heat capacity C.) The boundary condition at the top is perfect insulation. For simplicity, thermal resistances at the interfaces between the different materials are either assumed negligible or lumped into the k values. In Fig. 1, two temperature sensors are shown inside the plate, at locations L_1 and L_2. (In the IHCP, one or many measured temperature histories could be available.) These sensors provide measured temperatures at a series of discrete, uniformly spaced times. The objective is to estimate the surface heat flux, q(t), as a function of time.

A mathematical description of the thermal problem of the components such as shown in Fig. 1 is

$$\frac{\partial}{\partial x}[k_s \frac{\partial T_s}{\partial x}] = C_s \frac{\partial T_s}{\partial t}, \quad L_{s-1} < x < L_s, \; s = 1,2,3 \tag{1}$$

$$k_s \frac{\partial T_s}{\partial x}|_{L_s^-} = k_{s+1} \frac{\partial T_{s+1}}{\partial x}|_{L_s^+}, \quad s = 1,2 \tag{2}$$

$$T_s|_{L_s^-} = T_{s+1}|_{L_s^+}, \quad s = 1,2 \tag{3}$$

$$\frac{\partial T}{\partial x}|_{L_3} = 0 \tag{4}$$

$$T_s(x,0) = T_0, \quad s = 1,2,3 \tag{5}$$

$$T(x_j, t_i) = Y_j(t_i), \; i = 1,..,I; \; j = 1,..,J \tag{6}$$

The objective is to calculate the surface heat flux history (at x = 0),

$$-k_1 \frac{\partial T_1}{\partial x}|_{x=0} = q(t) = ? \tag{7}$$

(This experiment is anomalous because the heat flux is introduced electrically and is known, permitting a comparison of the measured and calculated values.) The temperature, $T_s(x,t)$, is for the sth plate and is calculated from the mathematical model and $Y_j(t_i)$ is the temperature measured by the jth sensor and at the ith time. In the experiment, temperature sensors are located at L_1 and L_2.

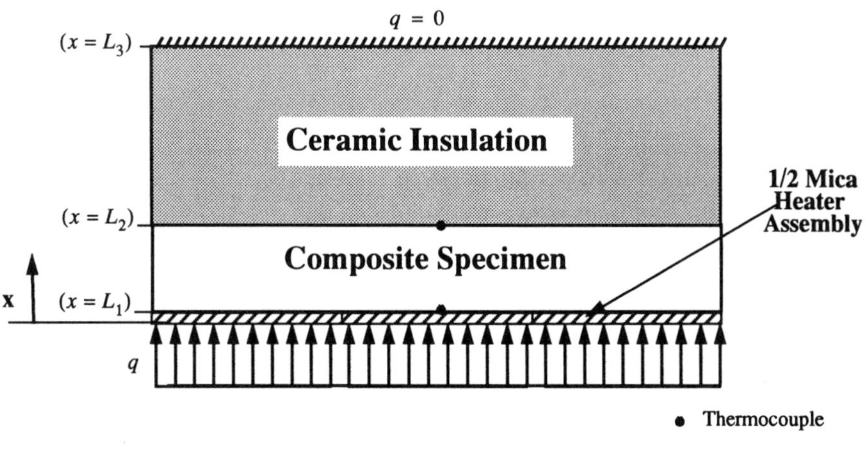

FIG. 1 *Schematic of geometry of experiment for inverse heat conduction problem*

In general, thermal properties are functions of temperature, T. However, for the temperature range in each experiment that was performed of the type shown in Fig. 1, the temperatures increased only about 25°C; this small increase permits an analysis assuming T-independent thermal properties for each test, although each temperature level required different thermal properties. For such analyses integral forms of the problems can be given.

One integral formulation is Duhamel's integral,

$$T(L,t) = T_0 + \int_0^t q(\lambda) \frac{\partial \phi(L,t-\lambda)}{\partial t} d\lambda \tag{8}$$

where $\phi(L,t)$ is the temperature rise at location L and time t for the heat flux q having a unit step change at time t = 0. The plate can be composed of several layers as shown in Fig. 1.

Eq. (8) can also represent a multi-dimensional body provided that there is a single heat flux history q(t) at a surface. Furthermore, it can represent the case of a sensor that is disturbing the local temperatures [3].

The integral equation can also be written without the derivative inside the integral. One way to do that is to use Green's functions [9]. For simplicity, assume that there is only a single material in Fig. 1 and write

$$T(L,t) = T_0 + \frac{1}{C}\int_{\tau=0}^{t} q(\tau)\, G(L,t|0,\tau)\, d\tau \tag{9}$$

where $G(L,t|0',\tau)$ is a Green's function representing the temperature rise at (L,t) caused by a plane instantaneous heat flux at $(0,\tau)$. Typically in the IHCP, eqs. (8) and (9) are evaluated numerically for many equally spaced time steps. Eqs. (8) and (9) give the same numerical results if each step is treated exactly; however, the numerical approximations may be different if approximate integrations for the individual steps are used. An accurate approximation for eq. (8) for the case of step basis functions is

$$T_M = T_0 + \sum_{n=1}^{M} q_n \Delta\phi_{M-n} \tag{10}$$

where q_n is considered to be at time $(n-1/2)\Delta t$ and $\Delta\phi_i = \phi_{i+1} - \phi_i$.

Any realistic method for solving the IHCP must be able to treat T-variable thermal properties. Exactly the same concepts used in the integral forms of the IHCP can be incorporated in finite difference, element or control volume methods for the FSM. The integral form of the IHCP is generally simpler and can be considered a "computational laboratory", since errors associated with finite space steps and finite time steps can be largely eliminated. Most of the results shown in this paper use finite difference solutions but the answers are virtually identical to those obtained using formulations based on eq. (10).

3 Desirable Features from an Engineering Viewpoint

From an engineering viewpoint, the FSM has several desirable features. Two mentioned earlier relate to the character of the problem being unchanged. The parabolic nature of the solution is retained. Limited future information is used for each q component, rather than all the measured times. Another feature is that the method remains linear if the direct problem is linear.

The function specification method is relatively easy for engineers to understand and to extend to multiple dimensions and heat fluxes. (Unfortunately while it is not difficult to understand the method, it is difficult to analyze mathematically but some work has been done, [15,21].) The numerical solution of the direct problem can be used as a building block since no special adjoint equations must be derived and programmed. More than one sensor can be used for one- and multi-dimensional problems. The sensors do not have to be located at special points. The method also has a statistical basis.

The method has wide applicability. The same basic concepts can be used for solving integral and differential forms of the IHCP. Various kinds of numerical methods can be used. It has been used for two-dimensional problems, linear and nonlinear. It can be extended to three-dimensional problems without fundamental difficulty.

The method also has growth potential. It can be used in a variety of hybrid techniques. For example, function specification can be used in the time dimension while Tikhonov or iterative regularization is used in space. For very small time steps, the mollification method with its gaussian filtering of the data is important.

One relatively straight forward application of some of the ideas in the FSM is to use the Tikhonov regularization method to eliminate the need to specify a particular function for several future time steps. Consider the case for which there is only one heat flux function to estimate. The sum of squares function for several future times can be written and a regularizing term can be added, such as first order regularization,

$$S = \sum_{i=M}^{M+r-1} (Y_i - T_i)^2 + \alpha \sum_{i=M}^{M+r-1} (q_i - q_{i-1})^2 \tag{11}$$

or in matrix notation

$$S = (\mathbf{Y} - \mathbf{T})^T(\mathbf{Y} - \mathbf{T}) + \alpha(\mathbf{H}_1\mathbf{q})^T(\mathbf{H}_1\mathbf{q}) \tag{12}$$

where α is the Tikhonov regularization parameter and $\mathbf{H}_1$ is first difference matrix with -1 on the main diagonal (except the last one) and 1's on the next upper diagonal. By using r steps into the future, and then using the first component of q, namely, q_M, sequential estimates of the heat flux can be found. If r is sufficiently large (about the same as used in the usual FSM), the q estimates are almost indistinguishable from those using the complete time domain in the estimation. The mathematical formulation to minimize S and to solve for the first estimated component of **q**, namely q_M, is not difficult and is given by the solution of eq. (5.4.7) of Beck et al. [8].

One of the aspects of the Tikhonov regularization method is that the regularization parameter α must be selected; in the FSM, the number of future time steps, r, is needed. If the errors are uncorrelated and gaussian with a standard deviation of σ, the α value is chosen so that the minimum sum of squares, S, is approximately equal to $n\sigma^2$ (which is related to the residual principle). (n is the number of measurements.) In the standard FSM, the value of r can be chosen in the same manner [6]. As shown in this paper, the measurement errors are not always uncorrelated with constant variance.

4 Sensitivity Coefficients

The concept of sensitivity coefficients gives considerable engineering insight. For simplicity assume that the heat flux can be approximated by a set of constant values, q_1, q_2, . . ., q_n where, for example

$$q = q_M = constant\ for\ t_{M-1} < t < t_M \tag{13}$$

This is consistent with what was discussed above but other more complex basis functions could be used such as B-splines. Two cases are discussed, a lumped body and a semi-infinite body.

A high-conductivity plate of thickness L on one side and insulated on the other can be modeled as a lumped body. That is the temperature is only a function of time. The describing problem is

$$C\frac{dT}{dt} = q(t), \quad T(0) = T_0 \tag{14}$$

Assume that q(t) is approximated by constant segments given above. The sensitivity coefficient for component q_M is found from the solution of the problem (with $X_M = \partial T/\partial q_M$),

$$C\frac{dX_M}{dt} = 1\ for\ t_{M-1} < t < t_M,\ otherwise = 0 \tag{15}$$

The solution of X_M is shown in Fig. 2a. It is zero until time t_{M-1}, rises linearly until t_M, and then is constant thereafter. The maximum value is $\Delta t/C$, indicating small Δt values have the effect of making the sensitivity coefficients smaller; both conditions result in making the estimation of the q components more difficult. Fig. 2b shows the first four sensitivity coefficients. Notice that all four sensitivity coefficients are equal after time t_4, indicating that measurements after that time can not distinguish between q_1,..,q_4. This has

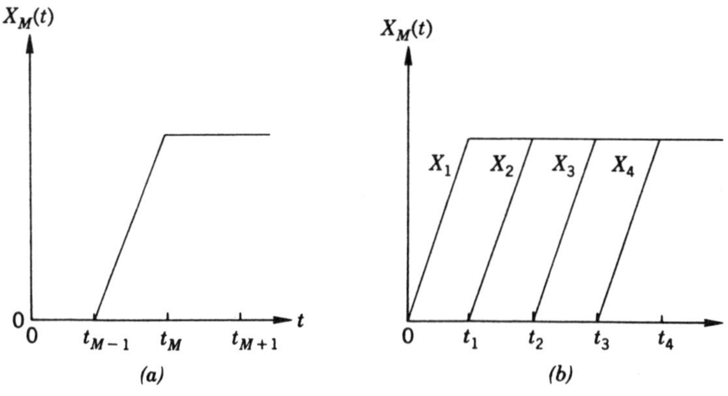

FIG. 2 . *Sensitivity coefficients for the lumped body case.*

implications for whole domain estimation in which all the components of q are simultaneously estimated. Let all the q components be estimated simultaneously from the time t = 0 to time t_n (with n much greater than 1). The perfect correlation between the q_M and q_{M+1} sensitivity coefficients after time t_{M+2} suggests that temperatures after time t_{M+2} are not effective in discerning distinctions between these two q's. The same statement can be made for many components. This suggests that a sequential method of estimating the q components would be appropriate in this problem, even though a whole domain approach of finding all the q components simultaneously can be used. Notice that the sensitivity coefficients for q_1 and q_2 are quite uncorrelated; in other words some future information for estimating q_1 could be helpful. More generally, future T's could be helpful even in the case of a lumped body.

For a semi-infinite body heated by a constant heat flux, q_c, the temperature distribution is given by

$$T(x,t) = T_0 + 2\frac{q_c}{k}(\alpha t)^{1/2} ierfc\left[\left(\frac{4\alpha t}{x^2}\right)^{-1/2}\right] \tag{16}$$

and for x = 0, this equation becomes

$$T(x,t) = T_0 + 2\frac{q_c}{k}\left(\frac{\alpha t}{\pi}\right)^{1/2} \tag{17}$$

Using superposition the results for T(x,t) and the sensitivity coefficients can be found. Again let the time step be denoted Δt. For the heated surface, the first four sensitivity coefficients are shown in Fig. 3. Each sensitivity coefficient has the same shape but there is no perfect correlation (after a short time) as there is in Fig. 2. However, there is increasing correlation as the time gets larger, as shown by the X_1 and X_2 curves. For an interior location *a*, dimensionless sensitivity coefficients are shown by Fig. 4 versus

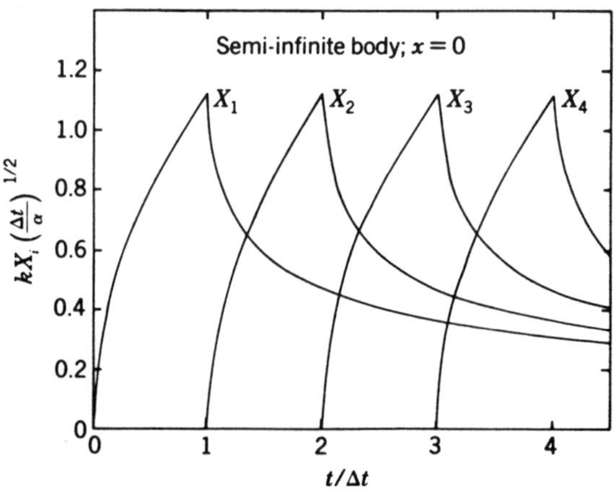

FIG. 3. *Sensitivity coefficients for heat flux components for surface of a semi-infinite body.*

dimensionless time. These dimensionless quantities are defined by

$$X_M^+ = \frac{k}{a}\frac{\partial T}{\partial q_M} \tag{18}$$

$$t^+ = \frac{\alpha t}{a^2} \tag{19}$$

Fig. 4a is for a dimensionless time step of 0.05 and Fig. 4b is for a dimensionless time step of 1.0. A comparison of these figures shows that the smaller time step has the smaller magnitude, about 5 % as large. Another consequence of the smaller time step is to have delayed effect of the heating. Notice in Fig. 4a that the first sensitivity coefficient has appreciable magnitude only beginning about dimensionless time about 0.1 (see the dot in Fig. 4a), which is 0.05 after the heat flux has dropped to zero. In other words, the heat

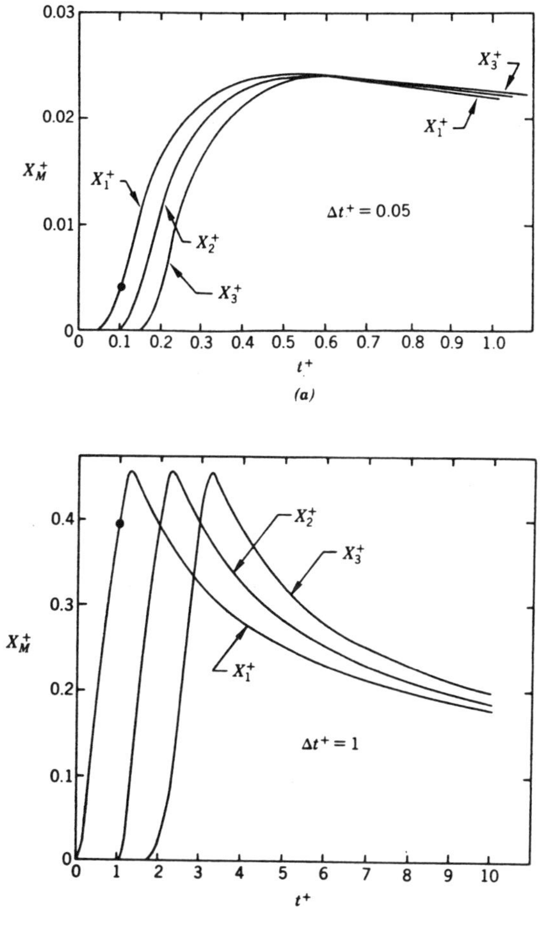

FIG. 4. *Dimensionless sensitivity coefficients for an interior point in a semi-infinite body, $\Delta t^+ = 0.05$ and 1.0.*

flux during the dimensionless time interval 0 to 0.05 has a significant effect upon the temperature at x only for times greater than 0.1 and the maximum effect is the dimensionless time about 0.5. Hence, the effect of a short heat pulse at the surface is both damped and lagged at an interior location. A dot in Fig. 4b for the first sensitivity coefficient is given at the end of the heat pulse. A comparison of Fig. 4a with Fig. 4b shows that both the damping and lagging are reduced for larger dimensionless time steps of heating. Note in both cases, however, as the dimensionless times increase that the sensitivity coefficients tend to become correlated; again we conclude that whole domain estimation is not as efficient as sequential estimation where only a limited window into the future is used.

5 Function Specification Method

Most methods for the IHCP assume that there is no or little prior information regarding the surface heat flux. That is also true for the FSM but it was developed with the capability of treating abruptly changing heat fluxes, such as calibrating plasma arcs and following rapid changes in quenching. A method that can treat rapid changes can also treat slowly varying conditions.

Other aspects relate to what is known about the measurement errors and also regarding the accuracy to which the thermal properties are known. In many methods the residual principle is used. These methods assume that we know the magnitude of the errors; usually it is assumed that the errors are uncorrelated and have a constant variance. Unfortunately real errors tend to be correlated or even worse, to be biased. In this paper actual data are used along with the residual principle.

The basic concept of specifying functions over time regions can be done two fundamentally different ways. One is to prescribe the heat flux over the total time domain and the other is to sequentially approximate the heat flux over a relatively few future times. For the total time domain, one can specify functions such as polynomials and B-splines. The emphasis of the FSM is upon the sequential aspects which can be used with many different specified functional forms. It can have one (described immediately below) or many parameters; see [5] and outlined below.

5.1 Single Unknown Parameter

The simplest functional form is that of a temporary constant for r future time steps,

$$q_{M+i-1} = q_M \quad for \quad i = 1, 2, \ldots, r \tag{20}$$

The unknown is q_M. This expression has been the most commonly used by the FSM and it is effective for abruptly changing heat fluxes. Another form is a linear variation with time, starting at time t_{M-1},

$$q_{M+i-1} = q_{M-1} + (q_M - q_{M-1})\tau_i \tag{21}$$

Again the unknown is q_M; and τ_i is

$$\tau_i = \frac{t_{M+i-1} - t_{M-1}}{\Delta t} = i \tag{22}$$

which can vary from 0 to r but does not have to be only integers. A parabolic approximation can be used by again starting at time t_{M-1} where q is known and also by having a zero slope in the q function at t_{M+r-1} (or equivalently, $\tau = r$),

$$q_{M+i-1} = q_{M-1} + (q_M - q_{M-1}) \frac{\tau_i(2r - \tau_i)}{2r - 1} \tag{23}$$

A similar expression for a cubic function in time which has the same constraints as the parabolic function (but with the additional condition of the second derivative at t_{M+r-1} equal to zero) is

$$q_{M+i-1} = q_{M-1} + (q_M - q_{M-1}) \frac{\tau_i(3r^2 - 3r\tau_i + \tau_i^2)}{3r^2 - 3r + 1} \tag{24}$$

The parabolic and cubic approximations are appropriate for smoothly varying surface heating conditions and actually are good for even cases with jumps.

Fig. 5 shows the q shape functions for these four approximations as a function of τ_i; the constant linear q curves are valid for any value of r while the parabolic and cubic curves are for r = 8 only. For $0 < \tau_i < 1$, the dimensionless q values are approximately linearly increasing from zero to one for each case except the q = c case which is one for all τ_i 's. However, all are larger than unity except the q = c case for $\tau_i > 1$.

By utilizing future time steps the sensitivity coefficients have different characters than for a single q_M component which are shown in Figures 2, 3, and 4. For the case of a semi-infinite body with a sensor at x = a, the sensitivity coefficients for these four cases

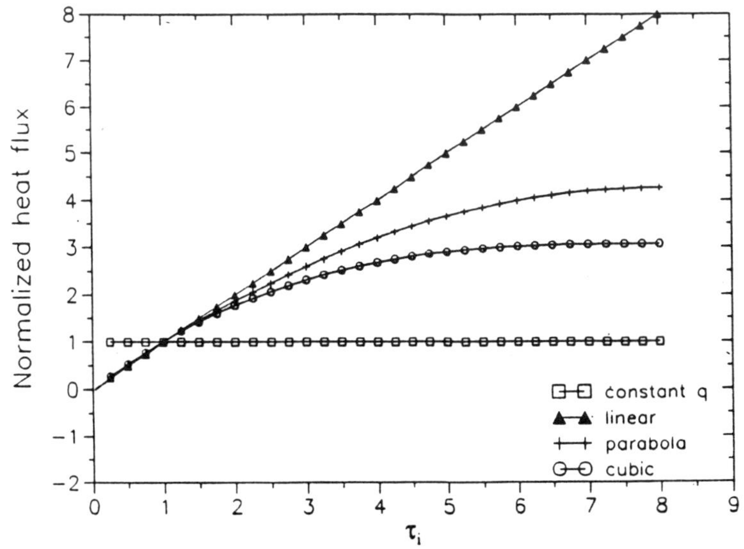

FIG. 5. *Normalized shape functions for heat flux for q equal to constant, linear, parbolic and cubic temporary approximations.*

functions are plotted in Fig. 6; each curve monotonically increases while for the single heat flux components shown in Figures 2 - 4 the curves approach a maximum, with some curves decreasing afterward. The sensitivity coefficients for q = c and q = ct cases are independent of r; the other two cases are plotted for r = 8. In all these cases the dimensionless time step is 0.05. Generally the q = c sensitivity coefficients are larger until four future time steps but for the later times the other sensitivity coefficients are larger, with the linear case being the largest.

The use of the linear, parabolic and cubic shape functions can be implemented in a similar way with the constant q temporary assumption. Assume that the heat flux is known until time t_{M-1} and that the heat flux is needed at the next experimental time step t_M. (If finite differences or elements are used to obtain the solution of the partial differential equation, intermediate calculational time steps might be needed to get the required accuracy, but that does not impact our derivation.) The temperature at time t_{M+i-1} is approximated by

$$T_{M+i-1} = T_{M+i-1}\Big|_{q_M=q_{M+1}=\cdots=q_{M+r-1}} + X_{M+i-1,M}(q_M - q_{M-1}) \tag{25}$$

where the sensitivity coefficient $X_{M+i-1,M}$ is defined by

$$X_{M+i-1,M} = \frac{\partial T_{M+i-1}}{\partial q_M}\Big|_{q_M=q_{M+1}=\cdots=q_{M+r-1}} \tag{26}$$

The shape function can be any of the four described above (eqs. (20, 21, 23 and 24)) but q_M is the unknown in each case. The notation in eqs. (25) and (26) does not explicitly denote the location of the sensor since only one is assumed; if there are two or more sensors, then the notation should include an index for sensors.

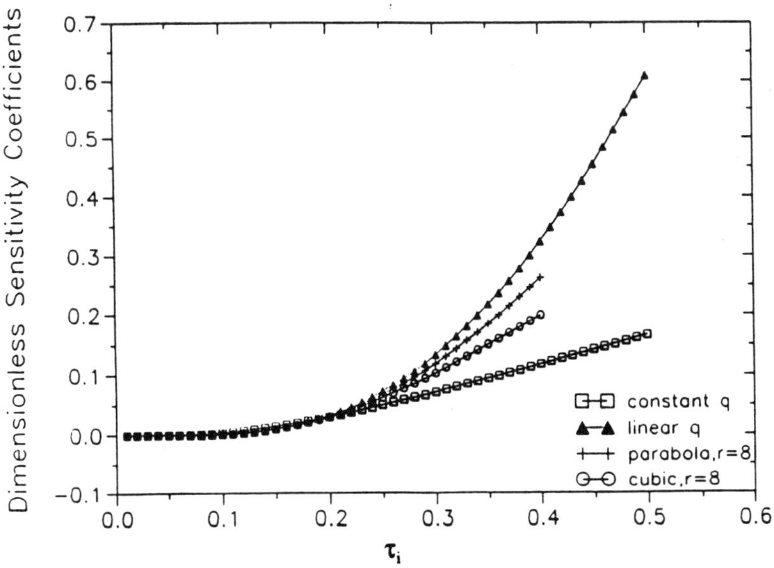

FIG. 6 *Sensitivity coefficients for a semi-infinite body with $\Delta t^+ = 0.05$ for q equal to constant, linear, parabola and cubic curves.*

As originally formulated, the regularization term (beginning with α in eq. (11) or (12)) was not present. If it is not wanted, α is simply set equal to 0. The regularization term in eq. (11) can be written for the above four cases by

$$S_\alpha = \alpha(q_M - q_{M-1})^2 \sum_{i=1}^{r} R_i^2 \tag{27}$$

where R_i is given by for the q = constant, linear, parabolic and cubic approximations, respectively,

$$R_i = 1 \text{ for } i = 1, \text{ otherwise } R_i = 0; \; q \text{ constant} \tag{28}$$

$$R_i = 1, \text{ for all } i; \; q \text{ linear} \tag{29}$$

$$R_i = \frac{2r - (2i-1)}{2r-1}; \; q \text{ parabolic} \tag{30}$$

$$R_i = \frac{3r^2 - 3(2i-1)r + 3i^2 - 3i + 1}{3r^2 - 3r + 1}; \; q \text{ cubic} \tag{31}$$

The procedure to find the estimate of q_M is to minimize eq. (11) with respect to q_M. For linear and nonlinear problems we use eq. (25); for convenience in notation, the subscript, $q_M = q_{M+1} = q_{M+2} = .. = q_{M-1}$, in eq. (25) is omitted. Using eq. (25) in eq. (11) gives

$$\frac{\partial S}{\partial q_M}\Big|_{\hat{q}_M} = 2\sum_{i=1}^{r} [Y_{M+i-1} - T_{M+i-1}| - X_{M+i-1,M}(\hat{q}_M - q_{M-1})][-X_{M+i-1,M}] + 2\alpha(\hat{q}_M - q_{M-1})\sum_{i=1}^{r} R_i^2 = 0 \tag{32}$$

which can be solved for the estimated q at time t_M which gives

$$\hat{q}_M = \hat{q}_{M-1} + \frac{\sum [Y_{M+i-1} - T_{M+i-1}|\,]}{\sum X_{M+i-1,M}^2 + \alpha \sum R_i^2} \tag{33}$$

where the summations are all from i =1 to i = r. For nonlinear cases iteration could be used for each q_M because the sensitivity coefficients are a function of temperature; our experience has shown that such iteration is rarely needed.

5.2 Multiple Unknown Heat Flux Components

A method to extend the FSM to multiple heat flux components using cubic splines is given below. Consider two cubics, one from t_{M-1} to t_{M+r-1} and the second one from t_{M+r-1} to t_{M+2r-1}. The first cubic is given by

$$q_i = a_1 + b_1\theta_i + c_1\theta_i^2 + d_1\theta_i^3 \qquad (34)$$

where for i = 0, 1, 2, ..., r

$$\theta_i = t_{M-1+i} - t_{M-1} \qquad (35)$$

The second cubic is given by

$$q_i = a_2 + b_2\eta_i + c_2\eta_i^2 + d_2\eta_i^3 \qquad (36)$$

where for i = 0, 1, 2, ..., r

$$\eta_i = t_{M+r-1+i} - t_{M+r-1} \qquad (37)$$

There are eight unknowns, $a_1,\ldots,d_2$ and eight conditions are

$$q|_{\theta_0} = q_{M-1},\ q|_{\theta_1} = q_M \qquad (38)$$

$$q|_{\theta_r} = q|_{\eta_0},\ q'|_{\theta_r} = q'|_{\eta_0} \qquad (39)$$

$$q|_{\theta_r} = q_{M+r-1},\ q''|_{\theta_r} = q''|_{\eta_0} \qquad (40)$$

$$q'|_{\eta_r} = 0,\ q''|_{\eta_r} = 0 \qquad (41)$$

These equations contain two unknowns, q_M and q_{M+r-1}. Using symbolic computation, the eight unknowns, $a_1,\ldots,d_2$ are solved in terms of these two unknowns. The future temperatures until time t_{M+2r-1} is solved in terms of these two variables and sensitivity coefficients for both parameters (q_M and q_{M+r-1}) are evaluated. Then S is minimized with respect to both parameters and only q_M is selected. Then M is increased by one and the procedure continues. This general procedure for two cubic splines can be extended to more cubic splines.

6 Results Using Experimental Data

In this section two similar (except for the temperature level) experiments are described, and the actual data generated was used both to estimate thermal properties (thermal conductivity and volumetric heat capacity) and the surface heat flux. The two estimation problems of estimating the thermal properties and surface heat flux were treated separately. The thermal properties were estimated first and then these values were used to estimate the surface heat flux histories given in this paper. The property estimation is discussed in [11].

6.1 Experiment

A schematic picture of the geometry of the specimens is shown in Fig. 1. A number of thermocouples were cemented in two identical carbon composite specimens. Figures 7 and 8 show temperatures at the mica/specimen interface and the insulated surface; actually

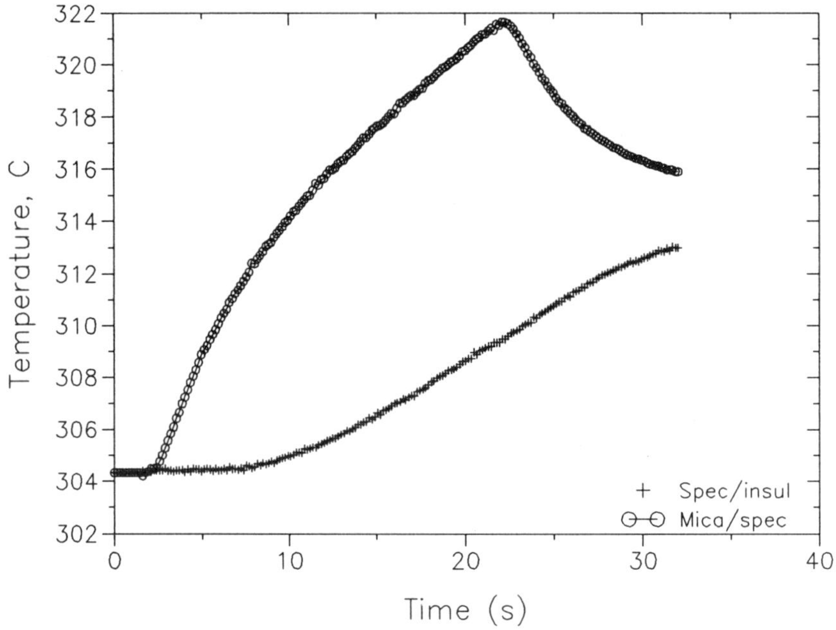

FIG. 7. ***Measured temperatures from test #30.***

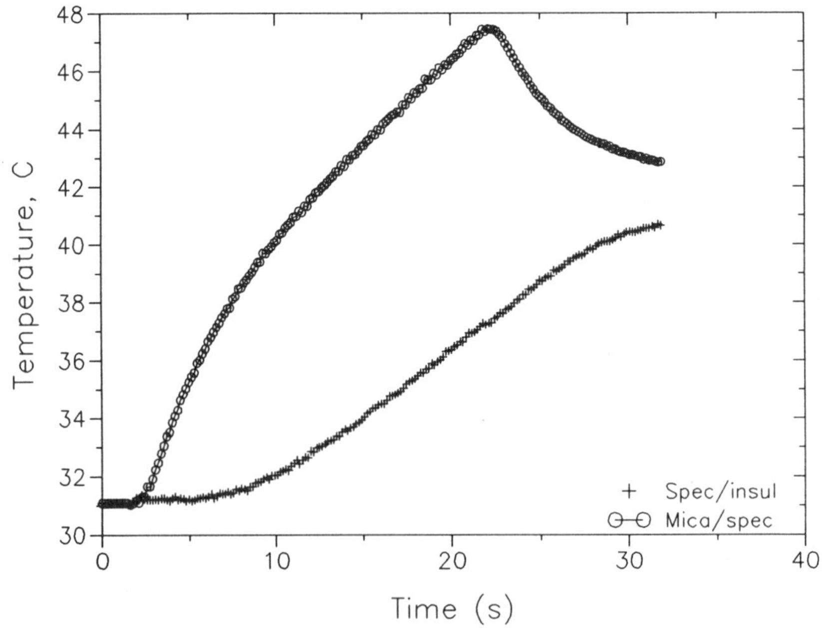

FIG. 8. ***Measured temperatures from test #304.***

these temperatures are averages of a number from seven thermocouples at mica interface and four at the other one. An electric heater is between the two specimens; it is mainly composed of mica. On the side opposite of the heater, there is ceramic insulation. The geometry is symmetric so that only one half is considered: from the center of the heater to the furthest side of the ceramic insulation. The temperature histories shown in Figures 7 and 8 for cases denoted #30 and #304, with the former having an initial temperature about 30°C and the second an initial T of about 304°C. In both cases the heating started at 2 s and at 22 s; the input heat flux is shown in Figures 9 and 10 by the circles. The input q increases and decreases abruptly. Because the thermocouple at the mica/specimen interface (the higher T in both cases) is not at the heated surface, there is a lag in the temperature response both in the response after the heating starts and after it stops. The lag is about 0.5 s; it is most visible when the heating stops at 22 s. The thermal conductivity of the specimen is about 100 times that of the mica heater assembly; this has the effect of making the IHCP more difficult than if the mica heater material was the only material present.

6.2 Inverse Heat Conduction Problem Results

Results for estimating the surface heat flux from the internal temperatures are now given. Three possibilities exist. Each temperature history can be used separately and both can be used together. Inspection of Figures 7 and 8 reveals that the temperature with the lowest rise has a very large lag; estimating the surface q with that temperature history is very difficult and has been considered. But it is not a practical case and is not discussed herein. We shall concentrate on using both temperature histories at the same time using a Crank-Nicolson finite difference-based program.

Estimated heat fluxes for the initial nominal temperature of 30°C are shown in Fig. 9 for the FSM with the constant future heat flux approximation with $r = 6$ and 9 as well as the measured heat flux input. Calculations for $r = 3$ were also run but were quite oscillatory, although stable. Both r values are satisfactory results but the $r = 9$ values are better. As expected the region of rapid change is the most difficult to reproduce; the nature of the algorithm is to anticipate changes. The amount of energy in the process (area under the curve) is accurately reproduced.

The q results for case #304 are shown in Fig. 10 for $r = 6$, 9 and 11. The results for the different r values are similar to those of case #30.

The surface temperature is estimated in the same calculational sequence as the surface heat flux. Results are shown in Fig. 11 for $r = 3$, 6 and 9. Notice that the estimated surface temperature rise (about 40 °C) is about twice that of the sensor nearest the heated surface (about 17°C, Fig. 9). The surface temperature for the latter two r values are indistinguishable in Fig. 11 but the $r = 3$ case has noticeable oscillations but much less than for the surface heat flux.

One of the problems with the FSM is the choice of the number of future times, r. (Other methods have analogous problems.) One method that has been successful for simulated data is to choose the parameter (r, α, or whatever) so that the sum of squares is reduced to the level that is indicated by the data [1,2,6,8,17]. In simulated data it is common to assume gaussian, additive, and uncorrelated errors but actual errors are more complicated. However, for the experiments mentioned here some estimates of the errors are available. For case #30, eleven thermocouples were used at the mica/heater interface but only seven were good (some had obvious problems such as negative temperatures). The temperature history used at the mica/specimen interface is the average of the seven thermocouples. The standard deviation of all the seven temperature histories is s = 0.2195 °C which leads to the value of $s\ n^{-1/2} = 0.083$ °C for the standard deviation of the average actually used at this interface. This value of 0.083 °C can then be used to estimate the appropriate value of r.

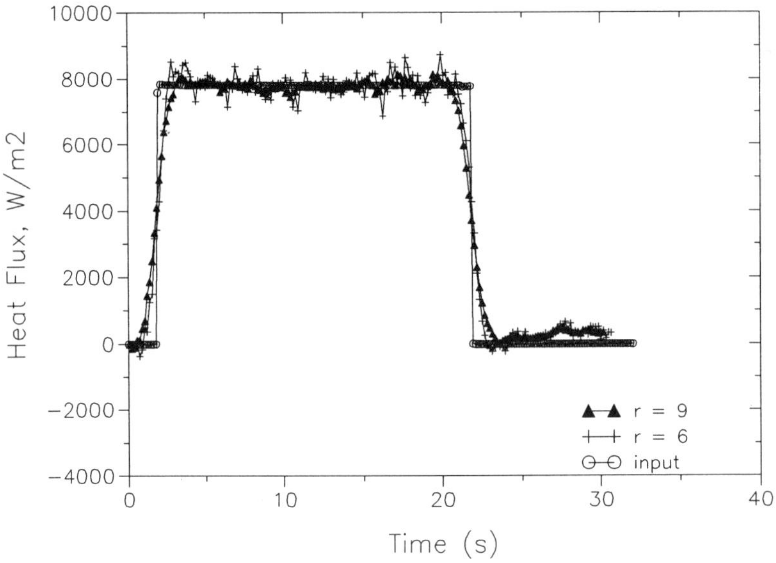

FIG. 9. ***Estimated heat fluxes for case #30.***

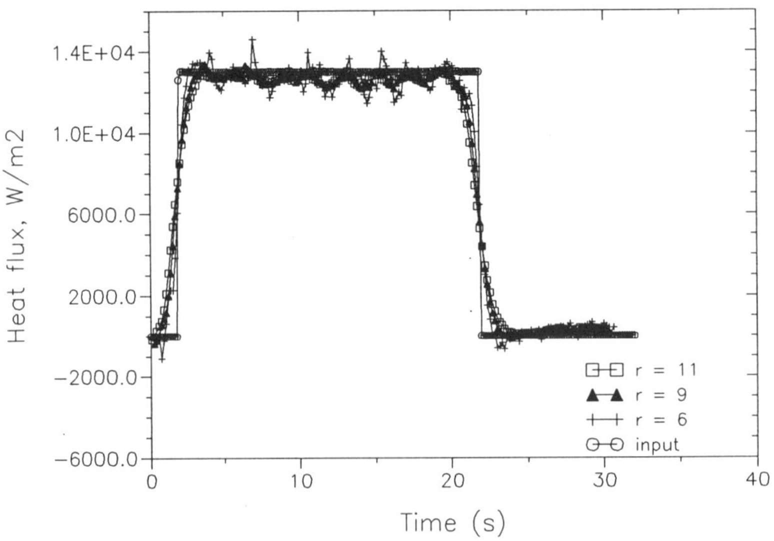

FIG. 10. ***Estimated heat fluxes for case #304.***

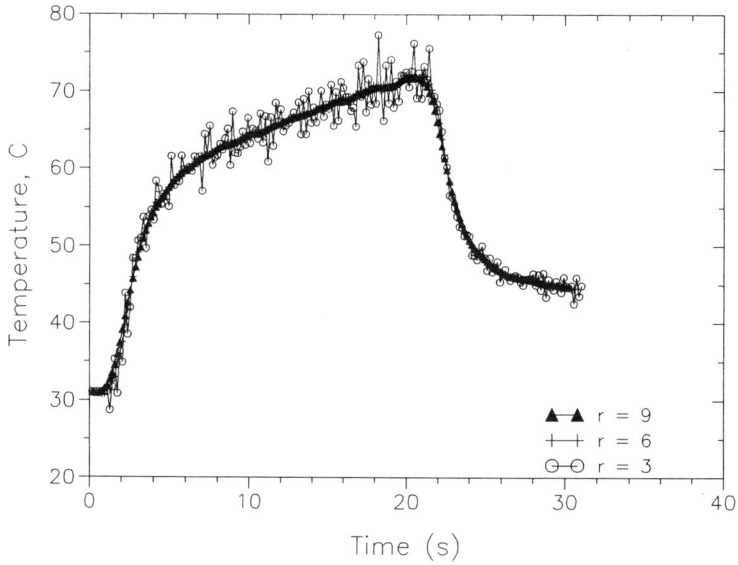

FIG. 11. *Estimated surface temperature for case #30.*

As the r value increases, the rms of the difference between the measured and calculated temperatures usually increases. This is particularly true for the sensor nearest the heated surface (mica/specimen interface). (The sensitivity coefficients are far larger there than further away from the heater, such as at the specimen/insulation interface.) Nevertheless three rms values for each r is shown in Table 1 for case #30. The first rms column (the most important) is at the mica/specimen interface, the second is at the specimen/insulation interface and the third is for both locations and is labeled "total". The rms at the mica/specimen interface increases with r while the specimen/insulation interface rms values slightly decrease. The total clearly increases in Table 1 with r. If either the mica/specimen interface or the total values are used, the r value associated with the rms of 0.083 is about r = 9. Hence this is an appropriate value to use. It is interesting to note that from an examination of the graphical results the same value of r was chosen, before the rms calculations were made. This is important because the appropriate rms values are not usually known, since replicated measurements are rarely obtained.

Table 2 displays the root mean square (rms) or equivalently, the estimated standard deviations, similar to those given in Table 1. Using the same average of seven thermocouples mentioned above, the estimated standard deviation of the errors in the average temperature at the mica/specimen interface is found to be 0.126 °C. Again using the mica/specimen interface, this value of 0.126 °C is between the r of 11 and 12 in Table 2. Hence the residual principle indicates that the optimal r is about 11. Table 2 does suggest a problem in using the residual principle for several temperature histories with one sensor being relatively near the heated surface and at least one being much further from the heated surface. The sensitivity coefficients are quite different and the temperature furthest from the heated surface is not greatly affected by the r value but yet this measured temperature might be included in the residual principle. If as in Table 2, the in-depth measurement is substantially biased high (or low), the chosen r value would be affected. This would be true for other solution methods as well.

6.3 Other specified functions and methods

Space does not permit detailed discussion of the use of the linear, parabola and cubic specified functions given by eqs. (21-24). The surprising aspect for this same data is that the seemingly crude approximation of constant specified heat fluxes produces results that are extremely similar to those for the others. The estimated q's using the same residual principle produce similar temperature rms values (such as shown in Tables 1 and 2) and also similar rms values for the difference between the measured and calculated q's. The major difference is that the cubic, for example, produces slightly smoother results but at the same time requires slightly larger r values.

Other methods have been examined for this same set of data. These include the whole domain methods of zeroth order Tikhonov regularization and iterative regularization, both implemented using adjoint equations and conjugate gradients. Again surprisingly, the results are very similar to those displayed in Figures 9 and 10. There were no marked improvements using these methods for this particular example. The computer time was substantially more using these whole domain methods. For two and three dimensional problems, the conclusions could be quite different.

7. Summary and Conclusions

The function specification method is reviewed and some specified functions (not previously used) are given. Some experimental data has been obtained and analyzed using the FSM. The estimated heat fluxes are very close to those that were measured. The residual principle is applied and works for the experimental data generated. However, for multiple temperature sensors with one much further from the heated surface than the others, application of the residual principle may be difficult if that (and other) temperature measurements are biased.

The various specified functions (when there is a single unknown) yield very similar answers for this data, which has abrupt variations.

REFERENCES

[1] O.M. Alifanov, *Identification of Heat Transfer Processes of Flight Vehicles, Mashinostroenie*, Moscow (in Russian),1979.

[2] O.M. Alifanov, E.A. Artyukhin and S.V. Rumyantsev, *Extreme Methods of Solving Ill-Posed Problems and Their Application to Inverse Heat Transfer Problems*, Nauka, Moscow (in Russian), 1988.

[3] J.V. Beck, *Correction of transient thermocouple temperature measurements in heat-conducting solids, Part II, The Calculation of Transient Heat Fluxes Using the Inverse Convolution*, AVCO Corp., Res. and Adv. Dev. Div., Wilmington, MA, Tech Report RAD-TR-7-60-38 (Part II), 1961.

[4] J.V. Beck, *Inverse problems in heat transfer with application to solidification and welding, Modeling of Casting, Welding and Advanced Solidification Processes*,- V, M. Rappaz, M.R. Ozgu and K. Mahin, Editors, pp. 503-514, TMS, Warrendale, PA, 1991.

[5] ____, *Surface heat flux determination using an integral method*, Nuclear Engineering and Design 7 (1968), pp. 170-178.

[6] ____, *Comparison of the iterative regularization and function specification algorithms for the inverse heat conduction problem*, Inverse Problems in Engineering: Theory and Practice, edited by N. Zabaras, K. Woodbury and M. Raynaud, ASME (1993), pp. 23-28.
[7] J.V. Beck and K.J. Arnold, *Parameter Estimation in Engineering and Science*, John Wiley and Sons, Inc., New York, 1977.
[8] J.V. Beck, B. Blackwell and C.R. St. Clair, Jr., *Inverse Heat Conduction: Ill-posed Problems*, John Wiley and Sons, Inc., New York, NY, 1985.
[9] J.V. Beck, K. Cole, A. Haji-Sheikh and B. Litkouhi, *Heat Conduction Using Green's Functions*, Hemisphere, Wash., D.C., 1992.
[10] J.V. Beck and A.M. Osman, *Analysis of quenching and heat treating processes using inverse heat conduction method*, Proceedings of the International Conference on Quenching, Chicago, Ill, Sept. 1992.
[11] K. Dowding, J. Beck, A. Ulrich, B. Blackwell and J. Hayes, *Estimation of thermal properties and surface heat flux in carbon-carbon composite*, to appear in Journal of Thermophysics and Heat Transfer, AIAA.
[12] E. Hensel, *Inverse Theory and Applications for Engineers*, Prentice-Hall, Advanced Reference Series Engineering, Englewood Cliffs, NJ, 1991.
[13] Y. Jarny, M.N. Ozisik and J.P. Bardon, *A general optimization method using adjoint equation for solving multidimensional inverse heat conduction*, Int. J. of Heat and Mass Transfer, 34, pp. 2911-2919, 1991.
[14] P.K. Lamm, *Local regularization methods for linear inverse problems*, Proceedings Sixth Inverse Problems in Engineering Seminar, D.A. Murio, editor, University of Cincinnati, 1994.
[15] P.K. Lamm, *Future-sequential regularization methods for ill-posed Volterra equations*, Proceedings of GAMM-SIAM Conference on Inverse Problems in Diffusion Processes, St. Wolfgang, Austria, June 27 - July 1, 1994, H. Engl, Editor
[16] A.V. Moultanovsky and Y.M. Matsevity, *A comprehensive method of adaptive filter for solution multiparameter inverse problems*, Proceedings Sixth Inverse Problems in Engineering Seminar, D.A. Murio, editor, University of Cincinnati, 1994.
[17] D.A. Murio, 1993, *The Mollification Method and the Numerical Solution of Ill-Posed Problems*, Wiley-Interscience, NY, 1993.
[18] A.M. Osman and J.V. Beck, *Nonlinear Inverse Problem for the Estimation of Time-and-Space-Dependent Heat-Transfer Coefficients*, J. of Thermophysics and Heat Transfer, 3 (1989), pp. 146-152.
[19] A.M. Osman and J.V. Beck, *Investigation of Transient Heat Transfer Coefficients in Quenching Experiments*, J. of Heat Transfer, 112 (1990), pp. 843-848.
[20] M. Raynaud and J. Bransier, *A New Finite Difference Method for Nonlinear Inverse Heat Conduction Problem*, Numer. Heat Transfer, 9 (1986), pp. 27-42.
[21] H. Reinhardt, *Towards a stability analysis and error analysis of sequential methods for the inverse heat conduction problem*, Inverse Problems in Engineering: Theory and Practice, edited by N. Zabaras, K. Woodbury and M. Raynaud, ASME (1993), pp. 11-16.

Table 1

Temperature standard deviations (rms values) for case #30 for various numbers of future temperatures

# future times, r	rms for mica/ /spec interface	rms for spec/ insul interface	overall rms
3	0.0380	0.0997	0.07543
6	0.0484	0.0976	0.07707
8	0.0674	0.0944	0.08206
9	0.0817	0.0924	0.08719
10	0.0986	0.0902	0.09452

Table 2

Temperature standard deviations (rms values) for case #304 for various numbers of future temperatures

# future times, r	rms for mica/ /spec interface	rms for spec/ insul interface	overall rms
3	0.025	0.169	0.1212
6	0.040	0.168	0.1222
9	0.079	0.162	0.1290
10	0.097	0.163	0.1344
11	0.119	0.162	0.1422
12	0.142	0.162	0.1521

UNIQUENESS AND STABILITY IN INVERSE PARABOLIC PROBLEMS

VICTOR ISAKOV

Wichita State University

ABSTRACT. We present some recent and some new results about identification of the coefficients and right sides of a second order parabolic equation with given extra filnal or lateral data.

0. INTRODUCTION

We are looking for the coefficients a_0, a, b, c of the second order parabolic equation
(0.1)

$$a_0 u_t - div(agradu) + b \cdot gradu + cu = f \text{ in } Q = \Omega \times (0, T)$$

where Ω is a bounded domain in the space $\mathbf{R}^n$ with the C^2-smooth boundary $\partial\Omega$. In one section we even consider the non-linear equation
(0.1_n)

$$a_0(x, u)u_t - \Delta u + c(x, t, u) = 0 \text{ in } Q$$

We prescribe the initial condition
(0.2)

$$u = 0 \text{ on } \Omega \times \{0\}$$

and the lateral boundary condition
(0.3)

$$u = g \text{ on } \partial\Omega \times (0, T)$$

Here a is generally a symmetric strictly positive matrix function and b is a vector function. We will denote by A the elliptic part of the operator in the left side of (0.1). Sometimes if certain coefficients are to be determined, we keep them outside of this operator.

To identify unknown a_0, a, b, c (or nonlinear terms) we will make use of the final overdetermination
(0.4)

$$u = h_T \text{ on } \Omega \times \{T\}$$

1991 *Mathematics Subject Classification.* 35R30.

Key words and phrases. Inverse Problems.

This work was partially supported by the NSF grant DMS 9101421

or of the lateral overdetermination

(0.5)
$$\partial_\nu u = h \text{ on } \Gamma \times (0,T), \Gamma \subset \partial\Omega$$

Here ν denotes the outward unit normal to $\partial\Omega$.

These inverse problems are of fundamental importance for oil and underground water search,chemical kinetics,and geophysics [E-T-F],[S-E-A],[Y].

It is known that if Ω is bounded and in $C^{2+\lambda}, 0 < \lambda < 1$, the functions $a_0, b, c, f \in C^{\lambda,0.5\lambda}(\overline{Q})$, the coefficients matrix $a \in C^{1+\lambda,0.5\lambda}(\overline{Q})$,then the initial boundary value problem (0.1)-(0.3) has a unique solution $u \in C^{2+\lambda,1+0.5\lambda}(\overline{Q})$ provided the boundary data $g \in C^{2+\lambda,1+0.5\lambda}(\partial\Omega \times [0,T])$ and the following natural compatibility conditions are satisfied : $g = 0, a_0 g_t = f$ on $\partial\Omega \times \{0\}$. When all coefficients are only in $C(\overline{Q}), a \in C^1(\overline{Q})$ and $f \in L^\infty(Q)$ then one can obtain an unique solution u in the anisotropic Sobolev spaces $H_p^{2,1}(Q), 1 < p < +\infty$. When the coefficients are in $L^\infty(Q)$ one has an unique generalized solution u in $H_2^{1,0}(Q) \cap C^{\mu,0.5\mu}(\overline{Q})$ for some $\mu \in (0,1)$. For these classical results we refer to the monographs [F] and [L-S-U].

While considering coefficients identification in sections 2-4 we will assume that a_0, a, b, c do not depend on t.

1.Final Overdetermination

First we assume that the additional data for our inverse problem are prescribed at the final moment of time $t = T$. This final overdetermination has been introduced in 1935 by Tikhonov who emphasized its geophysical interpretation (recovery of geological prehistory from contemporary data). In 1950-s Fritz John analysed stability and suggested numerical algorithms for this classically ill-posed problem. They both considered the problem without the initial data (0.2) where only u is to be found (backward parabolic equations). By adding the initial data and looking simulteneously for a solution u and the coefficient or the right side depending only on the space variable x one can obtain a well-posed inverse problem. We will consider two results in this direction.

Theorem 1.1. *Let us assume that $f = \alpha F$ where $\partial_t F = 0$ and the coefficients of the equation (0.1) and the weight function α are given and satisfy the following conditions*

$$a, b, c \text{ do not depend on } t$$

(1.1)

$$0 \leq a_0, \partial_t a_0 \leq 0, 0 \leq \alpha, 0 \leq \partial_t \alpha \text{ on } Q, \epsilon < \alpha \text{ on } \Omega \times \{T\}$$

Then there is a constant C depending only on $[]_k(Q)$-norms of the coefficients of the parabolic operator and of α and on the positive number ϵ such that any solution (u, F) to the inverse source problem (0.1)-(0.4) satisfies the inequality

(1.2)
$$[u]_{k+2}(Q) + |F|_k(\Omega) \leq C([g]_{k+2}(\partial\Omega \times [0,T]) + |h_T|_k(\Omega))$$

Moreover a solution (u, F) with the finite norm in the left side of (1.2) does exist for any (compatible) data g, h_T with the finite norms in the right side.

We observe that $[]_k$ denotes the norm in the anisotropic Hölder space $C^{k,k/2}$. Here k is a positive noninteger number. A proof of this result can be found in the paper [Is3]. In the same paper there is a conterexample which shows that the monotonicity condition on the weight function α is essential for uniqueness in this inverse source problem. We refer also to the papers [H-Y],[P-K],[P-S] where different functional spaces and generic results are discussed.

Now we will study identification of a time-independent coefficient c. We assume ,for simplicity, that $f = 0$ and a, b are given functions satisfying the conditions of Theorem 1.1.

Theorem 1.2. *Let $g \in C^{2+\lambda}(\partial\Omega \times [0,T])$*

i)(Uniqueness) If

(1.3)

$$0 \leq g, 0 \leq \partial_t g, g \text{ is not a zero function on } \partial\Omega \times [0,T]$$

then a solution $(u, c), c \in C^{\lambda}(\overline{\Omega})$ of the inverse coefficient problem (0.1)-(0.4) is unique.

ii)(Existence) If $h_T \in C^{2+\lambda}(\overline{\Omega})$

(1.4)

$$Ah_T + \partial_t v(;T) \leq 0, 0 < h_T \text{ on } \Omega, g = h_T \text{ on } \partial\Omega \times \{0\}$$

where v is a solution to the parabolic problem (0.1)-(0.3) with $c = 0$, then there is a solution (u, c) to the inverse problem (0.1)-(0.4)in the space $C^{2+k,1+k/2}(\overline{Q}) \times C^k(\overline{\Omega})$.

A proof is given in [Is3] as well.

To obtain uniqueness we consider two possible solutions (u_1, c_1) and (u_2, c_2) and subtract the equations for u_1 from the equations for u_2 to obtain

$$a_0\partial_t u + Au + c_2 u = u_1 F \text{ in } Q$$

where $F = c_1 - c_2$. Applying the maximum principles for parabolic equations for u_1 and for this equation differentiated with respect to t we conclude from the condition (1.3) that $0 \leq u_1, 0 \leq \partial_t u_1$ on Q. Then uniqueness will follow from Theorem 1.1 with $\alpha = u_1$.

To prove existence we consider the iterations $u(;j), c(;j-1)$ defined as follows

$$a_0\partial_t u(;j) + Au(;j) + c(;j-1)u(;j) = 0 \text{ on } Q$$

$$u(;j) = 0 \text{ on } \Omega \times \{0\}, u(;j) = g \text{ on } \partial\Omega \times (0,T)$$

$$c(;0) = 0, c(;j-1) = (-Ah_T - \partial_t u(,T;j-1))/h_T \text{ on } \Omega$$

Known results about parabolic boundary value problems guarantee existence and uniqueness of these iterations. Moreover it is proven that

$$u(;j+1) \leq u(;j), \partial_t u(;j+1) \leq \partial_t u(;j), c(;j-1) \leq c(;j)$$

so our iterations are monotone. Their convergence is proven in [Is3] as well.

2. Lateral Overdetermination: Single Measurements

In many important inverse problems it is quite unrealistic to have the final data h_T. On other hand, one can implement even all possible lateral boundary measurements. In this section we consider a simple but interesting case of such inverse problem where one is looking for the coefficient $c = c(x)$ given the additional Neumann data (0.5) of a solution of the parabolic problem (0.1)-(0.3) with $a_0 = 1, a = 1, b = 0, f = 0$. The single boundary measurement means that we are given (0.5) for only one g. In some sense, we have a minimal amount of data. This inverse problem is much better understood in the one-dimensional case. A uniqueness result is given by the following Theorem 2.1 where we assume that $n = 1$ and $\Omega = (0, 1)$.

Theorem 2.1. *Let $g(t) = t^{1/2}$ and let $\Gamma = \partial\Omega$.*
Then the coefficient $c = c(x) \in L^{\infty}(\Omega)$ is uniquely determined by the data (0.5) of the parabolic problem (0.1)-(0.3)

This result has been obtained by many authors (Lavrentiev and Reznitskaya, 1974;Pierce,1979; Rundell,1983;T.Suzuki,1984) in similar situations. The basic idea is to use analytic continuation with respect to t to obtain the data (0.5) for all positive t, and then via the Fourier(Laplace) transform to reduce our parabolic problem to the one-dimensional inverse Sturm-Liouville problem which is well understood.

In these notes we make use of a slightly different approach by exploiting the transform

(2.1)
$$u(x,t) = (\pi t)^{-1/2} \int_0^{\infty} exp(-\tau^2/(4t))u^*(x,\tau)d\tau$$

where u^* solves the hyperbolic problem

(2.2)
$$\partial_\tau^2 u^* - \partial_x^2 u^* + c(x)u^* = 0 \text{ in } Q^* = \Omega \times (0, T^*)$$
$$u^* = \partial_\tau u^* = 0 \text{ on } \Omega \times \{0\}$$
$$u^* = g^* \text{ on } \partial\Omega \times (0, T^*), (g^*(\tau) = 2\tau)$$

Integrating by parts in (2.1) one can conclude that

$$\partial_t u(x,t) = (\pi t)^{-1/2}(\partial_\tau u^*(x,0) + \int_0^{\infty} exp(-\tau^2/(4t))u^*(x,\tau)d\tau)$$

so if u^* solves (2.2) then the corresponding u will solve the parabolic problem (0.1)-(0.3).

The additional data

(2.3)
$$\partial_x u^* = g_1^* \text{ on } \partial\Omega \times (0, T^*), T^* = 1$$

to find c from (2.2) can be obtained by inverting the relation (2.1) where g_1 is the Neumann data (0.5) of the parabolic problem. We first will find the analytic continuation of g_1 from $(0, T)$ onto $(0, \infty)$ which is a conditionally stable operation discussed in more detail in section 3 (proof of Theorem 3.3). Let us fix x, then the formula (2.1) represents the solution $u(,t,s)$ to the heat equation $\partial_t u - \partial_s^2 u = 0$ in the (s,t)-half-plane $\{t > 0\}$ with the initial data $u(,0,s) = u^*(,s)$. So looking for u^* is equivalent to finding the initial data given the solution $u(,t,0)$ (i.e. at $\theta = 0$) for all $t > 0$.This second linear inversion is ill-posed, but quite well understood. The recovery of c from the data of the one-dimensional inverse problem (2.2),(2.3) is unique and stable. By using the Green's Functions for the string equation it is easy to reduce this last problem to a (non-linear) Volterra type equation of the second kind which is easy to solve numerically by iterations.

In the one-dimensional case there are even uniqueness and existence results for a inverse source problems [D].

When we turn to much more complicated multidimensional problems we can report only on some partial results. In the following Theorem we will replace the initial condition (0.2) by the condition $u = h_0$ on $\Omega \times \{0\}$ where h_0 is positive. Let P be a half-space and $\Gamma = \partial\Omega \cap P$.

Theorem 2.2. *Let $a_0 = a = 1, b = 0, f = 0$ and $g, h_0, c \in C^l$ where $(n+7)/2 \leq l$. Let the initial data satisfy the condition*

(2.4)

$$0 < \epsilon \leq h_0 \text{ on } \Omega_0 = \Omega \cap P$$

Then the coefficient $c = c(x)$ is uniquely determined on Ω_0 by the additional Neumann data (0.5).

A proof of this result is somehow similar to the outlined proof of Theorem 2.1. But now instead of the one-dimensional hyperbolic problem (2.2) we will use the similar problem with the equation

$$\partial_\tau^2 u^* - \Delta u^* + cu^* = 0$$

and T^* large enough. Then uniqueness in the inverse hyperbolic problem follows from Theorem 7.4.1 in the book [Is2] which has been obtained via Carleman estimates and Boukhgeim-Klibanov approach. It is quite a challenging question whether non-zero initial data are necessary in the parabolic inverse problem.

We observe that in the recent preprint of Engl,Scherzer,and Yamamoto [E-S-Y] they proved uniqueness of the source term

$$f(x,t) = \sum_{j=1}^{\infty} \chi((t_j, t_{j+1})((t-t_j)^{1/2} F_{1j}(x) + F_{0j}(x))$$

where (t_j, t_{j+1}) form a disjoint covering of the time interval $(0,T)$. On this Conference J.Gottlieb and T.Seidman reported about (independent) uniqueness proofs for coefficients when the boundary data have a similar form related to some complete family of functions on $\partial\Omega$.

3. Lateral Overdetermination: Many Boundary Measurements

In this section we will be given results of all possible lateral boundary measurements. In other words we are given the so-called lateral Dirichlet-to-Neumann operator $\Lambda_l : g \to h$ which we will consider on all smooth (C^3) Dirichlet data g on $\partial\Omega \times [0,T]$.

Theorem 3.1. *Let $a_0 = 1, b = 0$ and a be a scalar matrix. Let $f = 0$ and $h_0 = 0$. Let $2 \leq n$. Let $\Gamma = \partial\Omega$. Then Λ_l uniquely determines $a \in H^{2,\infty}(\Omega)$ and $c \in L^\infty(\Omega)$.*

Proof. The substitution

(3.1)
$$u = ve^{\lambda t}$$

transforms the equation (0.1) into the following equation

(3.2)
$$\partial_t v - div(agradv) + (c + \lambda)v = 0 \text{ in } Q$$

Choosing λ large we achieve that $0 \leq c + \lambda$. The Dirichlet-to-Neumann map for v is given as well.

Let $g = g^0\phi$ where $g^0 \in C^2(\partial\Omega)$ and $\phi \in C^2(\partial\Omega)$ satisfies the conditions $\phi = 0$ on $(-\infty, T/4), \phi = 1$ on $(3T/4, \infty)$. Extend g onto $\partial\Omega \times (T, \infty)$ as g^0. Since the coefficients of the equation (3.2) do not depend on t and so is g when $3T/4 \leq t$ the solution $v(x,t)$ is analytic with respect to such t, so $\partial_\nu v(x,t)$ on $\partial\Omega \times (3T/4, \infty)$ uniquely determines $\partial_\nu v$ on $\partial\Omega \times (3T/4, \infty)$.

The well known stabilization results ([F]) guarantee that

(3.3)
$$\|v(,t) - v^0\|_\infty(\Omega) \to 0 \text{ as } t \to \infty$$

where v^0 is an unique solution to the (stationary) Dirichlet Problem

(3.4)
$$-div(agradv^0) + (c + \lambda)v^0 = 0 \text{ in } \Omega, v^0 = g^0 \text{ on } \partial\Omega$$

The standard parabolic estimates for $v, \partial_t v$ give that $\|\partial_t v\|_{2,1;p}(\Omega\times(s, s+1)) \leq C$, so by Trace Theorems we have $\|v(,t)\|_{2,2}(\Omega) \leq C$, and (3.3) by interpolation implies that

(3.5)
$$\|v(,t) - v^0\|_{1,2}(\Omega) \to 0 \text{ as } t \to \infty$$

Here we are using anisotropic and isotropic Sobolev type norms, so that 2,1 denotes number of derivatives with respect to x and t,p is the index of the corresponding Lebesgue space, and then 2,2 and 1,2 stay for number of derivatives and the space L^2 and label isotropic spaces of functions on Ω. Since $\partial_\nu v(,t)$ is given for all t we obtain in the limit $\partial_\nu v^0$ on $\partial\Omega$ for any smooth Dirichlet data g^0. Finally we are given the Dirichlet-to-Neumann map for the elliptic equation (3.4).

The Riccati substitution $v^0 = a^{-1/2}w$ transforms the equation (3.4) into the Schrödinger equation

(3.6)
$$-\Delta w + a^{-1/2}(\Delta a^{1/2} + c + \lambda)w = 0 \text{ in } \Omega$$

Since a and its normal derivative on $\partial\Omega$ are uniquely determined by (stable) boundary reconstruction ([Is6],[U]) we are given the Dirichlet-to-Neumann map Λ for the equation (3.6). Due to the results of Sylvester and Uhlmann ($3 \leq n$, see[Is4],[S-U]) and to the recent result of Isakov and Nachman ($n = 2$, [Is-N]) the coefficient $c*$ of w in the equation (3.6) is uniquely determined by Λ. Therefore, $a^{-1/2}$ and therefore c are uniquely determined as the coefficients of powers of λ in c^*.

The proof is complete

A similar argument is valid for the vector-coefficient b of first order terms.

Theorem 3.2. *Let $a = 1$ and $3 \leq n$.*

There is positive ϵ such that if $\|curlb\|_\infty < \epsilon$, then the coefficients $a_0, c \in L^\infty(\Omega), b \in H_{2,\infty}(\Omega)$, are uniquely identified by Λ_l.

Proof. As in the proof of Theorem 3.1, Λ_l uniquely determines the Dirichlet-to-Neumann map Λ for the stationary (elliptic) equation

(3.7)
$$-\Delta v^0 + b \cdot gradv^0 + (c + a_0\lambda)v^0 = 0 \text{ on } \Omega$$

From the recent results of Isakov [Is8] and of Sun[Su] ($\Re b = 0$) we derive that $curlb, c + a_0\lambda$ are uniquely identified by Λ for all λ. So $curlb, c, a_0$ are unique. The proof is complete.

Recently Nakamura,Sun,and Uhlmann [N-S-U] obtained global uniqueness in the inverse elliptic problem (3.7) when $\Re b = 0$ and $3 \leq n$. We have little doubt that global uniqueness is valid in the real-valued case and also for lower dimensions of the space, because it is possible to obtain global uniqueness results for the elliptic equations when λ is large. However it has not been done yet.

The asymptotic analysis also allows us to derive stability estimates from available ones in the elliptic case.

Let $\Lambda_{l1}, \Lambda l2$ be two lateral Dirichlet-to-Neumann maps corresponding to the equations (0.1) with the coefficients $a_0 = a_{01}, c = c_1$ and $a_0 = a_{02}, c = c_2$ when $a = 1, b = 0$. Let ϵ be the operator norm of $\Lambda_{l1} - \Lambda_{l2}$ considered as the operator from the space $L^2((0,T); H_{(1/2)}(\partial\Omega)$ into $L^2((0,T); H_{(-1/2)}(\partial\Omega))$.

Theorem 3.3. *Assume that $n = 3$. Let*

(3.8)
$$\|a_{0j}\|_{1,\infty}(\Omega) + \|c_j\|_{1,\infty}(\Omega) \leq M$$

Then there are constants $C, \delta, 0 < \delta < 1$ such that

(3.9)
$$\|a_{01} - a_{02}\|_\infty(\Omega) + \|c_1 - c_2\|_\infty(\Omega) \leq C|log\epsilon|^{-\delta}$$

Proof. Using the substitution (3.1) we can achieve that $0 \leq (c_j + \lambda a_{0j})$. We will assume it henceforth.

Let $g^0 \in C^2(\partial\Omega), \|g^0\|_{(1/2)}(\partial\Omega) \leq 1$. Let $g = g^0\phi$ where ϕ is defined in the proof of Theorem 3.1 and let u_1, u_2 be solutions to the initial boundary value problem (0.1)-(0.3) with the coefficients $a_{01}, a_{02}, ...$ and with the described lateral Dirichlet data g for all positive times.

It is known ([Yo]) that the parabolic problem (0.1)-(0.3) generates an analytic semigroup (in $L^2(\Omega)$), so that its solution $u_j(x,t)$ admits the holomorphic extension onto the sector $S = \{t = t_1 + it_2 : |t_2| \leq 2(t_1 - T/2)\}$ of the complex t-plane.Moreover, we have the bound $\|u_j(,t)\|_{1,2}(\Omega) \leq C$ when $t \in S$ where C depends on Ω, ϕ,and M. By Trace Theorems,

(3.10)
$$\|\partial_\nu(u_1 - u_2)(,t)\|_{(-1/2)}(\partial\Omega) \leq C \text{ when } t \in S$$

A norm of an analytic function with the values in a Banach Space is a subharmonic function, so $s(t) = log\|\partial_\nu(u_1 - u_2)(,t)\|_{(-1/2)}(\partial\Omega)$ is subharmonic on S. Let $\mu(t)$ be the harmonic measure of the interval $I_T = (7T/8, T)$ in S, i.e. $\Delta\mu = 0$ on $S \setminus I_T, \mu = 0$ on ∂S, and $\mu = 1$ on I_T. By using conformal mappings (of $S \setminus \overline{I}_T$ onto an annulus) and maximum principles it is easy to see that μ does exist, is continuous on $\bar{S}$, and $0 \leq \mu \leq 1$ on $S \setminus I_T$. In addition, $t^{-2}/C < \mu(t)$ when $T < t$. The last claim can be checked by mapping S conformally onto the right half-plane $\{0 < \Re z\}$ as in [Is2,3.6]. Since $\partial_\nu u_j$ is analytic near $\bar{S}$ we have lower semicontinuity of $s(t)$ on $\bar{S}$.

We claim that

(3.11)

$$s(t) \leq C\mu(t)log\epsilon + (1 - \mu(t))logC, t \in \partial S \cup I_T$$

Indeed, when $t \in I_T$ the harmonic measure $\mu(t) = 1$ and then

$$s(t) = log\|\partial_\nu(u_1 - u_2)(,t)\|_{(-1/2)}(\partial\Omega) \leq Clog\|\Lambda_{l1} - \Lambda_{l2}\| = Clog\epsilon$$

When $t \in \partial S$ then $\mu(t) = 0$ and (3.11) follows from the bound (3.10). From the maximum principle for subharmonic functions we have the inequality (3.11) for $t \in S$, in particular, when $T < t < \infty$. Taking exponents of the both parts of (3.11) we obtain

(3.12)

$$\|\partial_\nu(u_1 - u_2)\|_{(-1/2)}(\partial\Omega) \leq C\epsilon^{\mu(t)}C^{1-\mu(t)} \leq C\epsilon^{t^{-2}/2}$$

On other hand,the exponential stabilization of solutions of parabolic problems with the maximum principle ([F]) guarantees that $\|u_j(,t) - u_{0j}\|_\infty(\Omega) \leq Ce^{-\theta t}$ where positive θ depends only on Ω. Here u_{0j} is the solution to the following stationary Dirichlet Problem

(3.13)

$$-\Delta u_{0j} + (c_j + \lambda a_{0j})u_{0j} = 0 \text{ in } \Omega, u_{0j} = g^0 \text{ on } \partial\Omega$$

By applying interior Schauder type estimates to $u_j(,t) - u_{0j}$ in the domains $\Omega \times (T_1, T_1 + 1)$ it is easy obtain the bounds of the $H_{(1)}(\Omega)$-norms of these differences by $Ce^{-\theta t}$. So by Traces Theorems we have

(3.14)

$$\|\partial_\nu(u_j(,t) - u_{0j})\|_{(-1/2)}(\partial\Omega) \leq Ce^{-\theta t}$$

From the triangle inequality we obtain

$$\|\partial_\nu(u_{01} - u_{02})\|_{(-1/2)}(\partial\Omega) \leq$$

$$\|\partial_\nu(u_1(,t) - u_2(,t))\|_{(-1/2)}(\partial\Omega) + \|\partial_\nu(u_1(,t) - u_{01})\|_{(-1/2)}(\partial\Omega)$$

$$+\|\partial_\nu(u_2(,t) - u_{02})\|_{(-1/2)}(\partial\Omega) \leq C(\epsilon^{t^{-2}/C} + e^{-\theta t})$$

when we make use of (3.12),(3.14). Choosing $t = (-log\epsilon)^{1/3}$ yields

$$\|\partial_\nu(u_{01} - u_{02})\|_{(-1/2)}(\partial\Omega) \leq Cexp(-(-log\epsilon)^{1/3}/C)$$

so denoting by ϵ_1 the norm of the difference of the elliptic Dirichlet-to-Neumann operators corresponding to the equations (3.13) and acting from $H_{(1/2)}(\partial\Omega)$ into $H_{(-1/2)}(\partial\Omega)$ we finally get
(3.15)
$$\epsilon_1 \leq Cexp(-(log\epsilon)^{1/3}/C)$$

The Alessandrini's stability estimates for the elliptic equation (3.13) [A] give
(3.16)
$$\|c_1^\dagger - c_2^\dagger\|_\infty(\partial\Omega) \leq C(-log\epsilon_1)^{-\delta}$$

where $c_j^\dagger = c_j + \lambda a_{0j}$. From ((3.15) it follows that ϵ_1 can be replaced by ϵ, only C, δ will be different. Observing that $c_1^\dagger - c_2^\dagger = (c_1 - c_2) + \lambda(a_{01} - a_{02})$, considering two different λ (say,λ and $\lambda + 1$), and solving the obtained system for $c_1 - c_2, a_1 - a_2$ we derive the estimate (3.9) from the estimate (3.16).

The proof is complete.

The stabilization method currently does not produce uniqueness results when measurements are implemented only on a part Γ of $\partial\Omega$. We suggest to handle this "local case" by using hyperbolic inverse problems for the function u^* related to u via (2.1). As in the proof of Theorem 2.1 if u^* solves the problem
(3.17)
$$a_0\partial_\tau^2 u^* - \Delta u^* + cu^* = 0 \text{ in } \Omega \times (0, \infty)$$
(3.18)
$$u^* = \partial_\tau u^* = 0 \text{ on } \Omega \times \{0\}$$
(3.19)
$$u^* = g^0\phi^* \text{ on } \partial\Omega \times (0, \infty)$$
Then u given by (2.1) will solve the parabolic problem (0.1)-(0.3) with $g = g^0\phi$.

Theorem 3.4. *Let $a_0 = 1, b = 0, a = 1$. Let P be a half-space, $2 \leq n$.Define $\Omega_0 = \Omega \cap P, \Gamma = \partial\Omega \cap P$. Then the local lateral Dirichlet-to-Neumann map $\Lambda_{l,\Gamma} : g \to \partial_\nu u$ on $\Gamma \times (0,T)$, $suppg \subset \Gamma \times (0,T)$ uniquely determines $c \in L^\infty(\Omega)$ on Ω_0.*

Proof. Let $u(x,t)$ be a solution to the parabolic problem (0.1)-(0.3) with the Dirichlet data $g = g^0\phi$.Consider ϕ^* supported in $(0, T/2)$ then ϕ is analytic with respect to t. The solution $u(x,t)$ to the parabolic problem will be analytic with respect to t as well, so $\partial_\nu u(x,t)$ on $\Gamma \times (0,T)$ uniquely determines $\partial_\nu u(x,t)$ on $\Gamma \times (0, \infty)$.

Consider the corresponding problem (3.17)-(3.19).Since $\partial_\nu u$ uniquely determines $\partial_\nu u^*$ via (2.1) we are given the later function on $\Gamma \times (0, \infty)$. Since functions $g^0\phi$ are complete in $C_0^2(\Gamma \times (0, \infty))$ we are given the local Dirichlet-to-Neumann map on Γ for all positive t. Applying the uniqueness results from [Is1] we obtain uniqueness of c in Ω_0.

The proof is complete.

We are quite sure that the same argument can prove that the local Dirichlet-to-Neumann map on $\Gamma = \partial\Omega \cap B$ where B is a ball around a boundary point uniquely recovers $a_0, curlb, c$ on the whole Ω. When $b = 0$ the needed hyperbolic results seemingly can be obtained by the BC-method of Belishev [B] augmented by the remarkable recent results of Tataru [T] on sharp uniqueness of the continuation for second order hyperbolic equations with time-independent coefficients. These results require a be in C^1 and lower order coefficients only measurable and bounded.

4. Recovery of discontinuous a

Of special applied importance is the problem of identification of the (scalar) coefficient

(4.1)
$$a = a^0 + k\chi(Q^*)$$

where a^0, k are C^2-smooth functions and $\chi(Q^*)$ is the characteristic function of un unknown domain $Q^* \subset Q$. We refer,say, to the review paper [Y].

In this section we will report on two new uniqueness results for Q^* and k: with a single (special) boundary measurement (then $Q^* = D \times (0,T)$) and with many measurements (then Q^* is a general domain whose time dependence is Lipschitz).

To prove the first result we combine uniqueness theorems for hyperbolic and elliptic equations by using the transform (2.1) and stabilization of solutions at infinity. To prove the second one we make use of singular solutions of parabolic equations as in the author's paper [Is6] on elliptic equations.

We call a domain D in the x-space Lipschitz if its boundary is locally the graph of a Lipschitz function. A domain is called x_n-convex if its intersections with straight lines parallel to the x_n-axis are intervals.

In the next result Γ is any open subset of the hyperplane $x_n = 0$.

Theorem 4.1. *Let $a^0 = 1$,D be a Lipschitz domain in the half-space $x_n < 0, 2 \leq n$ intersecting the strip $\Gamma \times \mathbf{R}^-$, and $Q^* = D \times (0,T)$.Assume that k is constant and*

(4.2)
$$-1 < k < 0$$

Let the Dirichlet data $g(x,t) = g^0(x)t^{1/2}$ where $g^0 = 1$ on an open subset of $\Gamma, g^0 \in C_0^2(\Gamma)$, and let u be a solution to the parabolic problem (0.1)-(0.3) (with $a_0 = 1, b = 0, c = 0$).

Then $\partial_\nu u$ on $\Gamma \times (0,T)$ uniquely determines D or,equivalently, the coefficient a given by (4.1).

Proof. Consider the hyperbolic initial boundary value problem

$$\partial_\tau^2 u^* - div((1 + k\chi(Q^*))gradu^*) = 0 \text{ in } \Omega \times \mathbf{R}^+$$

with zero initial data and with the lateral data $u^*(x,\tau) = g^0(x)\tau$ on $\partial\Omega \times \mathbf{R}^+$. Then $u(x,t)$ obtained from u^* via (2.1) solves the parabolic problem (0.1)-(0.3) with $a_0 = 1, a = 1 + k\chi(Q^*), b = 0, c = 0$. As above, the Dirichlet data for the parabolic problem are analytic with respect to t, so its solution $u(x,t)$ is t-analytic for all t greater than $3T/4$. Therefore,the data $\partial_\nu u(x,t)$ given on $\Gamma \times (0,T)$ are uniquely determined by the analytic continuation on $\Gamma \times (0,\infty)$. These data uniquely determine $\partial_\nu u*$ on $\Gamma \times (0,\infty)$. By the uniqueness theorem for the inverse hyperbolic problem [Is7] we conclude that $\partial D \cap (\Gamma \times \mathbf{R})$ is uniquely determined.

The function $v = \partial_t u^*$ solves the parabolic problem with t- independent data for large t, so it stabilizes to the solution v^0 to the conductivity problem

$$-div((1 + k\chi(D))gradv^0) = 0 \text{ in } \Omega, v^0 = g^0 \text{ on } \partial\Omega$$

Moreover $\partial_\nu v(x,t)$ converges to $\partial_\nu v^0(x)$ on $\partial\Omega$. So we are given the Cauchy data for v^0 on Γ. Two possible solutions D_1, D_2 of this inverse conductivity problem must have a common part of their boundaries, as we saw, so these domains are i-contact in the sense of the paper [F-I]. From the results of this paper we have $D_1 = D_2$. The proof is complete.

For an open subset Q^* of the layer $\mathbf{R}^n \times (0,T)$ the lateral boundary $\partial_x Q^*$ is defined as the closure of $\partial Q^* \cap \{0 < t < T\}$. We say that Q^* is x-Lipschitz if $\partial_x Q^*$ is Lipschitz (i.e. is locally the graph of a Lipschitz function).

Theorem 4.2. *Let $2 \leq n$. Suppose that Q^* is an open x-Lipschitz subset of Q and $\partial_x Q^* \cap \partial_x Q = \emptyset$. Suppose that*
(4.3)

$$\text{the sets } (Q \setminus \overline{Q}^*) \cap \{t = \tau\} \text{ are connected when } 0 < \tau < T$$

Then the lateral boundary data

$$h = \partial_\nu u \text{ on } \Gamma \times (0,T)$$

of a solution to (0.1)-(0.3) (with $a_0 = 1, a = 1+k\chi(Q^*), b = 0, c = 0, k \neq 0$ on $\partial_x Q^*$) given for any $g \in C_0^2(\Gamma \times (0,T))$ uniquely identify Q^* and t -independent $k \in C^2(\bar{\Omega})$ on Q^*.

A proof is given in the forthcoming paper [E-Is].We will outline its basic steps. Assume that there are two different domains Q_1^*, Q_2^* with the same data. Denote by $Q_{3\theta}$ the connected component of the open set $\Omega \setminus (\overline{Q}^*_{1\theta} \cup \overline{Q}^*_{2\theta})$ whose boundary contains Γ. Here $Q^*_{j\theta}$ is defined as $Q_j^* \cap \{t = \theta\}$. Let Q_3 be the union of $Q_{3\theta}$ over $0 < \theta < T$ and let $Q_4 = Q \setminus \overline{Q}_3$.

Lemma 4.3. *Under the conditions of Theorem 4.2*
(4.4)

$$\int_{Q_1^*} k_1 grad v_1 \cdot grad u_2^* dxdt = \int_{Q_2^*} k_2 grad v_1 \cdot grad u_2^* dxdt$$

for all solutions v_1 to the equation

$$\partial_t v_1 - div((1 + k_1\chi(Q_1)) grad v_1) = 0 \text{ near } \overline{Q}_4, v_1 = 0 \text{ when } t < 0$$

and all solutions v_2^* to the (adjoint) equation

$$-\partial_t u_2^* - div((1 + k_2\chi(Q_2)) grad u_2^*) = 0 \text{ near } \overline{Q}_4, u_2^* = 0 \text{ when } T < t$$

To prove this orthogonaly relation one subtract the equations

$$\partial_t u_j - div((1 + k_j\chi(Q_j)) grad u_j) = 0$$

to obtain

$$\partial_t u - div((1 + k_2\chi(Q_2)) grad u) = div(k_2\chi(Q_2) - k_1\chi(Q_1))$$

where $u = u_2 - u_1$. Since u_1, u_2 have the same lateral Cauchy data on $\Gamma \times (0,T)$ by known uniqueness results in the lateral Cauchy problem [Is2] we obtain that

$u = 0$ on Q_4. Then multiplying the equation for u by v_2^* and "integrating by parts" (or,more precisely, by using the definition of a weak solution) we obtain the orthogonality relations (4.4) for all needed u_2^* and for $v_1 = u_1$ which solve the parabolic problem (0.1)-(0.3) for the first equation.

By using uniqueness of the continuation for second order parabolic equations and the corresponding Runge Property we can approximate by such u_1 any v_1 and complete the proof of Lemma 4.3.

Returning to the proof of Theorem 4.2 we assume that

(4.5)
$$Q_1 \neq Q_2$$

Then we may assume that there is a point $(x_0, t_0) \in \partial_x Q_1 \setminus \overline{Q}_2$ which is contained in ∂Q_3 as well. By considering $g = 0$ when $t < t_0$ and using the translation $t \to t - t_0, x \to x - x_0$ we can also assume that $t_0 = 0, x_0 = 0$. Choose a ball $B \subset \mathbf{R}^n$ centered at 0 and a cylinder $Z = B \times (-\tau, \tau)$ so that $\overline{B} \subset \Omega, \overline{Z}$ is disjoint with $\overline{Q}_2$ and $(\partial_x Q_1) \cap \overline{Z}$ is a Lipschitz surface. By the Whitney Extension Theorems there is a $C^2(\overline{Q}_1 \cup \overline{Z})$-function a_3 coinciding with a_1 on Q_1. Extend a_3 onto $Q \setminus (\overline{Q}_1 \cup \overline{Z})$ as a^0.

Lemma 4.4. *Under the conditions of Theorem 4.2 we have*

(4.6)
$$\int_{Q_1} k_1 gradu_3 \cdot gradu_2^* = \int_{Q_2} k_2 gradu_3 \cdot gradu_2^*$$

for any solution u_3 to the equation

$$\partial_t u_3 - div(a_3 gradu_3) = 0 \text{ near } \overline{Q}_4, u_3 = 0 \text{ when } t < 0$$

and for any u_2^* from Lemma 4.3.

This result follows from Lemma 4.3 by approximating u_3 by solutions u_{3k} of the equations $\partial_t u_{3k} - div(a_{3k} gradu_{3k}) = 0$ with the same initial and lateral boundary data as u_3 and with a_{3k} equal to a_1 near $\overline{Q}_4$ and convergent pointwise to a_3.Lemma 4.3 is valid with $v_1 = u_{3k}$ and then one can pass to the limit.

To obtain a contradiction with the assumption (4.5) we will use as u_3 and u_3^* the fundamental solutions K^+ and K^- of the forward and backward Cauchy problems for the parabolic equations with the coefficients a_3 and a_2 with the poles at $(y, 0), (y, \tau)$, y is outside $\overline{Q}_2$ and close to the origin. To bound the integral (4.6) it is convenient to use new variables. We can assume that the direction e_n of the x_n-axis coincides with the interior unit normal to $\partial_x Q_1 \cap \{t = 0\}$. This surface near the origin is the graph of a Lipschitz function $x_n = q_1(x_1, ..., x_{n-1}, t)$. The substitution

$$x_k = x_k^\natural, k = 1, ..., n-1, x_n = x_n^\natural + q_1(0, 1), t = t^\natural$$

transforms the equations into similar ones (with additional first order terms in $x_n^\natural$). The domains Q_j are transformed into similar domains with the additional property that the points $(0, t), 0 < t < T$ belong to $\partial_x Q_1$. Further on we will drop the sign $\natural$.

The mentioned fundamental solutions have the following structure

(4.7)
$$K^+ = K_1^+ + K_0^+, K^- = K_1^- + K_0^-$$

where the first terms are so-called parametrices

$$K_1^+(x,t;y,\tau) = C(t-\tau)^{-n/2}exp(-|x-y|^2/(a_3(x)(t-\tau)))$$

$$K_1^-(x,t:y,\tau) = C(\tau-t)^{-n/2}exp(-|x-y|^2/(\tau-t))$$

and K_0^+, K_0^- are remainders, with weaker singularities [F]

Letting in Lemma 4.4 $u_3 = K^+, u_2^* = K^-$, splitting K^+ according to (4.7) and breaking the integration domain Q_1 into $Q_1 \cap Z$ and its complement we obtain

(4.8)
$$I_1 = I_2 + I_3$$

where

$$I_1 = \int_{Q_1 \cap Z} k_1 grad_x K_1^+ \cdot grad_x K_1^-$$

$$I_2 = -\int_{Q_1 \setminus Z} k_1 grad_x K^+ \cdot grad_x K^- + \int_{Q_2} k_2 grad_x K^+ \cdot grad_x K^-$$

and

$$I_3 = \int_{Q_1 \cap Z} k_1(grad_x K_1^+ \cdot grad_x K_0^- + grad_x K_0^+ \cdot grad K_1^- + grad_x K_0^+ \cdot grad_x K_0^-)$$

Observe that we let $K^+ = K^+(;0,0), K^- = K^-(;0,\tau)$ and we passed to the limit when $y \to 0$ which can be justified (when $0 < \tau$) by the Lebesgue Dominated Convergence Theorem using known bounds on $grad_x K^+, grad_x K^-$.

The basic analytic step of the proof is the following result

Lemma 4.5. *We have*

$$\tau^{-n/2}/C \leq |I_1|$$

$$|I_2| \leq C\tau^{-n/2+1}\epsilon^{-2}exp(-\epsilon^2/(M\tau))$$

$$|I_3| \leq C\tau^{-n/2+3/2}\epsilon^{-2}(1-exp(-\epsilon^2/(M\tau))$$

where M depends only on the upper bound of a_3, a^0 over Q.

A proof of these bounds is given in the forthcoming paper [E-Is]. It is based on known estimates of K_0^+, K_0^- and some technical tricks.

To finish the outline of the proof of Theorem 4.2 we let

(4.9)
$$\epsilon^2 = E\tau$$

where (large) E will be chosen later.From (4.8),(4.9), and Lemma 4.5 we have

$$\tau^{-n/2} \leq C^2(\tau^{-n/2+1}\epsilon^{-2}exp(-E/M) + \tau^{-n/2+3/2}\epsilon^{-2}(1 - exp(-E/M))$$

Using (4.9) again and multiplying the both parts by $\tau^{n/2}$ we obtain

$$1 \leq C^2E^{-1}exp(-E/M) + C^2E^{-1}\tau^{1/2}(1 - exp(-E/M))$$

Choose E so large that $E^{-1}e^{-E/M} \leq 1/(2C)^2$, then the first term in the right side is smaller than $1/4$. After that choose τ so small that the second term is smaller than $1/4$ as well.

We have a contradiction which shows that the assumption (4.5) is wrong, so $Q_1 = Q_2$.

When $Q_1 = Q_2$ it easy to show that $k_1 = k_2$. First, by similar argument we obtain that they are equal on $\partial_x Q_1$ which implies that they are the same on union of $\partial_x Q_{1\theta}$ over $0 < \theta < T$. Then we have unknown terms which does not depend on t. Their coincidence can be proven by using the methods of section 3.

5.Semi-linear equations

Now we are interested in recovery of the semi-linear equation

(5.1)
$$c(x,u)\partial_t u - \Delta u + a(x,u) = 0 \text{ in } Q$$

,with the zero initial data (0.2) and the lateral data (0.3), from the additional Neumann data on $\partial_x Q$. We assume that

$$0 < c \text{ on } \bar{\Omega} \times \mathbf{R}, 0 \leq a_u, a(x,0) = 0$$

(5.2)
$$c, c_u, c_{uu}, a, a_u, a_{uu} \text{ are in } C(\bar{\Omega} \times [-U, U]$$

for any finite U.

By using upper and lower solutions and standard estimates for parabolic equations ([F],[L-S-U]) one can show that for any Dirichlet lateral data $g \in C^2(\partial\Omega \times [0,\infty)), |g| \leq G, g = 0$ on $\partial\Omega \times \{0\}$ there is an unique solution $u \in H_{2,1;p}(Q), 1 < p$ of the problem (5.1),(0.2),(0.3), moreover, $|u| \leq G$. Therefore we have well-defined $\partial_\nu u$ on $\partial\Omega \times (0,\infty)$ and the lateral Dirichlet-to-Neumann map Λ_l.

Introduce the set $E = \{(x,u) : u_*(x) \leq u \leq u^*(x)\}$ where $u_*(x)(u^*(x))$ is $inf u(x)(sup u(x))$ over all admissible g.

Theorem 5.1. *Let $2 \leq n$. Then Λ_l uniquely determines a on E. If $3 \leq n$, then it uniquely determines also c on E.*

Until recently, only available uniqueness results for semi-linear equations were only in the case when c or a do not depend on x ([C-Y],[L-P],[P-R]),and even so in many important cases under additional "smallness assumptions". In the paper

[Is5] the author suggested a linearization technique to treat $a = a(x,u)$ which was transferred by the author and Sylvester [Is-S] onto elliptic operators when $3 \le n$ and by the author and Nachman [Is-N] onto elliptic operators in the plane. Here we derive Theorem 5.1 from the results of [Is-N] and [Is-S] by using the stabilization method described in section 3.

First we give necessary results about asymptotic behavior of $u(x,t)$ when $t \to \infty$. Let the Dirichlet data

(5.3)
$$g(x,t) = g^0(x) + t^{-1}g^1(x) + t^{-2}g^2(x), 1 \le t$$

where $g^0, g^1, g^2 \in C^2(\bar{\Omega}), g^1, g^2 \le 0$. Introduce solutions u^0, u^1, u^2 to the following (stationary) Dirichlet Problems

(5.4)
$$-\Delta u^0 + a(x,u^0) = 0 \text{ in } \Omega, u^0 = g^0 \text{ on } \partial\Omega$$

(5.5)
$$-\Delta u^1 + a_u(x,u^0)u^1 = 0 \text{ in } \Omega, u^1 = g^1 \text{ on } \partial\Omega$$

and

(5.6)
$$-\Delta u^2 + a_u(x,u^0)u^2 = -1/2a_{uu}(x,u^0)(u^1)^2 + c(x,u^0)u^1(x) \text{ in } \Omega, u^2 = g^2 \text{ on } \partial\Omega$$

From known results about parabolic problems we can derive that a solution $u(x,t)$ to (5.1),(0.2),(0.3) with g given in (5.3) admits the following representation

(5.7)
$$u(x,t) = u^0(x) + t^{-1}u^1(x) + t^{-2}u^2(x) + v(x,t)$$

where

$$t^2\|v(,t)\|_{1,p}(\Omega) \to 0$$

for any $1 < p$ when $t \to \infty$.

Proof of Theorem 5.1. First consider $g = g^0 (g^1 = g^2 = 0)$. From (5.7),(5.4) it follows that $u(,t) \to u^0$ in $H_{1,p}(\Omega)$ as t tends to ∞. So $\partial_\nu u(,t) \to \partial_\nu u^0$ in $H_{(-1/2)}(\partial\Omega)$. Since $\partial_\nu u(,t)$ is given, so is $\partial_\nu u^0$. In other words, we know the Dirichlet-to-Neumann map for the equation (5.4). Then our conditions (5.2) and the results of the recent papers [Is-N],[Is-S] imply that we are given $a(x,u)$ on E. The next step is to prove uniqueness of c and that is where we need the three terms of the asymptotic expansion (5.7).

Fix any $g^0 \in C^2(\partial\Omega)$ and the corresponding solution u^0 to the Dirichlet Problem (5.4). Given any $g^1 \in C^2(\partial\Omega)$ we know a solution u^1 to the Dirichlet problem (5.5). Since we are given $\partial_\nu u = \Lambda_l g$ and the asymptotic expansion (5.7) implies that

$$grad_x u = grad_x u^0 + t^{-1} grad_x u^1 + t^{-2} grad_x u^2 + grad_x v$$

with

$$t^2\|grad_x v(,t)\|_{(-1/2)}(\partial\Omega) \to 0$$

we are given $\partial_\nu u^2$ for any given g^0, g^1, g^2. Multiplying the both parts of the equations (5.6) by a $H_{2,2}(\Omega)$-solution v^2 to the linear equation

(5.8)
$$-\Delta v^2 + a_u(x, u^0)v^2 = 0 \text{ in } \Omega$$

and using the Green's Formula we obtain

$$\int_{\partial\Omega} (-(\partial_\nu u^2)v^2 + u^2\partial_\nu v^2) = \int_\Omega (-1/2a_{uu}(x, u^0)(u^1)^2 v^2 + \int_\Omega c(x, u^0)u^1 v^2$$

The left side and the first integral in the right side are known, so we are given

$$\int_{\partial\Omega} c(x, u^0(x))u^1(x)v^2(x)dx$$

for all solutions u^1, u^2 to the (linear) Schrödinger equations (5.5). It is known that when $3 \leq n$ the products $u^1 v^2$ are complete in $L^1(\Omega)$ [Is4], so these integrals uniquely determine $c(x, u^0(x))$ for any solution u^0 to the equation (5.4). This uniquely determines c on E.

The proof is complete.

6. Interior Sources

All previous results for coefficients identification were obtained when $f = 0$ (no interior sources) which is the case most important in applications. Unfortunately, these inverse problems are quite unstable. When it is possible to use boundary information from many interior sources one can obtain a simple a stable solution of an inverse problem. This was observed by Lowe and Rundell [L-R] in the one-dimensional case. We give manydimensional results removing many restrictive assumptions in [L-R] and using only sources localized around a point where we are recovering a coefficient.

Consider a solution u to the problem (0.1),(0.2),(0.3) with $a^0 = 1, a = 1, b = 0, g = 0, f = \delta(y, s)$ and unknown

(6.1)
$$c \in C^{\lambda,\lambda/2}(\overline{Q}), 0 \leq c$$

We denote this solution by $u(;y, s)$. So we are prescribing point sources (the Dirac delta-functions with the poles at (y, s)) Let Γ be a non-empty open part of $\partial\Omega$.

Theorem 6.1. *The Neumann data (0.5) of u given for all (y, s) in a neighborhhood of a point (y^0, s^0) uniquely (and in a stable way) determine $c(y^0, s^0)$.*

Proof. Let v be a solution to the following backward parabolic problem

(6.2)
$$\partial_t v + \Delta v = 0 \text{ in } Q$$

(6.3)
$$v = 0 \text{ on } \Omega \times \{T\}, v = g^* \text{ on } \partial\Omega \times (0, T)$$

where

(6.4)

$$0 \le g^* \text{ on } \Gamma \times (0,T), g^* \text{ is not identically zero on } \Gamma \times (0, T-\tau)$$

$$\text{for any positive } \tau, g^* = 0 \text{ on } (\partial\Omega \setminus \Gamma) \times (0,T), g^* \in C_0^2(\Gamma \times (0,T))$$

By the definition of a generalized solution of the problem (0.1)-(0.3) with $f = \delta(y,s)$ we have

$$v(y,s) = \int_Q u(-\partial_t v - \Delta v + cv) - \int_{\Gamma\times(0,T)} v\partial_\nu u$$

Using the equation (6.2) and the Neumann condition (0.5) we conclude that

(6.5)

$$U(y,s) = \int_Q u(;y,s)cv\,dx\,dt$$

is a known function

$$v(y,s) - \int_{\Gamma\times(0,T)} vh(;y,s)$$

when (y,s) is close to (y^0, s^0). Since $u(x,t:y,s)$ is the Green's Function for the equation $\partial_t u - \Delta u + cu = 0$ we have from (6.5)

(6.6)

$$-\partial_s U - \Delta_y U + cU = cv \text{ in } Q, U = 0 \text{ on } \partial Q \cap \{0 < t\}$$

According to (6.2) the function $v = v(y,s)$ solves the same parabolic equation and is zero on $\Omega \times \{T\}$ as U is. Due to the conditions (6.4) we have $U = 0 \le v$ on $\partial\Omega \times (0,T)$ and v is not identically zero there, so

$$U < v \text{ on } Q$$

and we can divide by $U - v$ to find $c(y,s)$ from the differential equation (6.6) as follows

(6.7)

$$c(y,s) = (\partial_s U + \Delta_y U)/(U - v)(y,s)$$

The proof is complete.

It is quite clear that the (local) reconstruction formula (6.7) gives a stable solution to the inverse problem.

Similarly, one can identify the leading coefficient.

Consider a solution u to the problem (0.1)-(0.3) with $a = 1, b = 0, c = 0, g = 0, f = \delta(y,s)$ and unknown

(6.8)

$$\text{positive } a_0 \in C^\lambda(\bar{\Omega})$$

Theorem 6.2. *The Neumann data (0.5) for (y,s) in a neighborhood of $(y^0,s^0), 0 < s^0 < T$ uniquely determine $a^0(y^0)$.*

Proof. Let v be defined as in the proof of Theorem 6.1, with the condition (6.4) replaced by

(6.9)

$$0 \leq \partial_t g^* \text{ on } \Gamma \times (0,T), g* \text{ is not identically equal to } 0$$

$$\text{on any } \Gamma \times (0, T-\tau), g^* = 0 \text{ on } (\partial\Omega \setminus \Gamma) \times (0,T), g^* \in C_0^3(\Gamma \times (0,T))$$

We modify the previous proof by replacing $-\partial_t v - \Delta v + cv$ by $-a_0\partial_t - \Delta v = (-a_0+1)\partial_t v$ to obtain the function

(6.10)

$$U(y,s) = \int_Q (1-a_0)u(;y,s)\partial_t v dx dt$$

known when (y,s) is close to (y_0,s_0).

Instead of (6.5) we will have

(6.11)

$$-a_0(y)\partial_s U - \Delta_y U = (1-a_0)\partial_s v \text{ in } Q, U = 0 \text{ on } \partial Q \cap \{0 < t\}$$

As above,from (6.2) it follows that

$$-a_0\partial_s v - \Delta v = (1-a_0)\partial_s v \text{ in } Q, v = 0 \text{ on } \Omega \times \{T\}$$

So $\partial_s(v-U)$ solves a homogeneous backward parabolic equation in Q and is zero on $\Omega \times \{T\}$. From (6.9),(6.11) we conclude that $\partial_s(v-U)$ has non-negative lateral Dirichlet data $\partial_s g^*$, so by maximum principles $\partial_s(v-U) > 0$ in Q and we can find $a_0(y)$ from (6.11) as follows

(6.12)

$$a_0(y^0) = (\partial_s v + \Delta_y U)/(\partial_s(v-U))(y^0,s^0)$$

The proof is complete.

7. Open Problems

We list here some challenging questions of theoretical and practical importance.

Problem 1. *(Two measurements at Final Moment of Time)*

Is it possible to use two different Dirichlet boundary data g in the situation of section 1 to guarantee global uniqueness of two coefficients $a_(x), c(x)$?

Problem 2. *(Global Uniqueness with Single Boundary Measurement and zero Initial Data)*

Does the single lateral data g, h,with a non-zero g (e.g.,$g(x,t) = t$), uniquely determine $c = c(x)$ in (0.1) ($a_0 = 1, a = 1, b = 0$) when the intial data is zero?

Problem 3. *(Global Uniqueness for variable principal part)*

Does the lateral Dirichlet-to-Neumann map uniquely determine $a_0 = a_0(x,t)$?

Problem 4. *(Restricted data for uniqueness of a domain)*
In the situation of Theorem 4.1 is it sufficient to know only $\partial_\nu u$ on $\partial\Omega \times \{T\}$ to uniquely determine D in the coefficient $a = 1 + \chi(D)$?

Problem 5. .
Does the lateral Dirichlet-to-Neumann map for the equation $\partial_t u - \Delta u + c(x, u) = 0$ uniquely determine $c(x, u)$ when T is finite?

These problem are generally interesting in the manydimensional case, but in some situations they are challenging even for dimension one.

ACKNOWLEDGEMENT

This research was partly supported by the NSF under the grant DMS-9101421 and CNR. The author is quite grateful to the referee of this paper for useful remarks and observing several misprints.

REFERENCES

[A] Alessandrini, G., *Determining conductivity by boundary measurements,the stability issue*, pp.317-324 In the book: Applied and Industrial Mathematics, R. Spiegler,Ed.,Kluwer,1991

[B] Belishev,M.I., *Wave bases in multidimensional inverse problems*, Mat.Sb. ,**180**(1989),584-602.

[C-Y] Cannon,J.,Yin,H., *A uniqueness result for a class of nonlinear parabolic inverse problems*,Inverse Problems,**4**(1988),411-416.

[D] Dinh Nho Hao, *A non-characteristic Cauchy Problem for linear parabolic equations and related inverse problems:I.Solvability*, Inverse Problems,**10**,(1994), 295-315.

[E-Is] Elayyan,A.,Isakov,V., *On Uniqueness of Recovery of the Discontinuous Conductivity Coefficient of a Parabolic Equation*, Preprint, 1994

[E-S-Y] Engl,H.W.,Scherzer,O.,Yamamoto,M., *Uniqueness of Forcing Terms in Linear Partial Differential Equations with Overspecified Boundary Data*, Inverse Problems (to appear)

[E-T-F] Ewing,R.,Tao Lin,Falk,R., *Inverse and ill-posed problems in reservoir simulation*, pp.483-492. In the book:Inverse and Ill-posed Problems, New York, Academic Press, 1990.

[F] Friedman,A., *Partial differential equations of parabolic type*, Prentice Hall, N.J.,1964.

[F-Is] Friedman,A.,Isakov,V., *On the uniqueness in the inverse conductivity problem with one measurement*, Indiana Univ.Math.J,**38**(1989), 553-580

[H-Y] Hoffman,K.H.,Yamamoto,M.,*Generic Uniqueness and Stability in Some Inverse Parabolic Problems* ,Proc. of the Lapland Conference on Inverse Problems, Eds. L.Päivärinta and E.Somersalo, Lecture Notes in Physics, Vol.422, Springer-Verlag, 1993.

[I] *Inverse Problems in PDE*, Eds.Colton,D.,Ewing,R.,Rundell,W. SIAM Proc. Series, Philadelphia,1990.

[Is1] Isakov,V., *Uniqueness for inverse parabolic problems with a lateral overdetermination* , Comm.Part.Diff.Equat.,**14**(1989),681-689.

[Is2] Isakov,V., *Inverse Source Problems* ,Math.Surveys and Monographs, Vol.**34**, AMS,Providence, R.I.,1990

[Is3] Isakov,V., *Inverse Parabolic Problems with Final Overdetermination*, Comm. Pure Appl.Math,**54**(1991),185-209.

[Is4] Isakov,V., *Completeness of products of solutions and some inverse problems for PDE* , J.Diff.Equat.,**92**(1991),305-317.

[Is5] Isakov,V., *On uniqueness in inverse problems for semilinear parabolic equations*, Arch.Rat.Mech.Anal., **124**(1993),1-13.

[Is6] Isakov,V., *Uniqueness and Stability in Multidimensional Inverse Problems* ,Inverse Problems,**9**(1993),579-621.

[Is7] Isakov,V., *On Uniqueness of Discontinuity Surface of the Speed of Propagation* ,preprint, 1994.

[Is8] Isakov,V., *Inverse Problems in PDE* , the book (in preparation).

[Is-N] Isakov,V.,Nachman,A., *A semiglobal uniqueness theorem for a two- dimensional nonlinear boundary value problem*, preprint, 1994.

[Is-S] Isakov,V.,Sylvester,J., *Global Uniqueness for a semilinear Elliptic Inverse Problem* , Comm.Pure Appl.Math.,**57**(1994),

[K] Klibanov,M.V., *Inverse Problems and Carleman Estimates*, Inverse Problems, **8**(1992),575-596.

[L-S-U] Ladyzenskaja,O.A.,Solonnikov,V.A.,Ural'ceva,N.N.,*Linear and quasilinear equations of parabolic type*, Transl. of Math.Monographs **23**,AMS,Providence, R.I.,1968.

[L-P] Lorenzi,A.,Paparoni,E., *An identification problem for a semilinear parabolic equation*, Ann.Mat.Pura Appl.,(4)**151**(1988),263-287.

[L-R] Lowe,B.,Rundell,W., *An Inverse Problem for a Sturm-Liouville Operator*, J.Math.Anal.Appl.,**181**(1994),188-200.

[N] Nachman,A.,*Global Uniqueness for a two-dimensional Inverse Boundary Value Problem*, Ann.Math. (in print).

[N-S-U] Nakamura,G.,Sun,Z.,Uhlmann,G., *Global Identifiability for an Inverse Problem for the Schrödinger Equation in a Magnetic Field*, preprint,1994.

[P-R] Pilant,M.,Rundell,W., *An inverse problem for a non-linear parabolic equation*, Comm.Part.Diff.Equat.,**11**(1986),445-457.

[P-K] Prilepko,A.I.,Kostin,A.B.,*Some inverse problems for parabolic equations with final and integral observations*, Mat.Sb., **183**(1992),49-68.

[P-S] Prilepko,A.I.,Soloviev,V.V., *Solvability theorems and the Rothe method in inverse problems for an equation of parabolic type*, II,Diff.Equat.,**23**(1987),1341-1349.

[S-E-A] Scherzer,O.,Engl,H.,Anderssen,R.S., it Parameter identification from boundary measurements in a parabolic equation arising from geophysics, Nonlin. Anal., **20** (1993),127-157.

[S] Sun,Z., *An inverse boundary value problem for Schrödinger operators with vector potentials*, Trans.AMS,**338**(1993),953-971.

[S-U] Sylvester,J.,Uhlmann,G., *Global Uniqueness Theorem for an Inverse Boundary Value Problem*, Ann.Math.,**41**(1987),153-169.

[U] Uhlmann,G., *Inverse boundary problems and applications, Methodes semi-classiques*,Vol.1(Nantes,91), Asterisque,**207**(1992),153-211.

[T] Tataru,D., *Unique continuation for solutions to PDE's:between Hörmander's theorem and Holmgren's theorem*, Comm.Part.Diff.Equat. (in print).

[Y] Yeh,W.W.-G., *Review of parameter identification procedures in groundwater hydrology.The inverse problem*, Water Res.Review,**22**(1986),95-108.

[Yo] Yosida,K., *Functional Analysis* ,Springer-Verlag,New York,1965.

WICHITA, KS 67260-0033, U.S.A., ISAKOV@ TWSUVM.UC.TWSU.EDU

Direct and inverse integrodifferential Maxwell problems for dispersive media related to cylindrical domains *

A. Lorenzi†

Abstract

As is well known, the propagation of electromagnetic waves in dispersive media is governed by integrodifferential equations. We consider here a linear model and assume that the medium is a *cylindrical* stratified rigid body. In this case all the physical characteristics, such as the dielectric coefficient, the magnetic permeability and the conductivity coefficient as well as the kernels accounting for memory effects, depend on one spatial variable, only.

This paper is devoted to solving both the *direct* problem, consisting in determining the electromagnetic field, when the physical characteristics are assumed to be known, and the *inverse* problem, consisting in determining, in addition to the electromagnetic field, also the physical characteristics, when such quantities are themselves unknown, by the means of additional measurements.

In both cases existence, uniqueness and continuous dependence results *in the large* are proved in the context of suitable functional spaces.

1 Introduction

This paper is concerned with the behaviour of some dielectric materials related to rigid bodies involving dispersive phenomena. As is well known, the corresponding constitutive equations contain memory terms (cf., e.g., [16], [9]) and are characterized by time convolution kernels which describe the main features of the medium. Since such kernels are unknown or scarcely known, their identification turns out to be of great importance for applications.

We note that in the last years the interest in wave propagation in complex media has been growing and, consequently, different approaches (cf. e.g. [7], [8], [11], [15], [17], [18], [20], [23]) have been developed to solve the related inverse problems.

In this paper we use some methods introduced in [7] and [3] and generalize them to the case of a nonhomogeneous medium related to a *generic* cylinder of the $\mathbf{R}^3$–space.

We now proceed first to derive the integrodifferential system of Maxwell equations from the constitutive relations and Maxwell laws and, then, to couple this system with some usual initial and boundary conditions and a set of additional information chosen according to the physical context. More explicitly, we assume that the medium under consideration occupies the cylinder $\Omega = (0, \ell) \times \omega$, where ω denotes either a bounded C^2–domain in $\mathbf{R}^2$ or

* Work partially supported by the Italian Ministero dell'Università e della Ricerca Scientifica e Tecnologica.

† Dipartimento di Matematica, Università degli Studi, via Saldini 50, 20133 Milan, Italy. Member of the National Group for Functional Analysis and Applications (G.N.A.F.A.) of the Italian Research National Council (C.N.R.).

the entire space $\mathbf{R}^2$, for any time $t \in [0,T]$, $T > 0$ being given.

Denoting by E, D, H, B, J the electric field, electric induction, magnetic field, magnetic induction and electric current vectors, respectively, we assume that the evolution of the electromagnetic field in our medium is described by the system of Maxwell equations

$$\text{rot}\, H(t,x) = \frac{\partial D}{\partial t}(t,x) + J(t,x) - f(t,x)Y_+(t) \qquad (t,x) \in \tilde{Q}_T \tag{1.1}$$

$$\text{rot}\, E(t,x) = -\frac{\partial B}{\partial t}(t,x) \qquad (t,x) \in \tilde{Q}_T \tag{1.2}$$

where $\tilde{Q}_T := (-\infty, T) \times \Omega$, Y_+ is the Heaviside function and $fY_+ : \tilde{Q}_T \to \mathbf{R}^3$ is a given electric current source vanishing in $(-\infty, 0) \times \Omega$.

Suppose that our medium consists of a dielectric material in which dispersion or absorption of electromagnetic waves is observed. Moreover, suppose that the medium is nonhomogeneous or, more precisely, that its electromagnetic characteristics change along one spatial direction.

Under these assumptions the electromagnetic wave propagation can be described, in the linear case, by the constitutive relations of hereditary type (cf., e.g., [4], [9])

$$D(t,x) = \varepsilon_0(x_1)E(t,x) + \int_0^{+\infty} \varepsilon(s,x_1)E(t-s,x)\,ds \tag{1.3}$$

$$B(t,x) = \mu_0(x_1)H(t,x) + \int_0^{+\infty} \mu(s,x_1)H(t-s,x)\,ds \tag{1.4}$$

$$J(t,x) = \sigma_0(x_1)E(t,x) + \int_0^{+\infty} \sigma(s,x_1)E(t-s,x)\,ds \tag{1.5}$$

where $\varepsilon_0, \sigma_0, \mu_0$ are the dielectric coefficient, the magnetic permeability and the conductivity coefficient, respectively. The functions $\varepsilon, \sigma, \mu : [0,+\infty) \times (0,\ell) \to \mathbf{R}$, accounting for the memory effects, are named relaxation kernels or susceptibility kernels. Here and in the sequel x_i $(i = 1,2,3)$ denotes the i-th coordinate in $\mathbf{R}^3$.

Assume that

$$\varepsilon_0(x_1) > 0\,,\ \mu_0(x_1) > 0\,,\ \sigma_0(x_1) \geq 0 \tag{1.6}$$

are prescribed smooth functions.

Moreover, assume (cf., e.g., [2], [6]), that the past histories of the electric and magnetic fields E and H are known and satisfy

$$E(t,x) = 0 \qquad (t,x) \in (-\infty, 0] \times \Omega \tag{1.7}$$

$$H(t,x) = 0 \qquad (t,x) \in (-\infty, 0] \times \Omega \tag{1.8}$$

Then, combining Maxwell equations (1.1)–(1.2) with the constitutive relations (1.3)–(1.5), and taking into account the assumptions (1.7)–(1.8), the evolution of the pair (E,H) is governed by the system of integrodifferential equations

$$\begin{aligned}\text{rot}\, H(t,x) = {} & \varepsilon_0(x_1)\frac{\partial E}{\partial t}(t,x) + \int_0^t \varepsilon(s,x_1)\frac{\partial E}{\partial t}(t-s,x)\,ds + \sigma_0(x_1)E(t,x) \\ & + \int_0^t \sigma(s,x_1)E(t-s,x)\,ds - f(t,x) \quad (t,x) \in Q_T := (0,T) \times \Omega\end{aligned} \tag{1.9}$$

$$\text{rot}\, E(t,x) = -\mu_0(x_1)\frac{\partial H}{\partial t}(t,x) - \int_0^t \mu(s,x_1)\frac{\partial H}{\partial t}(t-s,x)\,ds \quad (t,x) \in Q_T \tag{1.10}$$

As initial conditions we impose (cf. (1.7)–(1.8))

$$E(0,x) = 0 \qquad x \in \Omega \tag{1.11}$$

$$H(0,x) = 0 \qquad x \in \Omega \tag{1.12}$$

As far as the boundary conditions are concerned, we recall that they are given by

$$n(x) \times D(t,x) = 0 \qquad \forall (t,x) \in (0,T) \times \Omega \tag{1.13}$$

In terms of E condition (1.13) can be rewritten in the integral form

$$\varepsilon_0(x_1) n(x) \times E(t,x) + \int_0^t \varepsilon(s,x_1) n(x) \times E(t-s,x)\, ds = 0 \qquad \forall (t,x) \in (0,T) \times \Omega \tag{1.14}$$

Hence, from the bound $\varepsilon_0(x_1) \geq m > 0\ \forall x_1 \in [0,\ell]$ (due to (1.6) and the smoothness of ε_0) and the assumption $\varepsilon(\cdot, x_1) \in L^1((0,T))$ for a. e. $x_1 \in (0,\ell)$, we deduce that E must satisfy the boundary condition

$$n(x) \times E(t,x) = 0 \qquad \forall (t,x) \in (0,T) \times \Omega \tag{1.15}$$

Observe then that the combination of initial conditions with the constitutive relations imply the validity of another Maxwell law. For, from (1.4), (1.8) and (1.12) we deduce

$$B(0,x) = 0 \qquad x \in \Omega \tag{1.16}$$

Then, considering the chain of identities, where div denotes the *divergence* operator,

$$0 = \operatorname{div} \operatorname{rot} E(t,x) = -\operatorname{div} \frac{\partial B}{\partial t}(t,x) = -\frac{\partial \operatorname{div} B}{\partial t}(t,x) \qquad \text{in } Q_T \tag{1.17}$$

we obtain

$$\operatorname{div} B(t,x) = c(x) \qquad (t,x) \in Q_T \tag{1.18}$$

Finally, by (1.16), we get

$$\operatorname{div} B(t,x) = 0 \qquad (t,x) \in Q_T \tag{1.19}$$

The so-called *direct* problem, consisting in finding the pair (E,H) satisfying the equations (1.9)–(1.12), (1.15), $\varepsilon_0, \mu_0, \sigma_0, \varepsilon, \mu, \sigma, f, E_0, H_0$ being prescribed functions, is *well-posed.* Indeed, it can be solved, for instance, applying the technique used in section 3. Consequently, we can determine the vector fields E H, provided that the electromagnetic properties of the medium are known.

If this is not the case, we need some supplementary information in addition to the initial conditions (1.11–12). In order to determine the correct set of additional information we need to recover the kernels ϵ, μ, and σ it is necessary to transform equation (1.9), by an integration by parts, into the following equivalent formulation:

$$\begin{aligned} \operatorname{rot} H(t,x) &= \varepsilon_0(x_1)\frac{\partial E}{\partial t}(t,x) + \sigma_0(x_1)E(t,x) \\ &+ \int_0^t \nu(s,x_1)\frac{\partial E}{\partial t}(t-s,x)\, ds - f(t,x) \qquad (t,x) \in Q_T\,, \end{aligned} \tag{1.20}$$

where, for any $(t, x_1) \in [0, T] \times \mathbf{R}$, we have set

$$\nu(t, x_1) := \varepsilon(t, x_1) + \sigma_1(t, x_1) \qquad \sigma_1(t, x_1) := \int_0^t \sigma(s, x_1)\, ds \tag{1.21}$$

Assume now that we are given some additional information involving E and H to recover the three functions ε, μ, σ. Equation (1.20) shows that we cannot fully identify the triplet $(\varepsilon, \mu, \sigma)$. In other words, the structure of Maxwell equations allows to determine at most the pair $(\nu, \mu) = (\varepsilon + \sigma_1, \mu)$, whatever our additional information concerning the electromagnetic field (E, H) may be. This means that there exist *different conducting* dielectrics governed by the *same* set of Maxwell equations. According to this fact, we only need two additional information of scalar type depending on the variables t and x_1.

We now observe that from the physical viewpoint it would be reasonable to measure, for example, the following vector

$$\begin{aligned}(1.22) \qquad & (\Phi_1(E_i(t, x_1, \cdot, \cdot)), \Phi_2(H_j(t, x_1, \cdot, \cdot))) \\ &= \Big(\int_\omega E_i(t, x_1, x_2, x_3)\eta_1(x_2, x_3)\, dx_2 dx_3, \int_\omega H_j(t, x_1, x_2, x_3)\eta_2(x_2, x_3)\, dx_2 dx_3\Big)\end{aligned}$$

where $i, j \in \{1, 2, 3\}$ are fixed indices, v_i denotes the i-th component of vector v and $\eta_1, \eta_2 \in L^2(\omega; \mathbf{R})$.

We can now state our identification problem

Problem (I. P.). *Let ε_0, σ_0, μ_0, f, E_0, H_0, ψ_1, ψ_2 be prescribed smooth functions. Find two vector functions E, $H : Q_T \to \mathbf{R}^3$ and two scalar functions ν, $\mu : [0, T] \times \mathbf{R} \to \mathbf{R}$ satisfying the equations*

$$\begin{aligned}(1.23) \qquad & \varepsilon_0(x_1)\frac{\partial E}{\partial t}(t, x) - \operatorname{rot} H(t, x) + \sigma_0(x_1)E(t, x) \\ & + \int_0^t \nu(s, x_1)\frac{\partial E}{\partial t}(t - s, x)\, ds = f(t, x) \qquad (t, x) \in Q_T\end{aligned}$$

$$\mu_0(x_1)\frac{\partial H}{\partial t}(t, x) + \operatorname{rot} E(t, x) + \int_0^t \mu(s, x_1)\frac{\partial H}{\partial t}(t - s, x)\, ds = 0 \quad (t, x) \in Q_T \tag{1.24}$$

$$E(0, x) = 0 \qquad x \in \Omega \tag{1.25}$$

$$H(0, x) = 0 \qquad x \in \Omega \tag{1.26}$$

$$n(x) \times E(t, x) = 0 \quad (t, x) \in (0, T) \times \Omega \tag{1.27}$$

and the additional information

$$\Phi_1(E_i(t, x_1, \cdot, \cdot)) = \psi_1(t, x_1) \qquad (t, x_1) \in (0, T) \times (0, \ell) \tag{1.28}$$

$$\Phi_2(H_j(t, x_1, \cdot, \cdot)) = \psi_2(t, x_1) \qquad (t, x_1) \in (0, T) \times (0, \ell) \tag{1.29}$$

Φ_1 and Φ_2 being two given linear functionals continuous in $L^2(\omega; \mathbf{R})$.

Remark 1.1. In case of a non conducting medium, characterized by $\sigma_0 = 0$ and $\sigma = 0$, we can explicitly determine the remaining kernels ε and μ. For this purpose we observe that the additional requirement $\operatorname{div} E = 0$ in Q_T can be satisfied if we assume that f saitisfies the *same* equation in Q_T.

We conclude this section by observing that the direct problem (1.9)–(1.12), (1.15) can be given also the following equivalent formulation (of use in the sequel), which requires the differentiability of the kernels ε and μ with respect to t and makes use of an integration by parts and the following notation:

$$m_1(t,x_1) := \frac{\partial \varepsilon}{\partial t}(t,x_1) + \sigma(t,x_1), \qquad \lambda_1(x_1) := \sigma(x_1) + \varepsilon(0,x_1) \tag{1.30}$$

$$m_2(t,x_1) := \frac{\partial \mu}{\partial t}(t,x_1), \qquad \lambda_2(x_1) := \mu(0,x_1) \tag{1.31}$$

Problem (D.P.). *Let ε_0, μ_0, λ_j, m_j, $(j = 1,2)$, f, E_0, H_0 be prescribed smooth functions. Find two vector functions E, $H : Q_T \to \mathbf{R}^3$ satisfying the equations*

$$\begin{aligned} \varepsilon_0(x_1)\frac{\partial E}{\partial t}(t,x) - \operatorname{rot} H(t,x) + \lambda_1(x_1)E(t,x) \\ + \int_0^t m_1(s,x_1)E(t-s,x)\,ds = f(t,x) \qquad (t,x) \in Q_T \end{aligned} \tag{1.32}$$

$$\begin{aligned} \mu_0(x_1)\frac{\partial H}{\partial t}(t,x) + \operatorname{rot} E(t,x) + \lambda_2(x_1)H(t,x) \\ + \int_0^t m_2(s,x_1)H(t-s,x)\,ds = 0 \qquad (t,x) \in Q_T \end{aligned} \tag{1.33}$$

$$E(0,x) = 0 \qquad x \in \Omega \tag{1.34}$$

$$H(0,x) = 0 \qquad x \in \Omega \tag{1.35}$$

$$n(x) \times E(t,x) = 0 \quad (t,x) \in (0,T) \times \Omega \tag{1.36}$$

We conclude this introduction by noting that the existence, uniqueness and continuous dependence results related to problems **D. P.** and **I. P.** are reported in section 4, theorem 4.2 and section 6, theorem 6.1.

Finally, the plan of the paper is the following: section 2 is devoted to the analysis of functional spaces related to our problems; section 3 gives some basic results for the usual Maxwell equations; section 4 is devoted to the analysis of the solvability of problem **D. P.**, while sections 5 and 6 deals with the analogous question concerning problem **I. P.**.

2 Functional spaces related to Maxwell equations

This section is devoted to generalizing some well–known results concerning functional spaces related to smooth domains (cf. [5], [21]) to the case of C^2–cylindrical domains belonging to the class $\mathcal{N}^{0,1}$ introduced in [15, chpt. 3].
We now give the exact definition of a C^2–cylinder.

DEFINITION 2.1. A cylinder $\Omega = (0,\ell) \times \omega \in \mathcal{N}^{0,1}$, where ω is a $\mathcal{N}^{2,0}$–domain in $\mathbf{R}^2$ (briefly a C^2–bounded domain in $\mathbf{R}^2$), is a is called a bounded C^2–cylinder in $\mathbf{R}^3$.

We can now state some results concerning the following Hilbert spaces endowed with the graph–norms of rot or div :

$$H(\operatorname{rot};\Omega) = \{u \in L^2(\Omega;\mathbf{R}^3) : \operatorname{rot} u \in L^2(\Omega;\mathbf{R}^3)\} \tag{2.1}$$

$$(2.2) \qquad H(\mathrm{div}\,;\Omega) = \{u \in L^2(\Omega;\mathbf{R}^3) : \mathrm{div}\, u \in L^2(\Omega;\mathbf{R})\}$$

$$(2.3) \qquad H_0(\mathrm{rot}\,;\Omega) = \{u \in L^2(\Omega;\mathbf{R}^3) : \mathrm{rot}\, u \in L^2(\Omega;\mathbf{R}^3),\ n \times u = 0 \text{ a.e. in } \partial\Omega\}$$

$$(2.4) \qquad H_0(\mathrm{div}\,;\Omega) = \{u \in L^2(\Omega;\mathbf{R}^3) : \mathrm{div}\, u \in L^2(\Omega;\mathbf{R}),\ n \cdot u = 0 \text{ a.e. in } \partial\Omega\}$$

where n denotes the outward unit vector normal to $\partial\Omega$.

THEOREM 2.1. *Let Ω be a bounded $\mathcal{N}^{0,1}$-cylinder in $\mathbf{R}^3$ or the strip $(0,\ell) \times \mathbf{R}^2$. Then $C^\infty(\overline{\Omega};\mathbf{R}^3)$ is dense in $H(\mathrm{rot}\,;\Omega)$, $H(\mathrm{div}\,;\Omega)$ and $H(\mathrm{rot}\,;\Omega) \cap H(\mathrm{div}\,;\Omega)$.*

Proof. For the divergence case cf. [21, theorem 1.1], while for the rotor case cf. [19, lemma 2.2]. The remaining case can be dealt with in a similar way. □

THEOREM 2.2. *Let Ω be a bounded $\mathcal{N}^{0,1}$-cylinder in $\mathbf{R}^3$ or the strip $(0,\ell) \times \mathbf{R}^2$. Then $\mathcal{H}_0(\mathrm{rot}\,;\Omega) = \{u \in C^\infty(\overline{\Omega};\mathbf{R}^3) : n \times u|_{\partial\Omega} = 0\}$ is dense in $H_0(\mathrm{rot}\,;\Omega)$ and $\mathcal{H}_0(\mathrm{rot}\,;\Omega) \cap \mathcal{H}(\mathrm{div}\,;\Omega)$ is dense in $H_0(\mathrm{rot}\,;\Omega) \cap H(\mathrm{div}\,;\Omega)$, while $\mathcal{H}_0(\mathrm{div}\,;\Omega) = \{u \in C^\infty(\overline{\Omega};\mathbf{R}^3) : n \cdot u|_{\partial\Omega} = 0\}$ is dense in $H_0(\mathrm{div}\,;\Omega)$ and $\mathcal{H}(\mathrm{rot}\,;\Omega) \cap \mathcal{H}_0(\mathrm{div}\,;\Omega)$ is dense in $H(\mathrm{rot}\,;\Omega) \cap H_0(\mathrm{div}\,;\Omega)$.*

Proof. For the divergence case cf. [21, theorem 1.3], while for the rotor case adapt the proof of the quoted theorem 2.3. The remaining case can be dealt with in a similar way. □

THEOREM 2.3. *Let $\Omega = (0,\ell) \times \mathbf{R}^2$ or $\Omega = [0,\ell] \times \omega$, where $\omega \subset \mathbf{R}^2$ is a $\mathcal{N}^{0,1}$-open set in $\mathbf{R}^2$. Then the following properties hold true:*

i) the linear operator $u \to n \times u|_{\partial\Omega}$ defined on $C^\infty(\overline{\Omega};\mathbf{R}^3)$ can be continuously extended from $H(\mathrm{rot}\,;\Omega)$ to the fractional Sobolev space $H^{-1/2}(\partial\Omega;\mathbf{R}^3)$. Moreover, the bilinear map $\{u,v\} \to n \times u \cdot v$ from $C^\infty(\overline{\Omega};\mathbf{R}^3) \times C^\infty(\overline{\Omega};\mathbf{R}^3)$ into $C^{0,1}(\partial\Omega;\mathbf{R}^3)^*$ can be extended by continuity to a bounded bilinear map from $H(\mathrm{rot}\,;\Omega) \times H(\mathrm{rot}\,;\Omega)$ satisfying the formula

$$(2.5) \qquad < n \times u|_{\partial\Omega}, v|_{\partial\Omega} >= \int_\Omega [\mathrm{rot}\, u(x) \cdot v(x) - u(x) \cdot \mathrm{rot}\, v(x)]\, dx \qquad \forall u, v \in H(\mathrm{rot}\,;\Omega)$$

Here $< \cdot, \cdot >$ is defined as the pairing $< n \times u|_{\partial\Omega} \cdot v|_{\partial\Omega}, 1 >$ related to the spaces $C^{0,1}(\partial\Omega;\mathbf{R}^3)^*$ and $C^{0,1}(\partial\Omega;\mathbf{R}^3)$. In particular, $< \cdot, \cdot >$ takes on the usual meaning when $u \in H(\mathrm{rot}\,;\Omega)$ and $v \in H^1(\Omega;\mathbf{R}^3)$.

ii) the linear operator $u \to n \cdot u|_{\partial\Omega}$ defined on $C^\infty(\overline{\Omega};\mathbf{R}^3)$ can be continuously extended from $H(\mathrm{div}\,;\Omega)$ to the fractional Sobolev space $H^{-1/2}(\partial\Omega;\mathbf{R}^3)$ via the formula:

$$(2.6) \qquad < n \cdot u|_{\partial\Omega}, v|_{\partial\Omega} >= \int_\Omega [\mathrm{div}\, u(x) \cdot v(x) - u(x) \cdot Dv(x)]\, dx \qquad \forall v \in H^1(\Omega;\mathbf{R})$$

where Dv denotes the gradient of v.

Proof. i) The proof of the assertion is a consequence of theorem 1.1 and remark 1.2 in [19].
ii) We recall that under our assumption on Ω we get the following equations (cf. [15, chpt. 2]), where γ_0 denotes the trace operator on $\partial\Omega$:

$$(2.7) \qquad \gamma_0(H^1(\Omega;\mathbf{R}^3)) = H^{1/2}(\partial\Omega;\mathbf{R}^3), \quad H^1_0(\Omega;\mathbf{R}^3) = \{u \in H^1(\Omega;\mathbf{R}^3) : u|_{\partial\Omega} = 0\}$$

Consequently, according to theorem 1.1 and remark 1.3 in [21, chpt 1], we get the assertion. □

THEOREM 2.4. *Ω be a bounded C^2-cylinder in $\mathbf{R}^3$ or the strip $(0,\ell)\times\mathbf{R}^2$. Then the norms in the Hilbert spaces $H_0(\mathrm{rot}\,;\Omega)\cap H(\mathrm{div}\,;\Omega)$ and $H^1_* = \{u \in H^1(\Omega;\mathbf{R}^3) : n\times u|_{\partial\Omega} = 0\}$ are equivalent.*

THEOREM 2.5. *Let Ω be a bounded C^2-cylinder in $\mathbf{R}^3$ or the strip $(0,\ell)\times\mathbf{R}^2$. Then the norms in the Hilbert spaces $H(\mathrm{rot}\,;\Omega)\cap H_0(\mathrm{div}\,;\Omega)$ and $H^1_* = \{u \in H^1(\Omega;\mathbf{R}^3) : n\cdot u|_{\partial\Omega} = 0\}$ are equivalent.*

Before proving theorem 2.4 and 2.5 we introduce the following definition.

DEFINITION 2.2. A subset $E \subset \partial\Omega$ is ν–measurable if and only if the sets $E_1 = E\cap(\{0,\ell\}\times\omega)$ and $E_2 = E\cap[(0,\ell)\times\partial\omega]$ are measurable with respect to the positive measures $\delta_{\{0,\ell\}}\times m_2$ and $m_1\times\sigma$. Here $\delta_{\{0,\ell\}}$ and m_k denote the Dirac measure related to $\{0,\ell\}$ and the k–dimensional Lebesgue measure.
The measure $\nu(E)$ is defined by the following equation, where χ_F denotes the characteristic function of the set F and σ the Lebesgue measure induced on $\partial\omega$:

$$\nu(E) = \sum_{j=1}^{2}\int_\omega \chi_{E\cap(\{j\ell\}\times\omega)}(j\ell,x')\,dx' + \int_{(0,\ell)}dx_1\int_{\partial\omega}\chi_{E\cap[(0,\ell)\times\partial\omega]}(x_1,x')\,d\sigma(x') \tag{2.8}$$

Remark 2.1. We notice that the second integral in (2.13) vanishes, when $\partial\omega = \emptyset$.

LEMMA 2.1. *Let Ω be a bounded $\mathcal{N}^{1,1}$-cylinder in $\mathbf{R}^3$ or the strip $(0,\ell)\times\mathbf{R}^2$. Then the following formulas hold true for any $u \in C^1(\overline{\Omega};\mathbf{R}^3)$:*

$$\int_\Omega D_j u(x)\,dx = \int_{\partial\Omega} n_j u\,d\nu \qquad j = 1,2,3 \tag{2.9}$$

where n denotes the outward unit vector normal to $\partial\Omega$.

Proof. The assertion easily follows from definition 1.1 and the following equations:

$$\int_\Omega D_1 u(x)\,dx = \int_\omega dx'\int_{(0,\ell)} D_1 u(x_1,x')\,dx_1 \tag{2.10}$$

$$= \int_\omega [u(\ell,x') - u(0,x')]\,dx' = \sum_{j=0}^{1}\int_\omega n_1(j\ell,x')u(j\ell,x')\,dx' = \int_{\partial\Omega} n_1 u\,d\nu$$

$$\int_\Omega D_j u(x)\,dx = \int_{(0,\ell)} dx_1\int_\omega D_j u(x_1,x')\,dx' \tag{2.11}$$

$$= \int_{(0,\ell)} dx_1\int_{\partial\omega} n_j(x_1,x')u(x_1,x')\,d\sigma(x') = \int_{\partial\Omega} n_j u\,d\nu \qquad j = 2,3$$

LEMMA 2.2. *Let Ω be a bounded $\mathcal{N}^{0,1}$-cylinder in $\mathbf{R}^3$ or the strip $(0,\ell)\times\mathbf{R}^2$. Then the following identities hold true for any $u \in C^2(\overline{\Omega};\mathbf{R}^3)$:*

$$\int_\Omega |Du(x)|^2\,dx = \int_\Omega |\mathrm{rot}\,u(x)|^2\,dx + \int_\Omega |\mathrm{div}\,u(x)|^2\,dx \tag{2.12}$$

$$-\int_{\partial\Omega}\sum_{i,j=1}^{3}(n_iu_iD_ju_j-n_ju_iD_iu_j)\,d\nu$$

Proof. For any $u\in C^2(\overline{\Omega};\mathbf{R}^3)$ we consider the identity

$$\int_\Omega[\operatorname{div}u(x)]^2\,dx=\int_{\partial\Omega}(n\cdot u)\operatorname{div}u\,d\nu-\int_\Omega u(x)\cdot D\operatorname{div}u(x)\,dx \tag{2.13}$$

$$=\int_{\partial\Omega}\sum_{i,j=1}^{3}(n_iu_iD_ju_j-n_ju_iD_iu_j)\,d\nu+\int_\Omega\sum_{i,j=1}^{3}D_iu_j(x)D_ju_i(x)\,dx$$

$$=\int_{\partial\Omega}\sum_{i,j=1}^{3}(n_iu_iD_ju_j-u_in_jD_iu_j)\,d\nu+\int_\Omega\sum_{i=1}^{3}[D_iu_i(x)]^2\,dx$$

$$+2\int_\Omega\sum_{i=1}^{3}D_iu_{i+1}(x)D_{i+1}u_i(x)\,dx\qquad(u_4=u_1,\;D_4=D_1)$$

From (2.18) we derive the following equations

$$\int_\Omega[|\operatorname{rot}u(x)|^2+|\operatorname{div}u(x)|^2]\,dx-\int_{\partial\Omega}\sum_{i,j=1}^{3}(n_iu_iD_ju_j-n_ju_iD_iu_j)\,d\nu \tag{2.14}$$

$$=\sum_{i=1}^{3}[D_iu_{i+1}(x)-D_{i+1}u_i(x)]^2\,dx+\int_\Omega\sum_{i=1}^{3}[D_iu_i(x)]^2\,dx$$

$$+2\int_\Omega\sum_{i=1}^{3}D_iu_{i+1}(x)D_{i+1}u_i(x)\,dx=\int_\Omega\sum_{i,j=1}^{3}[D_iu_j(x)]^2\,dx$$

Proof of theorem 2.4. First we observe that for any $u\in\mathcal{H}_0(\mathrm{rot}\,;\Omega)$

$$\sum_{i,j=1}^{3}(n_iu_iD_ju_j-n_ju_iD_iu_j)=0\qquad\nu-\text{a.e. on }\{0,\ell\}\times\omega \tag{2.15}$$

since $u_2=u_3=0$ and $n_2=n_3=0$ on $\{0,\ell\}\times\omega$.
Hence, recalling that $n_1=0$ and $u_1=0$ on $(0,\ell)\times\partial\omega$, from lemma 2.2 we derive the equation

$$\int_\Omega|Du(x)|^2\,dx=\int_\Omega|\operatorname{rot}u(x)|^2\,dx+\int_\Omega|\operatorname{div}u(x)|^2\,dx-I(u) \tag{2.16}$$

where

$$I(u)=\int_{(0,\ell)\times\partial\omega}\sum_{i,j=1}^{3}(n_iu_iD_ju_j-n_ju_iD_iu_j)\,d\nu \tag{2.17}$$

Observe that

$$n\times u=0\text{ on }(0,\ell)\times\partial\omega\iff n_iu_j=n_ju_i\text{ on }(0,\ell)\times\partial\omega,\;i,j=1,2,3 \tag{2.18}$$

Assume now to have extended the coefficients n_2 and n_3 to the whole of $\overline{\Omega}$ so as to preserve the C^2-class (cf. lemma 3.1. in [12]). Hence we can deduce the chain of identities

$$\sum_{i,j=1}^{3}n_iu_iD_ju_j=\sum_{i,j=1}^{3}[u_iD_j(n_iu_j)-u_iu_jD_jn_i] \tag{2.19}$$

$$= \sum_{i,j=1}^{3} [u_i D_j(n_j u_i) - u_i u_j D_j n_i] = \sum_{i,j=1}^{3} [n_j u_i D_j u_i + u_i u_i D_j n_j - u_i u_j D_j n_i]$$

$$= \sum_{i,j=1}^{3} [n_j u_i D_i u_j + n_j u_i (D_j u_i - D_i u_j) + u_i u_i D_j n_j - u_i u_j D_j n_i]$$

$$= \sum_{i,j=1}^{3} [n_j u_i D_i u_j + u_i u_i D_j n_j - u_i u_j D_j n_i]$$

For, using (2.23) and interchanging indices, we get the following identities

$$\sum_{i,j=1}^{3} n_j u_i (D_j u_i - D_i u_j) = \sum_{i,j=1}^{3} n_i u_j (D_j u_i - D_i u_j) \tag{2.20}$$

$$= \sum_{i,j=1}^{3} n_j u_i (D_i u_j - D_j u_i) = - \sum_{i,j=1}^{3} n_j u_i (D_j u_i - D_i u_j)$$

which imply the vanishing of $\sum_{i,j=1}^{3} n_j u_i (D_j u_i - D_i u_j)$ on $(0,\ell) \times \partial\omega$.

Taking (2.22) and (2.24) into account, we obtain the estimates

$$|I(u)| = \left| \int_{(0,\ell)\times\partial\omega} \sum_{i,j=1}^{3} [u_i u_i D_j n_j - u_i u_j D_j n_i] \, d\nu \right| \tag{2.21}$$

$$\leq \int_{(0,\ell)\times\partial\omega} \sum_{i,j=1}^{3} |u_i u_i D_j n_j - u_i u_j D_j n_i| \, d\nu \leq C \|Dn\|_{C(\overline{\omega};\mathbf{R}^3)} \int_{(0,\ell)\times\partial\omega} |u|^2 \, d\nu$$

We use now the following compactness estimate, valid for any $\varepsilon \in (0,+\infty)$ and some positive constant $C_1(\varepsilon)$, depending also on ω :

$$\int_{(0,\ell)\times\partial\omega} |u|^2 \, d\nu \leq \varepsilon \int_\Omega |Du(x)|^2 \, dx + C_1(\varepsilon) \int_\Omega |u|^2 \, dx \tag{2.22}$$

Choosing $\varepsilon = (2C\|Dn\|_{C(\overline{\omega};\mathbf{R}^3)})^{-1}$, from (2.17), (2.20), (2.26), (2.27) we easily deduce the estimate

$$\int_\Omega [|Du(x)|^2 + |u(x)|^2] \, dx \tag{2.23}$$

$$\leq C_2 \int_\Omega [|\mathrm{rot}\, u(x)|^2 + |\mathrm{div}\, u(x)|^2 + |u(x)|^2] \, dx \quad \forall u \in \mathcal{H}_0(\mathrm{rot}\,;\Omega)$$

where C_2 depends on ω and $\|Dn\|_{C(\overline{\omega};\mathbf{R}^3)}$.

We now observe that the converse estimate

$$\int_\Omega [|\mathrm{rot}\, u(x)|^2 + |\mathrm{div}\, u(x)|^2 + |u(x)|^2] \, dx \tag{2.24}$$

$$\leq C_3 \int_\Omega [|Du(x)|^2 + |u(x)|^2] \, dx \quad \forall u \in \mathcal{H}_0(\mathrm{rot}\,;\Omega)$$

is trivial.

A density argument (which makes use of theorem 2.2) concludes the proof of the present theorem. □

Proof of theorem 2.5. First we observe that $n_2 = n_3 = 0$ on $\{0, \ell\} \times \omega$ and $u_1 = 0$ for any $u \in \mathcal{H}_0(\mathrm{rot}\,;\Omega) \cap \mathcal{H}(\mathrm{div}\,;\Omega)$. Such relations imply

$$\sum_{i,j=1}^{3} (n_i u_i D_j u_j - n_i u_j D_j u_i) = 0 \qquad \text{on } \{0, \ell\} \times \omega \tag{2.25}$$

Then, recalling that $n_1 = 0$ and $n_2 u_2 + n_3 u_3 = 0$ on $(0, \ell) \times \partial\omega$ imply $n_2 D_1 u_2 + n_3 D_1 u_3 = 0$ on $(0, \ell) \times \partial\omega$, we get

$$\sum_{i,j=1}^{3} (n_i u_i D_j u_j - n_j u_i D_i u_j) = - \sum_{i,j=2}^{3} n_j u_i D_i u_j \tag{2.26}$$

$$= - \sum_{i,j=2}^{3} u_i D_i (n_j u_j) + \sum_{i,j=2}^{3} u_i u_j D_i n_j = \sum_{i,j=2}^{3} u_i u_j D_i n_j \qquad \text{on } \{0, \ell\} \times \omega$$

For, operator $\sum_{i=2}^{3} u_i D_i$ is tangential on $\partial\omega$, since $n_2 u_2 + n_3 u_3 = 0$ on $\partial\omega$. This implies

$$\sum_{i,j=2}^{3} u_i D_i (n_j u_j) = \sum_{i=2}^{3} u_i D_i (n \cdot u) = 0 \qquad \text{on } \partial\omega \tag{2.27}$$

Then, reasoning as in the proof of theorem 2.4, we deduce that there exist two positive constants $C_1 < C_2$ (depending on ω and the C^1–norm of n) such that

$$C_1^2 \int_\Omega |Du(x)|^2\, dx \le \int_\Omega [|\mathrm{rot}\, u(x)|^2 + |\mathrm{div}\, u(x)|^2\, dx \tag{2.28}$$

$$\le \int_\Omega |Du(x)|^2\, dx \qquad \forall u \in \mathcal{H}_0(\mathrm{rot}\,;\Omega) \cap \mathcal{H}(\mathrm{div}\,;\Omega)$$

The density of $\mathcal{H}_0(\mathrm{rot}\,;\Omega) \cap \mathcal{H}(\mathrm{div}\,;\Omega)$ in $H^1_*(\Omega; \mathbf{R}^3)$ concludes our proof. □

3 The nonhomogeneous Maxwell system

Taking advantage of theorems 2.1 and 2.3 we can prove that the Hilbert spaces $H(\mathrm{rot}\,;\Omega)$ and $H_0(\mathrm{rot}\,;\Omega)$, defined in section 1, are separable. In fact, the vector spaces

$$Y = \{(u, \mathrm{rot}\, u) \in L^2(\Omega; \mathbf{R}^3) \times L^2(\Omega; \mathbf{R}^3)\} \tag{3.1}$$

$$Y_0 = \{(u, \mathrm{rot}\, u) \in L^2(\Omega; \mathbf{R}^3) \times L^2(\Omega; \mathbf{R}^3) : n \times u|_{\partial\Omega} = 0\} \tag{3.2}$$

are closed in the separable Hilbert space $L^2(\Omega; \mathbf{R}^6)$.

Likewise, we can show that the Hilbert spaces $H(\mathrm{div}\,;\Omega)$ and $H_0(\mathrm{div}\,;\Omega)$ are separable. Consequently, we can use the Faedo–Galerkin procedure to prove the existence of the solution to problem (1.32)–(1.36):

(P1) *find a pair of functions* (E, H) *such that*

$$E, H \in C^1([0, T]; L^2(\Omega; \mathbf{R}^3)) \cap C([0, T]; H(\mathrm{rot}\,;\Omega)) \tag{3.3}$$

$$l_1 D_t E - \mathrm{rot}\, H + \lambda_1 E = f, \qquad l_2 D_t H + \mathrm{rot}\, E + \lambda_2 H = g \tag{3.4}$$

$$E(0) = E_0, \qquad H(0) = H_0 \tag{3.5}$$

$$n(x) \times E(t,x) = 0 \quad \forall t \in [0,T], \ \textit{for a.e. } x \in \Omega \tag{3.6}$$

We assume that ε_0, μ_0, λ_1, λ_2 are scalar functions in $W^{1,\infty}(\Omega;\mathbf{R})$ and satisfy the following bounds for some positive constant γ :

$$l_j(x) \geq \gamma^{-1}, \qquad \textit{for a.e. } x \in \Omega, \ j = 1,2 \tag{3.7}$$

THEOREM 3.1. *Let Ω be any Lipschitz domain in $\mathbf{R}^3$. Assume further that the data satisfy the properties*

$$E_0 \in H_0(\mathrm{rot}\,;\Omega), \qquad H_0 \in H(\mathrm{rot}\,;\Omega), \qquad f,g \in W^{1,1}((0,T);L^2(\Omega;\mathbf{R}^3)) \tag{3.8}$$

Then problem (3.3)–(3.6) admits a unique solution (E,H) satisfying the estimate

$$\begin{aligned} &\left(\|D_tE(t)\|_0^2 + \|\mathrm{rot}\,E(t)\|_0^2 + \|E(t)\|_0^2 + \|D_tH(t)\|_0^2 + \|\mathrm{rot}\,H(t)\|_0^2 + \|H(t)\|_0^2\right)^{1/2} \\ &\leq C\left(\|\mathrm{rot}\,E_0\|_0^2 + \|E_0\|_0^2 + \|\mathrm{rot}\,H_0\|_0^2 + \|H_0\|_0^2 + \|f(0)\|_0^2 + \|g(0)\|_0^2\right)^{1/2} \\ &+C\int_0^t \left(\|D_sf(s)\|_0^2 + \|f(s)\|_0^2 + \|D_tg(s)\|_0^2 + \|g(s)\|_0^2\right)^{1/2} \qquad \textit{for a.e. } t \in (0,T) \end{aligned} \tag{3.9}$$

where

$$\|z\|_0 = \|z\|_{L^2(\Omega;\mathbf{R}^3)} \tag{3.10}$$

and C is a positive function depending upon γ and the norms of l_j, λ_j in $L^\infty(\Omega;\mathbf{R})$ $(j = 1,2)$.

Assume further that the data satisfy the properties

$$E_0 \in H_0(\mathrm{rot}\,;\Omega), \qquad H_0, \mathrm{rot}\,H_0, \mathrm{rot}\,E_0 \in H(\mathrm{rot}\,;\Omega) \tag{3.11}$$

$$f,g \in W^{2,1}((0,T);L^2(\Omega;\mathbf{R}^3)) \tag{3.12}$$

$$f'(0) - \mathrm{rot}\,g(0), \ g'(0) - \mathrm{rot}\,f(0) \in H(\mathrm{rot}\,;\Omega) \tag{3.13}$$

Then the strong solution (E,H) to problem (3.3)–(3.6) enjoys the property

$$E,H \in C^2([0,T];L^2(\Omega;\mathbf{R}^3)) \cap C^1([0,T];H(\mathrm{rot}\,;\Omega)) \tag{3.14}$$

and satisfies the estimate

$$\begin{aligned} &\big(\|D_t^2E(t)\|_0^2 + \|\mathrm{rot}\,D_tE(t)\|_0^2 + \|D_tE(t)\|_0^2 + \|D_t^2H(t)\|_0^2 + \|\mathrm{rot}\,D_tH(t)\|_0^2 \\ &+\|D_tH(t)\|_0^2\big)^{1/2} \leq C\Bigg(\sum_{j=0}^{2} [\|\mathrm{rot}^j E_0\|_0^2 + \|\mathrm{rot}^j H_0\|_0^2] \\ &+\|f(0)\|_0^2 + \|g(0)\|_0^2 + \|f'(0) + \mathrm{rot}\,g(0)\|_0^2 + \|g'(0) - \mathrm{rot}\,f(0)\|_0^2\Bigg)^{1/2} \\ &+C\int_0^t \left(\|D_t^2f(s)\|_0^2 + \|D_tf(s)\|_0^2 + \|D_t^2g(s)\|_0^2 + \|D_tg(s)\|_0^2\right)^{1/2} \end{aligned} \tag{3.15}$$

and C is a positive function depending upon γ and the norms of l_j, λ_j in $L^\infty(\Omega;\mathbf{R})$ $(j=1,2)$.

In order to solve our specific problem (3.3)–(3.6) we give it the following abstract formulation in a Hilbert space H : *determine two functions $v:[0,T]\to H$ and $w:[0,T]\to H$ such that*

$$L_1 v'(t) + A_1 w(t) + \Lambda_1 v(t) = f(t) \qquad \textit{for a.e. } t \in (0,T) \tag{3.16}$$

$$L_2 w'(t) + A_2 v(t) + \Lambda_2 w(t) = g(t) \qquad \textit{for a.e. } t \in (0,T) \tag{3.17}$$

$$v(0) = v_0, \qquad w(0) = w_0 \tag{3.18}$$

where

$$f, g \in L^1((0,T);H), \qquad v_0, w_0 \in H \tag{3.19}$$

As far as operators A_1, A_2, L_1 and L_2 are concerned, we assume the following:

H1 $W_j \hookrightarrow H \hookrightarrow W_j^*$ is a Hilbert triplet $(j=1,2)$, where W_j^* denotes the dual space to W_j;

H2 $W_2 \hookrightarrow W_1$;

H3 W_j $(j=1,2)$ are separable Hilbert spaces which are dense in H;

H4 $A_1 \in \mathcal{L}(W_1;H)$, $A_2 \in \mathcal{L}(W_2;H)$ and

$$(w, A_2 v) = -(A_1 w, v) + (\tilde{A} w, v) \qquad \forall v \in W_2, w \in W_1 \tag{3.20}$$

where $\tilde{A} \in \mathcal{L}(H)$;

H5 $(\psi, A_2\phi) = (f,\phi)$ $\forall \phi \in W_2$ and some $f \in H$ implies $\psi \in \mathcal{D}(A_1)$ and $A_1\psi = f$;

H6 $(\phi, A_1\psi) = (f,\psi)$ $\forall \psi \in W_1$ and some $f \in H$ implies $\phi \in \mathcal{D}(A_2)$ and $A_2\phi = f$;

H7 $\Lambda_j \in \mathcal{L}(H) \cap \mathcal{L}(W_{j+1})$ $(j=1,2;\ W_3 = W_1)$;

H8 $L_j \in \mathcal{L}(H)$ $(j=1,2)$ are positive definite and self–adjoint operators satisfying

$$(L_j v, v) \geq \gamma^{-1}\|v\|^2 \qquad \forall v \in H,\ j=1,2 \tag{3.21}$$

for some positive constant γ.

Remark 3.1. Conditions *H1–H8* are satisfied if we choose

$$H = L^2(\Omega;\mathbf{R}^3), \quad W_1 = H(\mathrm{rot}\,;\Omega), \quad W_2 = H_0(\mathrm{rot}\,;\Omega) \tag{3.22}$$

$$A_1 = -\mathrm{rot}\,, \quad A_2 = \mathrm{rot}\,, \quad \tilde{A} = 0 \tag{3.23}$$

$$L_j = l_j I, \quad \Lambda_j = \lambda_j I \quad j=1,2 \tag{3.24}$$

and take advantage of the inclusion $C_0^\infty(\Omega;\mathbf{R}^3) \subset W_1 \cap W_2$.

We note that *H4* is a consequence of property *i)* in theorem 2.3.

According to [5] we recall that by a *weak solution* to problem (3.16)–(3.18) we mean a pair $(v,w) \in L^\infty((0,T);H) \times L^\infty((0,T);H)$ satisfying the equation

$$\int_0^T [-(L_1 v(t), \phi'(t)) - (L_2 w(t), \psi'(t)) + (w(t), A_1\phi(t)) + (v(t), A_2\psi(t))]\,dt \tag{3.25}$$

$$+\int_0^T [(\Lambda_1 v(t), \phi(t)) + (\Lambda_2 w(t), \psi(t))]\,ds = \int_0^T [(f(t), \phi(t)) + (g(t), \psi(t))]\,dt$$

$$+(L_1 v_0, \phi(0)) + (L_2 w_0, \psi(0)) \qquad \forall(\phi,\psi) \in \Phi \times \Psi$$

where $(\cdot,\cdot)$ denotes the scalar product in H and the Banach spaces Φ and Ψ are defined by

$$\Phi = W^{1,1}((0,T);H) \cap L^1((0,T);W_2), \qquad \Psi = W^{1,1}((0,T);H) \cap L^1((0,T);W_1) \tag{3.26}$$

Conversely, by a *strong solution* we mean a pair (v,w) such that

$$v \in W^{1,\infty}((0,T);H) \cap L^\infty((0,T);W_1) \quad w \in W^{1,\infty}((0,T);H) \cap L^\infty((0,T);W_2) \tag{3.27}$$

satisfying (3.16)–(3.18).

In order to solve problem (3.16)–(3.18) under assumption (3.19) we follow the Faedo–Galerkin method. For this purpose we denote by $\{\overline{v}_n\}_{n=1}^{+\infty}$ and $\{\overline{w}_n\}_{n=1}^{+\infty}$ two bases for the separable spaces W_2 and W_1, respectively. Since $v_0, w_0 \in H$, from assumption *H3* we easily deduce that there exist two finite sequences of real numbers $\{\alpha_{0,m,i}\}_{1\le i\le m}$ and $\{\beta_{0,m,i}\}_{1\le i\le m}$ such that

$$\sum_{i=1}^m \alpha_{0,m,i}\overline{v}_i \to v_0 \qquad \sum_{i=1}^m \beta_{0,m,i}\overline{w}_i \to w_0 \quad \text{in } H \qquad \text{as } m \to +\infty \tag{3.28}$$

Then we introduce the functions

$$v_m(t) = \sum_{i=1}^m \alpha_{m,i}(t)\overline{v}_i \qquad w_m(t) = \sum_{i=1}^m \beta_{m,i}(t)\overline{w}_i \tag{3.29}$$

whose scalar coefficients are solutions to the Cauchy problem

$$\sum_{i=1}^m \alpha'_{m,i}(t)(L_1\overline{v}_i, \overline{v}_j) + \sum_{i=1}^m \beta_{m,i}(t)(\overline{w}_i, A_2\overline{v}_j) \tag{3.30}$$

$$+\sum_{i=1}^m \alpha_{m,i}(t)(\Lambda_1\overline{v}_i, \overline{v}_j) = (f(t), \overline{v}_j) \quad j = 1,\dots,m$$

$$\sum_{i=1}^m \beta'_{m,i}(t)(L_2\overline{w}_i, \overline{w}_j) + \sum_{i=1}^m \alpha_{m,i}(t)(\overline{v}_i, A_1\overline{w}_j) \tag{3.31}$$

$$+\sum_{i=1}^m \beta_{m,i}(t)(\Lambda_2\overline{w}_i, \overline{w}_j) = (g(t), \overline{w}_j) \quad j = 1,\dots,m$$

$$\alpha_{m,i}(0) = \alpha_{0,m,i}\,, \qquad \beta_{m,i}(0) = \beta_{0,m,i} \qquad i = 1,\dots,m \tag{3.32}$$

We observe that the differential system (3.30), (3.31) can be reduced to a *normal form*, since, according to assumption *H6*, we get

(3.33) $$\det\left((L_1\overline{v}_i,\overline{v}_j)\right)_{i,j=1,\ldots,m}\neq 0,\quad \det\left((L_1\overline{w}_i,\overline{w}_j)\right)_{i,j=1,\ldots,m}\neq 0\quad \forall m\in\mathbf{N}$$

For, we notice that the bilinear functionals $\{v,w\}\to(L_jv,w)$ $(j=1,2)$ are scalar products in H, which are equivalent to (v,w) and $\{\overline{v}_j\}_{j=1}^m$ and $\{\overline{w}_j\}_{j=1}^m$ are two systems of linearly independent vectors.
Multiply then equations (3.30) and (3.31) by $2\alpha_{m,j}(t)$ and $2\beta_{m,j}(t)$, respectively, and sum first with respect to j over $\{1,\ldots,m\}$ and then memberwise.
Taking advantage of definitions (3.29) and assumption *H6*, we easily derive the following equation for a.e. $t\in(0,T)$:

(3.34) $$\begin{aligned}2(f(t),v_m(t))+2(g(t),w_m(t)) &= D_t[(L_1v_m(t),v_m(t))+(L_2w_m(t),w_m(t))]\\ &+2(w_m(t),A_2v_m(t))+2(v_m(t),A_1w_m(t))+(\Lambda_1v(t),v(t))+(\Lambda_2w(t),w(t))\end{aligned}$$

Observe now that assumption *H4* implies

(3.35) $$(w_m(t),A_2v_m(t))+(v_m(t),A_1w_m(t))=(\tilde{A}w_m(t),v_m(t))\qquad \textit{for a.e } t\in(0,T)$$

Hence, from (3.34), (3.35) we deduce the integral inequalities

(3.36) $$\begin{aligned}&\gamma^{-1}[\|v_m(t)\|^2+\|w_m(t)\|^2]\le(L_1v_m(t),v_m(t))\\ &+(L_2w_m(t),w_m(t))=(L_1v_{0,m},v_{0,m})+(L_2w_{0,m},w_{0,m})\\ &-2\int_0^t[(\Lambda_1v(s),v(s))+(\Lambda_2w(s),w(s))+(w_m(s),\tilde{A}v_m(s))]\,ds\\ &+2\int_0^t[(f(s),v_m(s))+(g(s),w_m(s))]\,ds\le\|L_1\|_{\mathcal{L}(H)}\|v_{0,m}\|^2+\|L_2\|_{\mathcal{L}(H)}\|w_{0,m}\|^2\\ &+2\beta\gamma^{-1}\int_0^t[\|v_m(s)\|^2+\|w_m(s)\|^2]\,ds\\ &+2\int_0^t[\|f(s)\|^2+\|g(s)\|^2]^{1/2}[\|v_m(s)\|^2+\|w_m(s)\|^2]^{1/2}\,ds\quad \forall t\in(0,T)\end{aligned}$$

where

(3.37) $$\beta=\frac{1}{2\gamma}[\max(\|\Lambda_1\|_{\mathcal{L}(H)},\|\Lambda_2\|_{\mathcal{L}(H)}))+\|\tilde{A}\|_{\mathcal{L}(H)}]$$

$\|M\|_{\mathcal{L}(H)}$ denoting the norm of an operator M in $\mathcal{L}(H)$.
We use now the following simple variant of theorem 4.9 in [1].

LEMMA 3.1. *Let z be a nonnegative $C([0,T])$-function and let b, k be nonnegative $L^1((0,T))$-functions satisfying*

(3.38) $$z(t)\le a+\int_0^t b(s)z(s)\,ds+\int_0^t k(s)z(s)^p\,ds\qquad \forall t\in[0,T]$$

where $p>0$, $p\neq 1$, and $a\ge 0$ are given constants. Then for all $t\in[0,T]$

(3.39) $$z(t)\le\exp\left(\int_0^t b(s)\,ds\right)\left[a^{1-p}+(1-p)\int_0^t k(s)\exp\left((p-1)\int_0^s b(\sigma)\,d\sigma\right)ds\right]^{1/(1-p)}$$

Applying lemma 3.1 with $p = 1/2$ to our integral inequality (3.36), we easily obtain the estimate

(3.40) $$[||v_m(t)||^2 + ||w_m(t)||^2]^{1/2} \leq \gamma^{1/2} \max(||L_1||^{1/2}, ||L_2||^{1/2})[||v_{0,m}||^2 + ||w_{0,m}||^2]^{1/2}$$

$$\times \exp(\beta t) + \gamma \int_0^t \exp[\beta(t-s)][||f(s)||^2 + ||g(s)||^2]^{1/2}\, ds \qquad \textit{for a.e. } t \in (0,T)$$

According to (3.40) and corollary to theorem 1 in [24; chpt. 5, appendix], we can select two subsequences $\{v_{m_n}\}_{n=1}^{+\infty} \subset \{v_m\}_{m=1}^{+\infty}$ and $\{w_{m_n}\}_{n=1}^{+\infty} \subset \{w_m\}_{m=1}^{+\infty}$ *weakly star* converging in $L^\infty((0,T);H)$ to v and w, respectively. In particular, the pair (v,w) satisfies the estimate

(3.41) $$[||v(t)||^2 + ||w(t)||^2]^{1/2} \leq \gamma^{1/2} \max(||L_1||_{\mathcal{L}(H)}, ||L_2||_{\mathcal{L}(H)})[||v_0||^2 + ||w_0||^2]^{1/2}$$

$$\times \exp(\beta t) + \gamma \int_0^t \exp[\beta(t-s)][||f(s)||^2 + ||g(s)||^2]^{1/2}\, ds \qquad \forall t \in (0,T)$$

Following the same procedure as in theorem 4.1 in [5, chpt. 7], we can prove that (v,w) is the unique *weak* solution to problem (3.16)–(3.18).
Assume now that our data satisfy

(3.42) $$f, g \in W^{1,1}((0,T);H) \qquad v_0 \in W_2, w_0 \in W_1$$

Consequently, the solution to the Cauchy problem (3.30)–(3.32) belongs to $W^{2,1}((0,T);\mathbf{R}^{2m})$. Therefore, we can differentiate both members in equations (3.30), (3.31) with respect to time to obtain the following equations for $j = 1, \ldots, m$:

(3.43) $$\sum_{i=1}^{m} \alpha''_{m,i}(t)(L_1\overline{v}_i, \overline{v}_j) + \sum_{i=1}^{m} \beta'_{m,i}(t)(\overline{w}_i, A_2\overline{v}_j) + \sum_{i=1}^{m} \alpha'_{m,i}(t)(\Lambda_1\overline{v}_i, \overline{v}_j) = (f'(t), \overline{v}_j)$$

(3.44) $$\sum_{i=1}^{m} \beta''_{m,i}(t)(L_2\overline{w}_i, \overline{w}_j) + \sum_{i=1}^{m} \alpha'_{m,i}(t)(\overline{v}_i, A_1\overline{w}_j) + \sum_{i=1}^{m} \beta'_{m,i}(t)(L_2\overline{w}_i, \overline{w}_j) = (g'(t), \overline{w}_j)$$

Since $v_0 \in W_2$ and $w_0 \in W_1$, we can choose the bases $\{v_n\}_{n=1}^{+\infty}$ and $\{w_n\}_{n=1}^{+\infty}$ in such a way that $v_0 = \lambda v_1$ and $w_0 = \mu w_1$ for some real constants λ and μ. Consequently, from (3.30)–(3.32) and *H4* we easily deduce the equations

(3.45) $$(L_1 v'_m(0), v'_m(0)) = (f(0) - A_1 w_0 + \Lambda_1 v_0, v'_m(0))$$

(3.46) $$(L_2 w'_m(0), w'_m(0)) = (g(0) - A_2 v_0 + \Lambda_2 w_0, w'_m(0))$$

From (3.45), (3.46) and assumption *H6* we derive the estimates

(3.47) $$||v'_m(0)|| \leq \gamma ||f(0) - A_1 w_0 - \Lambda_1 v_0||, \qquad ||w'_m(0)|| \leq \gamma ||g(0) - A_2 v_0 - \Lambda_2 w_0||$$

Proceeding as above, we can show that the weak solution (v,w) to problem (3.16)–(3.18) is differentiable with respect to time, (v', w') belongs to $L^\infty((0,T);H) \times L^\infty((0,T);H)$ and satisfies

(3.48) $$[||v'(t)||^2 + ||w'(t)||^2]^{1/2} \leq \gamma^{3/2} \max(||L_1||_{\mathcal{L}(H)}^{1/2}, ||L_2||_{\mathcal{L}(H)}^{1/2})$$

$$\times[\|f(0)-A_1w_0-\Lambda_1v_0\|^2+\|g(0)-A_2v_0-\Lambda_2w_0\|^2]^{1/2}\exp(\beta t)$$
$$+\gamma\int_0^t \exp[\beta(t-s)][\|f'(s)\|^2+\|g'(s)\|^2]^{1/2}\,ds \quad \forall t\in(0,T)$$

Choose now $\phi(t)=\rho(t)z$, $\psi(t)=\sigma(t)q$ with $\rho,\sigma\in C_0^\infty((0,T);\mathbf{R})$ and $(z,q)\in W_2\times W_1$. Integrating by parts, from (3.25) we easily deduce the following equations

(3.49) $$(w(t),A_1z)=-(L_1v'(t)+\Lambda_1v(t)-f(t),z)\quad \forall z\in W_2,\ \textit{for a.e. } t\in(0,T)$$

(3.50) $$(v(t),A_2q)=-(L_2w'(t)+\Lambda_2w(t)-g(t),q)\quad \forall q\in W_1,\ \textit{for a.e. } t\in(0,T)$$

From assumptions *H5, H6* we conclude that $w(t)\in W_1$, $v(t)\in W_2$ for a.e. $t\in(0,T)$ and the following equations hold true:

(3.51) $$(w(t),A_1z)=(A_1w(t),z),\quad (v(t),A_2q)=(A_2v(t),q)\quad \forall z\in W_2,\ \textit{for a.e. } t\in(0,T)$$

Finally, from (3.49)–(3.51) it immediately follows that (v,w) is a strong solution to problem (3.16)–(3.18) belonging to $L^\infty((0,T);W_2)\times w\in L^\infty((0,T);W_1)$. In particular, from (3.16), (3.17), (3.41) and (3.48) we derive the estimate

(3.52) $$(\|A_2v(t)\|^2+\|A_1w(t)\|^2)^{1/2}\le(\|L_1v'(t)\|^2+\|L_2w'(t)\|^2)^{1/2}$$
$$+(\|f(t)\|^2+\|g(t)\|^2)^{1/2}\le C(\|L_1\|_{\mathcal{L}(H)},\|L_2\|_{\mathcal{L}(H)},\gamma)\{[\|A_1w_0\|^2+\|A_2v_0\|^2$$
$$+\|f(0)\|^2+\|g(0)\|^2]^{1/2}\exp(\beta t)+\int_0^t\exp[\beta(t-s)]\,(\|f'(s)\|^2+\|g'(s)\|^2)^{1/2}\,ds\Big\}$$

Combining estimates (3.41), (3.48), (3.52), we get the final estimate

(3.53) $$[\|v'(t)\|^2+\|w'(t)\|^2+\|v(t)\|^2_{W_2}+\|w(t)\|^2_{W_1}]^{1/2}$$
$$\le C(\|L_1\|_{\mathcal{L}(H)},\|L_2\|_{\mathcal{L}(H)},\gamma)\left\{[\|v_0\|^2_{W_2}+\|w_0\|^2_{W_1}+\|f(0)\|^2+\|g(0)\|^2]^{1/2}\exp(\beta t)\right.$$
$$\left.+\int_0^t\exp[\beta(t-s)]\,(\|f(s)\|^2+\|f'(s)\|^2+\|g(s)\|^2+\|g'(s)\|^2)^{1/2}\,ds\right\}$$

Assume now that our data satisfy

(3.54) $$f,g\in W^{2,1}((0,T);H)\qquad v_0\in W_2,w_0\in W_1$$

(3.55) $$f(0)-A_1w_0-\Lambda_1v_0\in W_2,\quad g(0)-A_2v_0-\Lambda_2w_0\in W_1$$

Using the previous regularity result, we can show that problem (3.16)–(3.18) admits a unique strong solution

(3.56) $$(v,w)\in[W^{2,\infty}((0,T);H)\cap W^{1,\infty}((0,T);W_2)]$$
$$\times[W^{2,\infty}((0,T);H)\cap W^{1,\infty}((0,T);W_1)]$$

satisfying the estimate

(3.57) $$[\|v''(t)\|^2+\|w''(t)\|^2+\sum_{j=0}^{1}\|v^{(j)}(t)\|^2_{W_2}+\sum_{j=0}^{1}\|w^{(j)}(t)\|^2_{W_1}]^{1/2}$$

$$\leq C(\|L_1\|_{\mathcal{L}(H)}, \|L_2\|_{\mathcal{L}(H)}, \|\Lambda_1\|_{\mathcal{L}(H)}, \|\Lambda_2\|_{\mathcal{L}(H)}, \gamma) \{[\|v_0\|_{W_2}^2 + \|w_0\|_{W_1}^2$$
$$+\|f(0)\|^2 + \|g(0)\|^2 + \|f(0) - A_1w_0 - \Lambda_1v_0\|_{W_2} + \|g(0) - A_2v_0 - \Lambda_2w_0\|_{W_1}]^{1/2} \exp(\beta t)$$
$$+\int_0^t \exp[\beta(t-s)] \left(\sum_{j=0}^{2} [\|f^{(j)}(s)\|^2 + \|g^{(j)}(s)\|^2] \right)^{1/2} ds \Bigg\}$$

By a density argument we can now prove that the weak solution (v, w) to problem (3.16)–(3.18) actually belongs to $C([0, T]; H) \times C([0, T]; H)$. For this purpose we consider four sequences $\{v_{0,m}\} \subset W_2$, $\{w_{0,m}\} \subset W_1$, $\{f_m\}$, $\{g_m\} \subset W^{1,1}((0, T); H)$ such that

(3.58) $$v_{0,m} \to v_0, \ w_{0,m} \to v_0 \quad \textit{in } H, \ f_m \to f, \ g_m \to g \quad \textit{in } L^1((0, T); H) \quad \textit{as } m \to +\infty$$

Taking advantage of the result just proved, we deduce that there exists a unique strong solution (v_m, w_m) to the problem

(3.59) $$(v, w) \in [W^{1,\infty}((0, T); H) \times L^\infty((0, T); W_2)] \times [W^{1,\infty}((0, T); H) \times L^\infty((0, T); W_1)]$$

(3.60) $$L_1v_m'(t) + A_1w_m(t) = f_m(t) \qquad \textit{for a.e. } t \in (0, T)$$

(3.61) $$L_2w_m'(t) + A_2v_m(t) = g_m(t) \qquad \textit{for a.e. } t \in (0, T)$$

(3.62) $$v_m(0) = v_{0,m}, \qquad w_0(0) = w_{0,m}$$

Multiplying scalarly equations (3.59) and (3.60) by $2v_m(t)$ and $2w_m(t)$, respectively, and summing memberwise, we obtain the energy equation

(3.63) $$(L_1v_m(t), v_m(t)) + (L_2w_m(t), w_m(t)) = (L_1v_{0,m}, v_{0,m})$$
$$+(L_2w_{0,m}, w_{0,m}) - 2\int_0^t [(\Lambda_1v(2), v(s)) + (\Lambda_2w(s), w(s)) + (\tilde{A}w_m(s), v_m(s))]\, ds$$
$$+2\int_0^t [(f(s), v_m(s)) + (g(s), w_m(s))]\, ds$$

From the energy estimate (3.41) (with (v, w, v_0, w_0, f, g) replaced by $(v_m, w_m, v_{0,m}, w_{0,m}, f_m, g_m)$) we deduce that (v_m, w_m) converges to a pair of functions (v^*, w^*) in $L^\infty((0, T);\ H) \times L^\infty((0, T); H)]$ as $m \to +\infty$. Since (v_m, w_m) satisfies (3.25), too, we deduce that (v^*, w^*) coincides with the weak solution (v, w) to problem (3.25). In particular, taking the limit as $m \to +\infty$, we obtain that (v, w) satisfies the equation

(3.64) $$(L_1v(t), v(t)) + (L_2w(t), w(t)) = (L_1v_0, v_0) + (L_2w_0, w_0)$$
$$-2\int_0^t [(\Lambda_1v(s), v(s)) + (\Lambda_2w(s), w(s)) + (\tilde{A}w(s), v(s))]\, ds$$
$$+2\int_0^t [(f(s), v(s)) + (g(s), w(s))]\, ds$$

This implies that (possibly after correction on a null set) the function

(3.65) $$t \to (L_1v(t), v(t)) + (L_2w(t), w(t)) \textit{ is continuous in } [0, T]$$

Choose now $\phi(t) = \rho(t)u$, $\psi(t) = \sigma(t)z$, where $\rho, \sigma \in C^\infty([0,T];\mathbf{R})$, $\rho(0) = \sigma(0) = 0$ and $u \in W_2$, $z \in W_1$. From (3.25) we easily deduce that the functions $t \to (L_1 v(t), u)$ and $t \to (L_2 w(t), z)$ belong to $W^{2,1}((0,T);\mathbf{R})$. Hence the functions $t \to (L_1 v(t), u)$ and $t \to (L_2 w(t), z)$ are continuous in $[0,T]$ for any pair $(u,z) \in W_2 \times W_1$. From lemma 1.4 in [21, chpt. 3] and assumption *H1* we conclude that the function

$$(3.66) \qquad t \to (L_1 v(t), u) + (L_2 w(t), z) \textit{ is continuous in } [0,T], \forall u, z \in H$$

Then, from assumption *H6* we derive the following inequalities valid for any pair $t, t_0 \in [0,T]$:

$$(3.67) \qquad \gamma^{-1}[||v(t) - v(t_0)||^2 + ||w(t) - w(t_0)||^2] \leq (L_1[v(t) - v(t_0)], v(t) - v(t_0))$$

$$+(L_2[w(t) - w(t_0)], w(t) - w(t_0)) = (L_1 v(t), v(t)) + (L_2 w(t), w(t))$$

$$+(L_1 v(t_0), v(t_0)) + (L_2 w(t_0), w(t_0)) - 2[(L_1 v(t), v(t_0)) + (L_2 w(t), w(t_0))]$$

Finally, from (3.65)–(3.67) we deduce the weak solution (v,w) to problem (3.16)–(3.18) actually belongs to $C([0,T];H) \times C([0,T];H)$. Consequently, we can prove that also the first (the second) derivative of the strong solution (v,w) to problem (3.16)–(3.18) is continuous, if assumptions (3.42) ((3.42) and (3.54), (3.55)) are satisfied.

Finally, we note that under (3.42), (3.54), (3.55) and the additional assumption

$$(3.68) \qquad f, g \in C([0,T]; W_1)$$

from equations (3.16), (3.17), assumption *H9* and estimate (3.57) we deduce that $A_2 v$, $A_1 w \in C([0,T]; W_1)$ and (v,w) satisfies the additional estimate

$$(3.69) \qquad [||A_2 v(t)||^2 + ||A_1 w(t)||^2]^{1/2} \leq [||f(t)||^2_{W_2} + ||g(t)||^2_{W_1}]^{1/2}$$

$$+C(||L_1||_{\mathcal{L}(H)}, ||L_2||_{\mathcal{L}(H)}, ||\Lambda_1||_{\mathcal{L}(H)}, ||\Lambda_2||_{\mathcal{L}(H)}, \gamma)\{[||v_0||^2_{W_2} + ||w_0||^2_{W_1}$$

$$+||A_1 w_0||_{W_2} + ||A_2 v_0||_{W_1} + ||f(0)||^2_{W_2} + ||g(0)||^2_{W_1}]^{1/2} \exp(\beta t)$$

$$+ \int_0^t \exp[\beta(t-s)] \left(\sum_{j=0}^{2} [||f^{(j)}(s)||^2 + ||g^{(j)}(s)||^2] \right)^{1/2} ds \right\}$$

We summarize the previous results in the following theorem 3.2.

THEOREM 3.2. *Assume that the data f, g, v_0, w_0 satisfy properties (3.19). Then problem (3.16)–(3.18) admits a unique weak solution*

$$(3.70) \qquad (v,w) \in C([0,T];H) \times C([0,T];H)$$

satisfying estimate (3.41).

Further, if f, g, v_0, w_0 satisfy properties (3.42), then problem (3.16)–(3.18) admits a unique strong solution

$$(3.71) \qquad (v,w) \in [C^1([0,T];H) \cap C([0,T];W_2)] \times [C^1([0,T];H) \cap C([0,T];W_1)]$$

satisfying estimate (3.53).

Finally, if f, g, v_0, w_0 satisfy properties (3.42), (3.54), (3.55), then the strong solution (v, w) to problem (3.16)–(3.18) belongs to

$$[C^2([0,T];H) \cap C^1([0,T];W_2)] \times [C^2([0,T];H) \cap C^1([0,T];W_1)] \tag{3.72}$$

and satisfies estimate (3.57).

If, in addition, the functions f, g belong to $C([0,T];W_1)$, then $A_2 v, A_1 w \in C([0,T];\, W_1)$ and (v, w) satisfies estimate (3.54), too.

Proof of theorem 3.1. When our data satisfy property (3.8), the existence of a unique solution (E, H) to problem (3.3)–(3.6) satisfying estimate (3.9) is an immediate consequence of theorem 3.2 according to remark 3.1.

A further application of theorem 3.2, under assumptions (3.11)–(3.13), implies that (E, H) satisfies relationships (3.14)–(3.16). $\square$

THEOREM 3.3. *Let Ω be a bounded C^2-cylinder in $\mathbf{R}^3$ or the strip $(0,\ell) \times \mathbf{R}^2$. Assume further that l_j, λ_j $(j = 1,2)$ belong to $W^{1,\infty}(\Omega;\mathbf{R})$ and the data (E_0, H_0, f, g) satisfy the properties*

$$E_0, H_0 \in H^1(\Omega;\mathbf{R}^3), \qquad n \times E_0 = 0 \qquad n \cdot H_0 = 0 \quad \text{on } \partial\Omega \tag{3.73}$$

$$f, g \in W^{1,1}((0,T);L^2(\Omega;\mathbf{R}^3)) \cap L^1((0,T);H(\operatorname{div},\Omega)) \tag{3.74}$$

$$n \cdot g(t) = 0 \qquad \forall t \in [0,T] \tag{3.75}$$

Then the solution (E, H) to problem (3.3)–(3.6) belongs to $C^1([0,T];L^2(\Omega;\mathbf{R}^3)) \cap C([0,T];$ $H^1(\Omega;\mathbf{R}^3))$ and satisfies

$$n \times E = 0, \qquad n \cdot H = 0 \qquad \text{on } [0,T] \times \partial\Omega \tag{3.76}$$

Moreover, the following estimate holds true for some positive constant C depending on ω, γ, the norms of l_j, λ_j $(j = 1,2)$ in $L^\infty((0,\ell);\mathbf{R})$ and for any $t \in [0,T]$:

$$\begin{aligned} &(\|D_t E(t)\|_0^2 + \|E(t)\|_1^2 + \|D_t H(t)\|_0^2 + \|H(t)\|_1^2)^{1/2} \\ &\le C\,(\|E_0\|_1^2 + \|H_0\|_1^2 + \|f(0)\|_0^2 + \|g(0)\|_0^2)^{1/2} \\ &+ C \int_0^t (\|D_s f(s)\|_0^2 + \|\operatorname{div} f(s)\|_0^2 + \|D_t g(s)\|_0^2 + \|\operatorname{div} g(s)\|_0^2)^{1/2} \end{aligned} \tag{3.77}$$

where

$$\|z\|_1 = \|z\|_{H^1(\Omega;\mathbf{R}^3)} \tag{3.78}$$

Proof. Integrating both members in (3.5) with respect to t and using the initial conditions (3.6), we obtain the equations

$$l_1 E(t) = l_1 E_0 - \int_0^t \operatorname{rot} H(s)\,ds + \lambda_1(x_1) \int_0^t E(s,x)\,ds + \int_0^t f(s)\,ds \qquad \forall t \in [0,T] \tag{3.79}$$

By virtue of the density of $C^\infty(\bar{\Omega};\mathbf{R}^3)$ in $H(\mathrm{rot}\,;\Omega)$ and $H(\mathrm{div}\,;\Omega)$ (cf. theorem 2.1) and the membership of H and f in $C([0,T];H(\mathrm{rot}\,;\mathbf{R}^3)$ and $C([0,T];H(\mathrm{div}\,;\mathbf{R}^3)$, respectively, we easily deduce the following relations for any $\phi \in C_0^\infty(\Omega;\mathbf{R})$:

$$(3.80)\qquad \left(D\phi, \int_0^t \mathrm{rot}\, H(s,x)\,ds\right) = 0$$

$$(3.81)\qquad \left(D\phi,\, l_1 E_0 + \lambda_1 \int_0^t E(s)\,ds + \int_0^t f(s)\,ds\right) = -\left(\phi,\, \mathrm{div}\,(l_1 E_0) + \int_0^t \mathrm{div}\, f(s)\,ds\right)$$

From formula (3.81) we deduce that the function

$$(3.82)\qquad \tilde{E}(t,x) = l_1(x_1)E(t,x) + \lambda_1(x_1)\int_0^t E(s)\,ds \qquad (t,x)\in[0,T]\times\Omega$$

belongs to $C([0,T];H(\mathrm{div}\,;\mathbf{R}^3))$ and satisfies

$$(3.83)\qquad \mathrm{div}\,\tilde{E}(t) = \mathrm{div}\,(l_1 E_0) + \int_0^t \mathrm{div}\, f(s)\,ds$$

Solving equation (3.82) for E, we derive the representation

$$(3.84)\qquad E(t,x) = \frac{1}{l_1(x_1)}\tilde{E}(t,x) + \frac{\lambda_1(x_1)}{l_1(x_1)}\int_0^t \tilde{E}(s,x)\,ds \qquad (t,x)\in[0,T]\times\Omega$$

Recalling that $l_1, \lambda_1 \in W^{1,\infty}((0,\ell))$, from (3.84) we deduce that $E \in C([0,T];H(\mathrm{div}\,;\mathbf{R}^3))$. An explicit computation shows that $\mathrm{div}\, E$ solves the equation

$$(3.85)\qquad \mathrm{div}\,(l_1 E)(t,x) = \mathrm{div}\,(l_1 E_0)(x) - \lambda_1(x_1)\int_0^t \mathrm{div}\,\tilde{E}(s,x)\,ds$$

$$-\lambda_1'(x_1)\int_0^t \tilde{E}(s,x)\,ds + \int_0^t \mathrm{div}\, f(s,x)\,ds \qquad \forall t\in[0,T],\ \textit{for a.e. } x\in\Omega$$

Hence $E \in C^1([0,T],L^2(\Omega;\mathbf{R}^3)$ and satisfies the Cauchy problem

$$(3.86)\qquad l_1(x_1)\frac{\partial \mathrm{div}\, E}{\partial t}(t,x) + \lambda_1(x_1)\mathrm{div}\, E(t,x)$$

$$= -[\varepsilon_0'(x_1) + \lambda_1'(x_1)]E(t,x) + \mathrm{div}\, f(t,x)\,ds \qquad \forall t\in[0,T],\ \textit{for a.e. } x\in\Omega$$

$$(3.87)\qquad \mathrm{div}\, E(0,x) = \mathrm{div}\,(E_0)(x) \qquad \textit{for a.e. } x\in\Omega$$

From (3.86), (3.87) and lemma 3.1. we deduce the following estimate for some positive constant C depending only on γ and the norms $||l_1||_{W^{1,\infty}((0,\ell))}$, $||\lambda_1||_{W^{1,\infty}((0,\ell))}$

$$(3.88)\qquad ||\mathrm{div}\, E(t)||_0 \le C(||\mathrm{div}\, E_0||_0 + \int_0^t (||E(s)||_0 + ||\mathrm{div}\, f(s)||_0)\,ds$$

A similar reasoning shows that $H \in C^1([0,T];H(\mathrm{div}\,;\mathbf{R}^3))$ and satisfies an estimate like (3.88). In conclusion we have proved that $E, H \in C^1([0,T];H(\mathrm{div}\,;\mathbf{R}^3))$ and, owing to (3.9), the following estimate holds true

$$(3.89)\qquad (||\mathrm{div}\, E(t)||_0^2 + ||\mathrm{div}\, H(t)||_0^2)^{1/2} \le C\Big[(||E_0||^2_{H(\mathrm{rot}\,;\Omega)} + ||H_0||^2_{H(\mathrm{rot}\,;\Omega)})^{1/2}$$

$$+\int_0^t (\|f(s)\|^2_{H(\mathrm{div}\,;\Omega)} + \|g(s)\|^2_{H(\mathrm{div}\,;\Omega)})^{1/2}\,ds\Big]$$

In order to conclude the proof of the theorem it remains to show that $n \cdot H = 0$ on $\partial\Omega$. First we note that the trace $n \cdot H(t)$ is a. e. defined on $\partial\Omega$ for any $t \in [0,T]$, since we have proved that $H \in C([0,T]; H(\mathrm{div}\,;\Omega))$. Actually $n \cdot H \in C([0,T]; H^{-1/2}(\partial\Omega))$.
Then we observe that from the equation

(3.90)
$$\mu_0 H(t,x) = \mu_0 H_0(x) + \int_0^t \mathrm{rot}\, E(s,x)\,ds$$

$$-\lambda_2(x_1)\int_0^t H(s,x)\,ds + \int_0^t g(s,x)\,ds \qquad \forall t \in [0,T],\ \textit{for a.e. } x \in \Omega$$

we derive

(3.91)
$$\mu_0 n \cdot H(t,x) = \mu_0 n \cdot H_0(x) + \int_0^t n \cdot \mathrm{rot}\, E(s,x)\,ds$$

$$-\lambda_2(x_1)\int_0^t n \cdot H(s,x)\,ds + \int_0^t n \cdot g(s,x)\,ds = 0 \qquad \forall t \in [0,T],\ \textit{for a.e. } x \in \Omega$$

In fact, the first and the third term in the righthand side in (3.91) vanish according to (3.73) and (3.75), while the second one equals 0 owing to the following relationships:

(3.92)
$$e_1 \times E(t) = 0 \ \textit{on}\ \{0,\ell\} \times \omega \iff E_2(t) = E_3(t) = 0 \quad \textit{on}\ \{0,\ell\} \times \omega$$

$$\Longrightarrow e_1 \cdot \mathrm{rot}\, E(t) = D_2E_3(t) - D_3E_2(t) + D_1(n_2E_3 - n_3E_2) = 0 \quad \textit{a.e. on}\ \{0,\ell\} \times \omega$$

(3.93)
$$n \times E(t) = 0 \ \textit{on}\ (0,\ell) \times \partial\omega \iff E_1(t) = n_2E_3(t) - n_3E_2(t) \ \ \textit{on}\ (0,\ell) \times \partial\omega$$

$$\Longrightarrow n \cdot \mathrm{rot}\, E(t) = (n_2D_3 - n_3D_2)E_1(t) + D_1(n_2E_3 - n_3E_2) = 0 \quad \textit{a.e. on}\ (0,\ell) \times \partial\omega$$

since the differential operators $(n_2D_3 - n_3D_2)$ and D_1 are tangential to $(0,\ell) \times \partial\omega$.
From (3.7) and (3.91) we conclude that $n \cdot H(t) = 0$ a.e. on $\partial\Omega$ for any $t \in [0,T]$.
Summing up, we have proved that $E \in C([0,T]; H_0(\mathrm{rot}\,;\Omega) \cap H(\mathrm{div}\,;\Omega))$ and $H \in C([0,T]; H(\mathrm{rot}\,;\Omega) \cap H_0(\mathrm{div}\,;\Omega))$. Whence and from theorems 3.1, 2.4, 2.5 we derive

(3.94)
$$E, H \in C^1([0,T]; L^2(\Omega;\mathbf{R}^3)) \cap C([0,T]; H^1(\Omega;\mathbf{R}^3))$$

Finally, from (3.9) and (3.89) we easily obtain estimate (3.77). □

THEOREM 3.4. *Let Ω be a bounded C^2–cylinder in $\mathbf{R}^3$ or the strip $(0,\ell) \times \mathbf{R}^2$. Assume further that $l_j, \lambda_j \in W^{1,\infty}(\Omega;\mathbf{R})$ $(j = 1,2)$ and assume that the data (E_0, H_0, f, g) satisfy the properties*

(3.95)
$$f, g \in W^{2,1}((0,T); L^2(\Omega;\mathbf{R}^3)) \cap W^{1,1}((0,T); H(\mathrm{div}\,;\Omega))$$

(3.96)
$$n \cdot g(t) = 0 \qquad \forall t \in [0,T]$$

(3.97)
$$E_0, H_0 \in H^1(\Omega;\mathbf{R}^3), \qquad -\mathrm{rot}\, E_0 + g(0), \mathrm{rot}\, H_0 + f(0) \in H^1(\Omega;\mathbf{R}^3),$$

$$(3.98)\qquad n\times E_0=0,\quad n\cdot H_0=0,\quad n\times[\mathrm{rot}\,H_0+f(0)]=0,\quad n\cdot[-\mathrm{rot}\,E_0+g(0)]=0\quad on\ \partial\Omega$$

Then the solution (E,H) to problem (3.3)–(3.6) belongs to $C^2([0,T];L^2(\Omega;\mathbf{R}^3))\cap C^1([0,T];$ $H^1(\Omega;\mathbf{R}^3))$ and satisfies

$$(3.99)\qquad n\times E=0,\qquad n\cdot H=0\qquad on\ [0,T]\times\partial\Omega$$

Moreover, the following estimate holds true for some positive constant C depending on ω, γ, the norms of l_j, λ_j $(j=1,2)$ in $W^{1,\infty}((0,\ell);\mathbf{R})$ and for any $t\in[0,T]$:

$$(3.100)\qquad \left(\|D_t^2E(t)\|_0^2+\|D_t^2H(t)\|_0^2+\sum_{j=0}^{1}\|D_t^jE(t)\|_1^2+\sum_{j=0}^{1}\|D_t^jH(t)\|_1^2\right)^{1/2}$$

$$\le C\,(\|E_0\|_1^2+\|H_0\|_1^2+\|\mathrm{rot}\,E_0\|_0^2+\|\mathrm{rot}\,H_0\|_0^2+\|f'(0)\|_0^2+\|g'(0)\|_0^2$$

$$+\|f(0)\|_0^2+\|g(0)\|_0^2+\|\mathrm{rot}\,f(0)\|_0^2+\|\mathrm{rot}\,g(0)\|_0^2)^{1/2}$$

$$+C\int_0^t\left[\|D_s^2f(s)\|_0^2+\|D_s^2g(s)\|_0^2+\sum_{j=0}^{1}\left(\|\mathrm{div}\,D_t^jf(s)\|_0^2+\|\mathrm{div}\,D_t^jg(s)\|_0^2\right)\right]^{1/2}ds$$

Proof. We observe that under assumptions (3.95)–(3.98) the pair (D_tE,D_tH) is a strong solution to the problem (3.4)–(3.6) with (f,g,E_0,H_0) replaced by $(D_tf,D_tg,\ f(0)+\mathrm{rot}\,H_0,$ $g(0)-\mathrm{rot}\,E_0)$. This, in turn, implies that $D_tE\in H_0(\Omega;\mathbf{R}^3)$, $D_tH\in H(\Omega;\mathbf{R}^3)$ and $n\cdot D_tH=0$ on $(0,T)\times\partial\Omega$. From assumption (3.95) and Cauchy problem (3.85), (3.86) we conclude that $D_tE\in H_0(\mathrm{rot}\,;\Omega)\cap H(\mathrm{div}\,;\Omega)$ and $D_tH\in H(\mathrm{rot}\,;\Omega)\cap H_0(\mathrm{div}\,;\Omega)$. Consequently, from theorems 2.4 and 2.5 we deduce that $D_tE,D_tH\in H^1(\Omega;\mathbf{R}^3)$. Moreover, from (3.16) in theorem 3.1 and the inequalities $\|D_tE\|_1\le C\|\mathrm{rot}\,D_tE\|_0$, $\|D_tE\|_1\le C\|\mathrm{rot}\,D_tE\|_0$ we get (in particular) the estimate

$$(3.101)\qquad (\|D_tE(t)\|_1^2+\|D_tH(t)\|_1^2)^{1/2}\le C\left(\sum_{j=0}^{2}[\|\mathrm{rot}^j\,E_0\|_0^2+\|\mathrm{rot}^j\,H_0\|_0^2]\right.$$

$$\left.+\|f(0)\|_0^2+\|g(0)\|_0^2+\|f'(0)+\mathrm{rot}\,g(0)\|_0^2+\|g'(0)-\mathrm{rot}\,f(0)\|_0^2\right)^{1/2}$$

$$+C\int_0^t\left[\|D_s^2f(s)\|_0^2+\|D_s^2g(s)\|_0^2+\sum_{j=0}^{1}\|\mathrm{div}\,D_t^jf(s)\|_0^2+\|\mathrm{div}\,D_t^jg(s)\|_0^2\right]^{1/2}\qquad\forall t\in(0,T)$$

4 Integrodifferential Maxwell equations

In this section we consider the following integrodifferential problem related to the cylinder $\Omega=(0,\ell)\times\omega$, which is a slight generalization of problem (1.32)–(1.36): *determine a pair of functions $E,H:\Omega\to\mathbf{R}^3$ belonging to the space*

$$(4.1)\qquad X(T):=C^2([0,T];L^2(\Omega;\mathbf{R}^3))\cap C^1([0,T];H^1(\Omega;\mathbf{R}^3))$$

and satisfying the equations

$$(4.2)\qquad l_1(x_1)D_tE(t,x)-\mathrm{rot}\,H(t,x)+\lambda_1(x_1)E(t,x)$$

$$= -\int_0^t m_1(t-s,x_1)H(s,x)\,ds + f(t,x) \qquad (t,x) \in [0,T]\times\Omega$$

(4.3)
$$l_2(x_1)D_tH(t,x) + \operatorname{rot} E(t,x) + \lambda_2(x_1)H(t,x)$$

$$= -\int_0^t m_2(t-s,x_1)H(s,x)\,ds + g(t,x) \qquad (t,x) \in [0,T]\times\Omega$$

(4.4)
$$E(0,x) = E_0(x), \qquad H(0,x) = H_0(x) \qquad x \in \Omega$$

(4.5)
$$n \times E(t,x) = 0, \qquad n \cdot H(t,x) = 0 \qquad (t,x) \in [0,T]\times\partial\Omega$$

Here the scalar functions l_j, λ_j, $(j = 1,2)$ are assumed to satisfy

(4.6)
$$l_j, \lambda_j \in W^{1,\infty}((0,\ell);\mathbf{R}) \qquad j = 1,2$$

and condition (3.7).
We now introduce the following Banach spaces (when equipped with their natural norms) depending on the parameter $t \in (0,T]$:

(4.7)
$$X(t) = C^2([0,t];H) \cap C^1([0,t];V), \qquad Y(t) = W^{2,1}((0,t);H) \cap W^{1,1}((0,t);F)$$

(4.8)
$$Z(t;u_0) = \{M \in L^1((0,t);H^1((0,\ell))) : Mu_0 \in W^{1,1}((0,T);H)\} \qquad u_0 \in L^2((\Omega;\mathbf{R}^3)$$

(4.9)
$$\mathcal{E}(T) = \{(f,g,E_0,H_0) \in Y(T)\times Y(T)\times V\times V : \operatorname{rot} E_0, \operatorname{rot} H_0 \in V,$$
$$\operatorname{rot}^2 E_0, \operatorname{rot}^2 H_0 \in H, \quad n\cdot g(t) = 0 \ \forall t\in[0,T], \quad n\cdot H_0 = 0,$$
$$n\cdot[-\operatorname{rot} E_0 + g(0)] = 0, \quad n\times E_0 = 0, n\times[\operatorname{rot} H_0 + f(0)] = 0, \quad on\ \partial\Omega\}$$

where we have set

(4.10)
$$V = H^1(\Omega;\mathbf{R}^3), \qquad F = H(\operatorname{div};\Omega), \qquad H = L^2(\Omega;\mathbf{R}^3)$$

Remark 4.1. We note that $Z(t;0) = L^1((0,t);H^1((0,\ell)))\ \forall t \in (0,T]$.

Finally, we make the following assumptions concerning our data $(m_1, m_2, f, g, E_0, H_0)$:

(4.11)
$$(f,g,E_0,H_0) \in \mathcal{E}(T) \qquad m_1 \in Z(T;E_0), \qquad m_2 \in Z(T;H_0)$$

As usual, we reformulate our explicit problem in an abstract form in a Hilbert space H. For this purpose we now assume that V and F are two general Hilbert spaces such that

H9 $V \hookrightarrow F \hookrightarrow H$.

Moreover, we introduce the following Banach spaces:

(4.12)
$$V_1 = \{w \in V : A_w \in V\}, \quad W = \mathcal{L}(V;F) \cap \mathcal{L}(H)$$

(4.13)
$$Z(t;x) = \{M \in L^1((0,t);W) : Mx \in W^{1,1}((0,T);H)\} \quad \forall t \in (0,T], \ x \in H$$

(4.14) $X(T,J)=\{w\in X(T):Jw(t)=0\forall t\in[0,T]\},\ Y(T,J)=\{g\in Y(T):Jg(t)=0\forall t\in[0,T]\}$

$$\mathcal{E}(T)=\{(f,g,v_0,w_0)\in Y(T)\times Y(T;J)\times V_2\times V_1: f(0)\in V \tag{4.15}$$
$$\mathcal{C}_j(f(0),g(0),v_0,w_0)=0\quad j=1,\dots,n_1+n_2\}$$

In (4.14) and (4.15) we assume that the linear operators J and $\mathcal{C}_j$ $(j=1,\dots,N-1+n_2)$ enjoy the properties

(4.16) $J\in\mathcal{L}(V;\mathbf{R}),\ \ \mathcal{C}_j\in\mathcal{L}(F^2\times V^2;\mathbf{R})\ \ j=1,\dots,n_1,\ \ \mathcal{C}_j\in\mathcal{L}(F^2\times V^2;H)\ \ j=n_1,\dots,n_1+n_2$

Finally, we add to *H1–H9* the following basic assumptions:

H10 $L_k\in\mathcal{L}(H)$ $(k=1,2)$;

H11 $A_k\in\mathcal{L}(V;H)$ $(k=1,2)$;

H12 $\Lambda_j\in W$ $(j=1,2)$;

H13 there exist a linear functional $J\in\mathcal{L}(V;\mathbf{R})$ such that for any quadruplet $(f,g,v_0,w_0)\in\mathcal{E}(T)$ problem (3.16)–(3.18) admits a unique solution $(v,w)=R(f,g,v_0,w_0)$ for some linear operator $R\in\mathcal{L}(\mathcal{E}(T);X(T)\times X(T;J))$;

H14 operator R satisfies the following estimate for β defined by (3.37) and some positive constant C_1 depending only on γ, $\|L_k\|_{\mathcal{L}(H)}$ $(k=1,2)$:

$$\|D_t^2R(f,g,v_0,w_0)(t)\|_{H\times H}+\sum_{j=0}^{1}\|D_t^jR(f,g,v_0,w_0)(t)\|_{V\times V} \tag{4.17}$$
$$\le C_1(\|v_0\|_{V_1}+\|w_0\|_{V_1}+\|f'(0)\|+\|g'(0)\|+\|f(0)\|_F+\|g(0)\|_F)e^{\beta t}$$
$$+C_1\int_0^t e^{\beta(t-s)}\left(\sum_{j=1}^{2}[\|f^{(j)}(s)\|+\|g^{(j)}(s)\|+\|f(s)\|_F+\|g(s)\|_F]\right)ds\quad\forall t\in[0,T]$$

H15 $M_1\in Z(T;v_0)$, $M_2\in Z(T;w_0)$ $(j=1,2)$;

H16 $J[M_2*w]=M_2[Jw]$ $\forall M_2\in Z(T;w_0)$, $\forall w\in X(T,J)$ such that $w(0)=w_0$.

Remark 4.2. From assumptions (4.6), (4.10) and equations (3.23), (3.24) we easily deduce that the linear operators L_k, A_k, Λ_k $(k=1,2)$ in our explicit problem (4.2)–(4.5) satisfy assumptions *H10-H12*. Moreover, the operators J and $\mathcal{C}_j$ $(j=1,2,3,4)$ are nothing but the left hand sides in definition (4.9). Finally, assumptions *H13* and *H14* are a reformulation of theorem 2.4.

We can now state our abstract problem: *for any given quadruplet* $(f,g,v_0,w_0)\in\mathcal{E}(T)$ *and any given pair* $(M_1,M_2)\in Z(T;v_0)\times Z(T;w_0)$ *determine a pair of functions* $(v,w)\in X(T)\times X(T;J)$ *such that*

$$L_1v'(t)+A_1w(t)+\Lambda_1(t)v(t)+\int_0^t M_1(t-s)v(s)\,ds+f(t)\qquad t\in[0,T] \tag{4.18}$$

$$L_2w'(t)+A_2v(t)+\Lambda_2(t)w(t)+\int_0^t M_2(t-s)w(s)\,ds+g(t)\qquad t\in(0,T) \tag{4.19}$$

$$v(0) = v_0 \qquad w(0) = w_0 \tag{4.20}$$

Theorem 4.1. *Let $(f, g, v_0, w_0) \in \mathcal{E}(T)$ and $(M_1, M_2) \in Z(T; v_0) \times Z(T; w_0)$ satisfy assumptions H1–H16. Then problem (4.18)–(4.20) admits a unique solution $(v, w) \in X(T) \times X(T; J)$ satisfying the following estimates for any $t \in [0, T]$:*

$$\|v''(t)\| + \|w''(t)\| + \sum_{j=0}^{1} [\|v^{(j)}(t)\|_V + \|w^{(j)}(t)\|_V] \tag{4.21}$$

$$\leq C_2(\|M\|_t, T)(\|v_0\|_{V_1} + \|w_0\|_{V_1} + \|f'(0)\| + \|g'(0)\| + \|f(0)\|_F + \|g(0)\|_F)e^{\beta t}$$

$$\leq C_2(\|M\|_t, T) \int_0^t e^{\beta(t-s)} \left(\sum_{j=1}^{2} [\|f^{(j)}(s)\| + \|g^{(j)}(s)\| + \|f(s)\|_F + \|g(s)\|_F] \right) ds$$

where

$$\|M\|_t = \|M_1\|_{Z(T;v_0)} + \|M_2\|_{Z(T;w_0)} \tag{4.22}$$

and C_2 is a positive function, which is continuous and increasing in each of the variables pointed out and depends also on γ, $\|L_k\|_{\mathcal{L}(H)}$, $\|\Lambda_k\|_W$, $(k = 1, 2)$ and $\|\tilde{A}\|_{\mathcal{L}(H)}$.
Assume further that $(f_j, g_j, v_{0,j}, w_{0,j})$ $(j = 1, 2)$ and $(M_{1,j}, M_{2,j}) \in Z(T; v_{0,j}) \times Z(T; w_{0,j})$ are given. Let $(\tilde{v}, \tilde{w}) \in X(T) \times X(T; J)$ be the solution to problem (4.18)–(4.20) with $(M_1, M_2, f, g, v_0, w_0)$ replaced by $(\tilde{M}_1, \tilde{M}_2, \tilde{f}, \tilde{g}, \tilde{v}_0, \tilde{w}_0)$. Then the following estimate holds true for all $t \in [0, T]$:

$$\sum_{j=0}^{1} [\|v^{(j)}(t) - \tilde{v}^{(j)}(t)\|_V + \|w^{(j)}(t) - \tilde{w}^{(j)}(t)\|_V] + \|v''(t) - \tilde{v}''(t)\| \tag{4.23}$$

$$+\|w''(t) - \tilde{w}''(t)\| \leq e^{\beta t/2} C_3(\|M\|_t, T)(\|v_0 - \tilde{v}_0\|_{V_1} + \|w_0 - \tilde{w}_0\|_{V_1} + \|f'(0) - \tilde{f}'(0)\|$$

$$+\|g'(0) - \tilde{g}'(0)\| + \|f(0) - \tilde{f}(0)\|_F + \|g(0) - \tilde{g}(0)\|_F)$$

$$+C_3(\|M\|_t, T) \int_0^t e^{\beta(t-s)}(\|M - \tilde{M}\|_s + \sum_{j=1}^{2} [\|f^{(j)}(s) - \tilde{f}^{(j)}(s)\| + \|g^{(j)}(s) - \tilde{g}^{(j)}(s)\|]$$

$$+\|f(s) - \tilde{f}(s)\|_F + \|g(s) - \tilde{g}(s)\|_F) \, ds \qquad \forall t \in (0, T)$$

where C_3 enjoys the same properties as C_2.

Remrk 4.4. The following more explicit estimates (of use in the sequel to solve our identification problem) can be proved:

$$\|v''(t)\| + \|w''(t)\| + \sum_{j=0}^{1} [\|v^{(j)}(t)\|_V + \|w^{(j)}(t)\|_V] \tag{4.24}$$

$$\leq \Theta(M_1, M_2, f, g, v_0, w_0)(t) + \int_0^t k_1(M_1, M_2)(t - s)\Theta(M_1, M_2, f, g, v_0, w_0)(s) \, ds$$

$$\sum_{j=0}^{1} [\|v^{(j)}(t) - \tilde{v}^{(j)}(t)\|_V + \|w^{(j)}(t) - \tilde{w}^{(j)}(t)\|_V] + \|v''(t) - \tilde{v}''(t)\| \tag{4.25}$$

$$\leq \Xi(M_1, M_2, \tilde{M}_1, \tilde{M}_2, f, g, v_0, w_0, \tilde{f}, \tilde{g}, \tilde{v}_0, \tilde{w}_0)(t)$$

$$+ \int_0^t k_2(M_1, M_2)(t-s)\Xi(M_1, M_2, \tilde{M}_1, \tilde{M}_2, f, g, v_0, w_0, \tilde{f}, \tilde{g}, \tilde{v}_0, \tilde{w}_0)(s)\, ds$$

Operators Θ, Ξ are defined by formulas (4.75), (4.82), respectively, while k_j $(j = 1, 2)$ is a solution to equation (4.84).

We can now state the result related to our explicit problem (4.2)–(4.5).

THEOREM 4.2. *Let $(f, g, v_0, w_0) \in \mathcal{E}(T)$, $(m_1, m_2) \in Z(T; E_0) \times Z(T; H_0)$ and let l_j, λ_j $(j = 1, 2)$ be functions in $W^{1,\infty}((0, \ell); \mathbf{R})$ satisfying (3.7). Then problem (4.2)–(4.5) admits a unique solution $(E, H) \in X(T) \times X(T)$ satisfying the following estimate for all $t \in [0, T]$:*

$$\left(\|D_t^2 E(t)\|_0^2 + \|D_t^2 H(t)\|_0^2 + \sum_{j=0}^{1} [\|D_t^j E(t)\|_1^2 + \|D_t^j H(t)\|_1^2] \right)^{1/2} \tag{4.26}$$

$$\leq C_4(\|m\|_t, T)\, (\|E_0\|_1^2 + \|H_0\|_1^2 + \|\text{rot}\, E_0\|_0^2 + \|\text{rot}\, H_0\|_0^2 + \|D_t f(0)\|_0^2 + \|D_t g(0)\|_0^2$$

$$+ \|f(0)\|_0^2 + \|g(0)\|_0^2 + \|\text{rot}\, f(0)\|_0^2 + \|\text{rot}\, g(0)\|_0^2)^{1/2}$$

$$+ C_4(\|m\|_t, T) \int_0^t \left(\sum_{j=0}^{2} [\|D_t^j f(s)\|_0^2 + \|D_t^j g(s)\|_0^2] + \|\text{div}\, f(s)\|_0^2 + \|\text{div}\, g(s)\|_0^2 \right)^{1/2}$$

where for all $t \in [0, T]$ we have set (cf. (4.8))

$$\|m\|_t = \|M_1\|_{Z(t;E_0)} + \|M_2\|_{Z(t;H_0)} \tag{4.27}$$

and C_4 is a positive function, which is continuous and increasing in each of the variables pointed out and depends also on ω, γ and the norms of μ_0, λ_1, λ_2 in $W^{1,\infty}(\Omega; \mathbf{R})$. Assume further that $(f_j, g_j, E_{0,j}, H_{0,j})$ and $(m_{1,j}, m_{2,j}) \in Z(T; E_{0,j}) \times Z(T; H_{0,j})$ are given $(j = 1, 2)$. Let $(\tilde{E}, \tilde{H}) \in X(T) \times X(T)$ be the solution to problem (4.2)–(4.5) with $(m_1, m_2, f, g, E_0, H_0)$ replaced by $(\tilde{m}_1, \tilde{m}_2, \tilde{f}, \tilde{g}, \tilde{E}_0, \tilde{H}_0)$. Then the following estimates hold true for all $t \in [0, T]$:

$$\Big(\|D_t^2 E(t) - D_t^2 \tilde{E}(t)\|_0^2 + \|D_t^2 H(t) - D_t^2 \tilde{H}(t)\|_0^2 \tag{4.28}$$

$$+ \sum_{j=0}^{1} [\|D_t^j E(t) - D_t^j \tilde{E}(t)\|_1^2 + \|D_t^j H(t) - D_t^j \tilde{H}(t)\|_1^2] \leq C_5(\|m\|_t, T) \Big(\|E_0 - \tilde{E}_0\|_1^2$$

$$+ \|H_0 - \tilde{H}_0\|_1^2 + \|\text{rot}^2 E_0 - \text{rot}^2 \tilde{E}_0\|_0^2 + \|\text{rot}\, H_0 - \text{rot}\, \tilde{H}_0\|_0^2$$

$$+ \|D_t f(0) - D_t \tilde{f}(0)\|_0^2 + \|D_t g(0) - D_t \tilde{g}(0)\|_0^2 + \|f(0) - \tilde{f}(0)\|_0^2 + \|g(0) - \tilde{g}(0)\|_0^2$$

$$+ \|\text{rot}\, f(0) - \text{rot}\, \tilde{f}(0)\|_0^2 + \|\text{rot}\, g(0) - \text{rot}\, \tilde{g}(0)\|_0^2 \Big)^{1/2} + C_5(\|m\|_t, T)$$

$$\times \int_0^t \Big(\|m - \tilde{m}\|_s + \sum_{j=1}^{2} [\|D_t^j f(s) - D_t^j \tilde{f}(s)\|_1^2 + \|D_t^j g(s) - D_t^j \tilde{g}(s)\|_1^2]$$

$$+ \|\text{div}\, f(s) - \text{div}\, \tilde{f}(s)\|_0^2 + \|\text{div}\, g(s) - \text{div}\, \tilde{g}(s)\|_0^2 \Big)^{1/2}$$

where C_5 enjoys the same properties as C_4.

Remark 4.5. Estimates analogous to (4.24) and (4.25) hold, of course, in the present explicit case, too.

In order to simplify the proof of theorem 4.1 we need to prove an abstract fixed–point theorem. For this purpose we introduce the function Banach space $C^m([0,T];\mathbf{X})$ related to an m–tuple of Banach spaces X_j $(j=0,\ldots,m)$ satisfying

$$X_{j+1} \hookrightarrow X_j \qquad j=0,\ldots,m-1 \tag{4.29}$$

$$C^m([0,T];\mathbf{X}) = \{u \in C^m([0,T];X_0) : u^{(j)} \in C([0,T];X_j),\ j=0,\ldots,m\} \quad m \in \mathbf{N} \tag{4.30}$$

THEOREM 4.3. *Let $\{X_j\}_{0\le j\le n}$ be a finite sequences of Banach spaces satisfying (4.29), let Q be a metric space and let $\Theta : Q \to C([0,T];[0,+\infty))$ be a (possibly) nonlinear operator. Assume that $N : C^m([0,T];\mathbf{X}) \times Q \to C^m([0,T];\mathbf{X})$ be a (possibly) nonlinear operator satisfying the following estimates for some functions $h_1 : Q \to L^p((0,T);\mathbf{R}_+)$ and $h_2 : Q \times Q \to L^p((0,T);\mathbf{R}_+)$ $(p \in (1,+\infty])$:*

$$\sum_{j=0}^{m} \|D_t^j N(z,q)(t)\|_{X_j} \le \Theta(q)(t) + \int_0^t h_1(q)(t-s) \sum_{j=0}^{m} \|z^{(j)}(s)\|_{X_j}\, ds \tag{4.31}$$

$$\forall t \in [0,T];\ \forall z \in C^m([0,T];\mathbf{X}),\ \forall q \in Q$$

$$\sum_{j=0}^{m} \|D_t^j N(z_2,q)(t) - D_t^j N(z_1,q)(t)\|_{X_j} \le \int_0^t h_2(q,q)(t-s) \sum_{j=0}^{m} \|z_2^{(j)}(s) - z_1^{(j)}(s)\|_{X_j}\, ds \tag{4.32}$$

$$\forall t \in [0,T];\ \forall z_1, z_2 \in E(q),\ \forall q \in Q$$

where

$$E(q) = \Big\{ z \in C^m([0,T];\mathbf{X}) : \sum_{j=0}^{m} \|z^{(j)}(t)\|_{X_j} \le \Theta(q)(t) + \int_0^t k_1(q)(t-s)\Theta(q)(s)\, ds \qquad \forall t \in [0,T] \Big\} \tag{4.33}$$

$$k_1(q) = \sum_{j=0}^{+\infty} [h_1(q)*]^j h_1(q) \tag{4.34}$$

Then the fixed–point equation

$$z = N(z,q) \tag{4.35}$$

admits, for any $q \in Q$, a unique solution $z \in C^m([0,T];\mathbf{X})$, which belongs to $E(q)$. Assume further that N satisfies the estimate

$$\sum_{j=0}^{m} \|D_t^j N(z,q)(t) - D_t^j N(\tilde{z},\tilde{q})(t)\|_{X_j} \tag{4.36}$$

$$\leq \Xi(q,\tilde{q})(t) + \int_0^t h_2(q,\tilde{q})(t-s) \sum_{j=0}^{m} \|z^{(j)}(s) - \tilde{z}^{(j)}(s)\|_{X_j}\, ds$$

$$\forall t \in [0,T];\ \forall q, \tilde{q} \in Q,\ \forall z \in E(q),\ \forall \tilde{z} \in E(\tilde{q})$$

where $\Xi : Q^2 \to C([0,T]; \mathbf{R}_+$ *is a (possibly) nonlinear operator. Then, if* z *and* $\tilde{z}$ *denote the solutions to (4.35) and the equation*

$$\tilde{z} = N(\tilde{z}, \tilde{q}) \tag{4.37}$$

where $q, \tilde{q} \in Q$*, then*

$$\sum_{j=0}^{m} \|z^j(t) - \tilde{z}^j(t)\|_{B_j} \leq \Xi(q,\tilde{q})(t) + \int_0^t k_2(q,\tilde{q})(t-s)\Xi(q,\tilde{q})(s)\, ds \quad \forall t \in [0,T] \tag{4.38}$$

where

$$k_2(q,\tilde{q}) = \sum_{j=0}^{+\infty} [h_2(q,\tilde{q})*]^j h_2(q,\tilde{q}) \tag{4.39}$$

Remark 4.3. If operator N satisfies, in addition to the properties listed in theorem 4.3, also the following

$$N(z,q)(0) \textit{ is independent of } z \in C^m([0,T];\mathbf{X}).$$

then every solution $z \in C^m([0,T];\mathbf{X})$ to (4.35) satisfies the initial condition $z(0) = N_0(q)$, where $N_0(q) = N(z,q)$. Consequently, we can restrict ourselves to solving the fixed–point equation (4.35) in the complete metric space

$$E_0(q) = \{z \in E(q) : z(0) = N_0(q)\} \tag{4.40}$$

$E(q)$ being defined by (4.33).
We can, therefore, replace $E(q)$ by $E_0(q)$ in the statement of theorem 4.3. Moreover, in (4.31) we can add the restriction $z(0) = N_0(q)$.

Remark 4.4. Assume that the nonlinear operator Ξ is continuous on $Q \times Q$ and satisfies the additional property

$$\Xi(q,q)(t) = 0 \qquad \forall q \in Q,\ \forall t \in [0,T] \tag{4.41}$$

Then from estimate (4.38) we immediately deduce that the solution z to equation (4.35) continuously depends on q.

We premise with the proof of theorem 4.3 the following lemma 4.1, which is a generalization of lemma 1.1 in [13].

LEMMA 4.1. *Let* $f, g \in L^1((0,T))$ *and* $h \in L^p((0,T))$ $(p \in (1,+\infty])$ *be three nonnegative functions satisfying the inequality*

$$f(t) \leq \int_0^t h(t-s) f(s)\, ds + g(t) \qquad \textit{for a.e. } t \in (0,T) \tag{4.42}$$

Then f *can be estimated by*

$$f(t) \leq g(t) + \int_0^t k(t-s) g(s)\, ds \qquad \textit{for a.e. } t \in (0,T) \tag{4.43}$$

where

$$k = \sum_{j=0}^{+\infty} (h*)^j h \tag{4.44}$$

Proof of lemma 4.1. The linear operator $L \in \mathcal{L}(L^1((0,T)))$ defined by the equation

$$Lf(t) = \int_0^t h(t-s)f(s)\,ds \qquad \textit{for a.e. } t \in (0,T) \tag{4.45}$$

maps the closed cone $L^1_+((0,T)) = \{f \in L^1((0,T)) : f(t) \geq 0 \textit{ for a.e. } t \in (0,T)\}$ into itself. The iterates L^n of L are convolution linear operators with kernels h_n defined by

$$h_n = (h*)^{n-1}h \qquad n \in \mathbf{N}\backslash\{0\} \tag{4.46}$$

where $(h*)^{n-1}h$ denotes the convolution $h * \ldots * h$ (n times).
According to lemma 2.5 in [15], we get the estimates

$$\|h_n\|_{L^1((0,t))} \leq 2^{-1/p}(n!)^{-1/p'}t^{n/p'}\|h\|^n_{L^p((0,T))} \qquad \forall t \in (0,T],\ n \in \mathbf{N}\backslash\{0,1\} \tag{4.47}$$

Hence, estimates (4.47) imply the convergence of the series $\sum_{n=0}^{+\infty}\|L^n\|_{\mathcal{L}(L^1((0,T)))}$. We have thus shown that operator $I-L$ admits an inverse bounded operator $(I-L)^{-1}$ given by the formula

$$(I-L)^{-1} = \sum_{n=0}^{+\infty} L^n \tag{4.48}$$

Consequently, $(I-L)^{-1}$ maps the cone $L^1_+((0,T))$ into itself.
Finally, we observe that from inequality (4.42), rewritten in the form

$$(I-L)f \leq g \tag{4.49}$$

we easily derive the desired inequality

$$f \leq (I-L)^{-1}g = \sum_{n=0}^{+\infty} L^n g = \sum_{n=0}^{+\infty} h_n * g = k * g \tag{4.50}$$

Proof of theorem 4.3. First we observe that for any fixed $q \in Q$, the set $E(q)$ defined by (4.33) is closed in $C^m([0,T];\mathbf{X})$.
Then we note that, for any fixed $q \in Q$, operator $N(\cdot,q)$ maps $E(q)$ to itself. For this purpose we observe that the kernel $k_1(q)$ solves the integral equation

$$k_1(q) = h_1(q) + h_1(q) * k_1(q) \tag{4.51}$$

where $*$ denotes convolution with respect to time. Hence, for any $z \in E$, from (4.31), (4.33), (4.51) we derive the chain of bounds proving the assertion:

$$\sum_{j=0}^{m} \|D_t^j N(z,q)(t)\|_{X_j} \leq \Theta(q)(t) + \int_0^t [h_1(q) + h_1(q) * k_1(q)](t-s)\Theta(q)(s)\,ds \tag{4.52}$$

$$= \Theta(q)(t) + \int_0^t k_1(q)(t-s)\Theta(q)(s)\,ds$$

In order to show that operator $N(\cdot,q)$ is a contraction mapping on $E(q)$, we note that from estimate (4.32) we easily deduce (by recursion) the following estimates, where $N^k(\cdot,q)$ denotes the k–th iterate of $N(\cdot,q)$:

$$\text{(4.53)} \qquad \sum_{j=0}^{m} \|D_t^j N^k(z,q)(t) - D_t^j N^k(\tilde{z},q)(t)\|_{X_j}$$

$$\leq \int_0^t [h_2(q,q)*]^{k-1}h_2(q,q)(t-s)\sum_{j=0}^{m} \|z^{(j)}(s) - \tilde{z}^{(j)}(s)\|_{X_j}\,ds$$

$$\leq \|z-\tilde{z}\|_{C^m((0,T);\mathbf{X})}\|[h_2(q,q)*]^{k-1}h_2(q,q)\|_{L^1((0,T))} \quad \forall t \in [0,T], \forall z, \tilde{z} \in C^m([0,T];\mathbf{X})$$

From estimate (4.53) and (4.47) we conclude that $N(\cdot,q)$ is a contraction mapping on $E(q)$. Consequently, equation (4.35) admits a unique solution $z \in E(q)$.
We also observe that equation (4.35) cannot admit any else solution in $C^m([0,T];\mathbf{X})$. For, if $z \in C^m([0,T];\mathbf{X})$ solves (4.35), it must satisfy the integral inequality

$$\text{(4.54)} \qquad \sum_{j=0}^{m} \|z^{(j)}(t)\|_{X_j} = \sum_{j=0}^{m} \|D_t^j N(z,q)(t)\|_{X_j}$$

$$\leq \Theta_1(\|q\|_{Q(t)}) + \int_0^t h_1(q)(t-s)\sum_{j=0}^{m} \|z^{(j)}(s)\|_{X_j}\,ds \qquad \forall t \in [0,T]$$

Owing to lemma 4.1, we conclude that $z \in E(q)$.

Finally, from (4.36) and (4.37) we obtain the integral inequality

$$\text{(4.55)} \qquad \sum_{j=0}^{m} \|z_2^{(j)}(t) - z_1^{(j)}(t)\|_{B_j} \leq \Xi(q_1,q_2)(t)$$

$$+ \int_0^t h_2(q,\tilde{q})(t-s)\sum_{j=0}^{m} \|z^{(j)}(s) - \tilde{z}_1^{(j)}(s)\|_{X_j}\,ds \qquad \forall t \in [0,T]$$

Applying again lemma 4.1, we easily derive estimate (4.38). □

Proof of theorem 4.2. First we introduce the following operator $\mathcal{M}$ defined on $X(T) \times Z(T) \times Y(T)$ by

$$\text{(4.56)} \qquad \mathcal{M}(u,M,f)(t) = -\int_0^t M(t-s)u(s)\,ds + f(t)$$

where the linear operator M satisfies assumption *H15*.
From (4.56) and the following formulas, valid for any $u \in X(T) \times Z(T) \times Y(T)$ such that $Mu(0) \in W^{1,1}((0,T);H)$ and any $t \in [0,T]$,

$$\text{(4.57)} \qquad D_t\mathcal{M}(u,M,f)(t) = -M(t)u(0) - \int_0^t M(t-s)u'(s)\,ds + f'(t)$$

$$\text{(4.58)} \qquad D_t^2\mathcal{M}(u,M,f)(t) = -D_t[M(t)u(0)] - M(t)u'(0) - \int_0^t M(t-s)u''(s)\,ds + f''(t)$$

we derive the estimates

$$\|\mathcal{M}(u,M,f)(t)\|_F \le \int_0^t \|M(t-s)\|_{\mathcal{L}(V;F)}\|u(s)\|_V\,ds + \|f(t)\|_F \tag{4.59}$$

$$\|D_t\mathcal{M}(u,M,f)(t)\|_F \le \|M(t)\|_{\mathcal{L}(V;F)}\|u(0)\|_V + \int_0^t \|M(t-s)\|_{\mathcal{L}(V;F)}\|u'(s)\|_V\,ds + \|f'(t)\|_F \tag{4.60}$$

$$\|D_t^2\mathcal{M}(u,M,f)(t)\| \le \|D_t[M(t)u(0)]\| + \|M(t)\|_{\mathcal{L}(V;H)}\|u'(0)\|_V + \int_0^t \|M(t-s)\|_{\mathcal{L}(H)}\|u''(s)\|\,ds + \|f''(t)\| \tag{4.61}$$

Likewise, for any $u_1, u_2 \in X(T)$, $M_1 \in Z(T;u_1(0))$, $M_2u \in Z(T;u_2(0))$, $f_1, f_2 \in Y(T)$, $t \in [0,T]$ we derive the estimates

$$\|\mathcal{M}(u_2,M_2,f_2)(t) - \mathcal{M}(u_1,M_1,f_1)(t)\|_F \le \int_0^t \|M_1(t-s)\|_{\mathcal{L}(V;F)}\|u_2(s)-u_1(s)\|_V\,ds + \int_0^t \|M_2(t-s)-M_1(t-s)\|_{\mathcal{L}(V;F)}\|u_2(s)\|_V\,ds + \|f_2(t)-f_1(t)\|_F \tag{4.62}$$

$$\begin{aligned}\|D_t\mathcal{M}(u_2,M_2,f_2)(t) - D_t\mathcal{M}(u_1,M_1,f_1)(t)\|_F &\le \|M_2(t)-M_1(t)\|_{\mathcal{L}(V;F)}\|u_2(0)\|_V \\ &+\|M_1(t)\|_{\mathcal{L}(V;F)}\|u_2(0)-u_1(0)\|_V + \int_0^t \|M_1(t-s)\|_{\mathcal{L}(V;F)}\|u_2'(s)-u_1'(s)\|_V\,ds \\ &+ \int_0^t \|M_2(t-s)-M_1(t-s)\|_{\mathcal{L}(V;F)}\|u_2'(s)\|_V\,ds + \|f_2'(t)-f_1'(t)\|_F\end{aligned} \tag{4.63}$$

$$\begin{aligned}&\|D_t^2\mathcal{M}(u_2,M_2,f_2)(t) - D_t^2\mathcal{M}(u_1,M_1,f_1)(t)\| \\ &\le \|D_t[M_2(t)u_2(0)-M_1(t)u_1(0)]\| + \|M_2(t)-M_1(t)\|_{\mathcal{L}(V;H)}\|u_2'(0)\|_V \\ &+\|M_1(t)\|_{\mathcal{L}(V;H)}\|u_2'(0)-u_1'(0)\|_V + \int_0^t \|M_1(t-s)\|_{\mathcal{L}(H)}\|u_2''(s)-u_1''(s)\|\,ds \\ &+ \int_0^t \|M_2(t-s)-M_1(t-s)\|_{\mathcal{L}(H)}\|u_2''(s)\|\,ds + \|f_2''(t)-f_1''(t)\|\end{aligned} \tag{4.64}$$

Since the linear convolution operators in (4.18), (4.19) vanish at $t=0$, according to remark 4.2 we can restrict our considerations to the following complete metric spaces

$$\mathcal{X}_1 = \{v \in X(T) : v(0) = v_0,\ v'(0) = L_1^{-1}[f(0) - A_1w_0 - \Lambda_1v_0]\} \tag{4.65}$$

$$\mathcal{X}_2 = \{w \in X(T;J) : w(0) = w_0,\ w'(0) = L_1^{-1}[g(0) - A_2v_0 - \Lambda_2w_0]\} \tag{4.66}$$

In fact, every solution $(v,w) \in X(T)\times X(T)$ to the Cauchy problem (4.18)–(4.20) necessarily belongs to $\mathcal{X}_1(T)\times\mathcal{X}_2(T)$.

Since assumption *H9* implies the continuous embedding $\mathcal{L}(V;F) \hookrightarrow \mathcal{L}(V;H)$, from (4.59)–(4.61) we deduce that operator $\mathcal{M}$ belongs to $\mathcal{L}(X(T) \times Z(T) \times Y(T); Y(T))$ and satisfies the following estimates, where C_6 and C_7 are absolute positive constants:

(4.67)
$$\sum_{j=0}^{1} \|D_t^j \mathcal{M}(u,M,f)(t)\|_F + \|D_t^2 \mathcal{M}(u,M,f)(t)\|$$
$$\leq C_6 \int_0^t \|M(t-s)\|_{\mathcal{L}(H)\cap\mathcal{L}(V;F)}(\|u''(s)\| + \|u'(s)\|_V + \|u(s)\|_V)\,ds + \|D_t[M(t)u(0)]\|$$
$$+\|M(t)\|_{\mathcal{L}(V;F)}(\|u'(0)\|_V + \|u(0)\|_V) + \|f''(t)\| + \|f'(t)\|_F + \|f(t)\|_F$$
$$\forall u \in X(T),\ \forall M \in Z(T;u(0)),\ \forall f \in Y(T),\ \forall t \in [0,T]$$

(4.68)
$$\sum_{j=0}^{1} \|D_t^j \mathcal{M}(u_2,M_2,f_2)(t) - D_t^j \mathcal{M}(u_1,M_1,f_1)(t)\|_F$$
$$+\|D_t^2 \mathcal{M}(u_2,M_2,f_2)(t) - D_t^2 \mathcal{M}(u_1,M_1,f_1)(t)\|$$
$$\leq C_7 \int_0^t \|M_1(t-s)\|_{\mathcal{L}(H)\cap\mathcal{L}(V;F)}(\|u_2''(s) - u_1''(s)\| + \|u_2'(s) - u_1'(s)\|_V + \|u_2(s) - u_1(s)\|_V)\,ds$$
$$+C_7 \int_0^t \|M_2(t-s) - M_1(t-s)\|_{\mathcal{L}(H)\cap\mathcal{L}(V;F)}(\|u_2''(s)\| + \|u_2'(s)\|_V + \|u_2(s)\|_V)\,ds$$
$$+\|D_t[M_2(t)u_2(0) - M_1(t)u_1(0)]\| + \|M_1(t)\|_{\mathcal{L}(V;F)}(\|u_2'(0) - u_1'(0)\|_V + \|u_2(0) - 2_1(0)\|_V)$$
$$+\|M_2(t) - M_1(t)\|_{\mathcal{L}(V;F)}(\|u_2'(0)\|_V + \|u_2(0)\|_V)$$
$$+\|f_2''(t) - f_1''(t)\| + \|f_2'(t) - f_1'(t)\|_F + \|f_2(t) - f_1(t)\|_F$$
$$\forall u_1,u_2 \in X(T),\ \forall M_1 \in Z(T;u_1(0)),\ \forall M_2 \in Z(T;u_2(0)),\ \forall f_1,f_2 \in Y(T),\ \forall t \in [0,T]$$

Moreover, from assumption *H16* it immediately follows (cf. (4.15))

(4.69)
$$J[M_2 * w] = M_2 * [Jw] = 0 \qquad \forall w \in X(T;J)$$

Finally, since $M_1 * v(0) = M_2 * w(0) = 0$, we easily deduce

(4.70)
$$\mathcal{C}_j(M_1 * v(0), M_2 * w(0), 0, 0) = 0 \qquad \forall (v,w) \in X(T) \times X(T;J)$$

We have thus shown that the mapping $\{v,w\} \to (M_1 * v, M_2 * w, 0, 0)$ maps $X(T) \times X(T,J)$ to $\mathcal{E}(T)$.

We now observe that, according to assumptions *H13* and *H16* the integrodifferential problem (4.18)–(4.20) can be rewritten in the fixed–point form (4.35), where we have set

(4.71)
$$z = (v,w), \qquad q = (M_1, M_2, f, g, v_0, w_0)$$

(4.72)
$$N(z,q) = -R(M_1 * v, M_2 * w, 0, 0) + R(f, g, v_0, w_0)$$

We introduce also the Banach spaces

(4.73)
$$X_0 = H^2, \qquad X_1 = X_2 = V^2$$

$$Q = \{q : (f, g, v_0, w_0) \in \mathcal{E}(T),\ M_1 \in Z(T; v_0),\ M_2 \in Z(T; w_0)\} \qquad \forall t \in (0, T] \tag{4.74}$$

Then we observe that our data $(M_1, M_2, f, g, v_0, w_0)$ actually belong to Q.
Since $N(z, q)(0) = (v_0, w_0)$ for any $z \in C^2([0, T]; \mathbf{X})$, according to remark 4.2 we can restrict ourselves to solving the fixed–point equation (4.35) in the complete metric space $E_0(q)$ defined by (4.40) with $m = 2$.
Then, from assumptions *H13* ,*H14*, *H16*, definition (4.15) and estimate (4.62)–(4.68) we easily derive that N satisfies the integral inequality (4.31) with $z(0) = N_0(q)$ and

$$\begin{aligned}\Theta(q)(t) = \Theta(M_1, M_2, f, g, v_0, w_0)(t) &= C_8 \Theta_0(f(0), g(0), v_0, w_0) e^{\beta t} \\ &+ C_8 \int_0^t e^{\beta(t-s)} \{\|D_t[M_1(s)v_0]\| + \|D_t[M_2(s)w_0]\| \\ &+ \sum_{j=1}^{2} \|M_j(s)\|_W \Theta_j(f(0), g(0), v_0, w_0) + \Theta_3(f, g)(s)\Big\}\, ds\end{aligned} \tag{4.75}$$

$$h_1(M_1, M_2)(t) = C_9 \int_0^t e^{\beta(t-s)} \sum_{j=1}^{2} \|M_j(s)\|_W\, ds \tag{4.76}$$

where

$$\Theta_0(f(0), g(0), v_0, w_0)(t) = \|v_0\|_{V_1} + \|w_0\|_{V_1} + \|f'(0)\|_F + \|g'(0)\|_F + \|f(0)\|_V + \|g(0)\|_V \tag{4.77}$$

$$\Theta_1(f(0), g(0), v_0, w_0) = \|L_1^{-1}[f(0) - A_1 w_0 - \Lambda_1 v_0\|_V + \|v_0\|_V \tag{4.78}$$

$$\Theta_2(f(0), g(0), v_0, w_0) = \|L_2^{-2}[g(0) - A_2 v_0 - \Lambda_2 w_0\|_V + \|w_0\|_V \tag{4.79}$$

$$\Theta_2(f, g)(t) = \|f''(t)\| + \|f'(t)\|_F + \|f(t)\|_F + \|g''(t)\| + \|g'(t)\|_F + \|g(t)\|_F \tag{4.80}$$

Moreover, N satisfies (4.32) and (4.33), with $m = 2$, $E(q)$, $E(q_1)$, $E(q_2)$ replaced by $E_0(q)$, $E_0(q_1)$, $E_0(q_2)$, respectively, and

$$h_2(M_1, M_2)(t) = C_{10} \int_0^t e^{\beta(t-s)} \sum_{j=1}^{2} \|M_j(s)\|_W\, ds \tag{4.81}$$

$$\begin{aligned}\Xi(q, \tilde{q})(t) &= \Xi(M_1, M_2, \tilde{M}_1, \tilde{M}_2, f(0), g(0), v_0, w_0, \tilde{f}(0), \tilde{g}(0), \tilde{v}_0, \tilde{w}_0) \\ &= C_{11} \Theta_2(f(0) - \tilde{f}(0), g(0) - \tilde{g}(0), v_0 - \tilde{v}_0, w_0 - \tilde{w}_0) e^{\beta t} \\ &+ C_{11} \int_0^t e^{\beta(t-\tau)} \Big\{ \int_0^\tau + \|D_t[M_1(\tau)v_0 - \tilde{M}_1(\tau)\tilde{v}_0]\| + \|D_t[M_2(\tau)w_0 - \tilde{M}_2(\tau)\tilde{w}_0]\| \\ &+ \sum_{j=1}^{2} \|\tilde{M}_j(\tau)\|_W \Theta_j(f(0) - \tilde{f}(0), g(0) - \tilde{g}(0), v_0 - \tilde{v}_0, w_0 - \tilde{w}_0) \\ &+ \sum_{j=1}^{2} \|M_j(\tau) - \tilde{M}_j(\tau)\|_W \Theta_j(f(0), g(0), v_0, w_0)\end{aligned} \tag{4.82}$$

$$+C_8\int_0^\tau \sum_{j=1}^{2} \|M_j(\tau-s)-\tilde{M}_j(\tau-s)\|_W \Big(\Theta(M_1,M_2,f,g,v_0,w_0)(s)$$

$$+\int_0^s k_1(M_1,M_2)(s-\sigma)\Theta(M_1,M_2,f,g,v_0,w_0)(\sigma)\,d\sigma\Big) + \Theta(f-\tilde{f},g-\tilde{g})(t)\Big\}\,d\tau$$

In fact, owing to (4.33), the following inequalities hold true:

$$\int_0^t \sum_{j=1}^{2} \|M_j(t-s)-\tilde{M}_j(t-s)\|_W \sum_{j=1}^{2} \|z^{(j)}\|_{X_j}(s)\,ds \tag{4.83}$$

$$\le C_8\int_0^t \sum_{j=1}^{2} \|M_j(t-s)-\tilde{M}_j(t-s)\|_W \Big\{\Theta(M_1,M_2,f,g,v_0,w_0)(s)$$

$$+\int_0^s k_1(M_1,M_2)(s-\sigma)\Theta(M_1,M_2,f,g,v_0,w_0)(\sigma)\,d\sigma\Big\}\,ds$$

where $k_1(M_1,M_2)$ solves the Volterra integral equation

$$k_1(M_1,M_2) = h_1(M_1,M_2) + h_1(M_1,M_2) * k_1(M_1,M_2) \tag{4.84}$$

From remark 4.2 and theorem 4.3, we deduce that, for any $q \in Q$, the fixed–point equation (4.32) admits a unique solution $z \in E_0(q)$ satisfying (4.30) and (4.35). These latter, via standard computations, easily imply estimates (4.20) and (4.22). □

Proof of theorem 4.1. It is an immediate consequence of definitions (4.7)–(4.10), theorem 4.2, remark 4.2 and the embedding $H^1(\Omega;\mathbf{R}^3) \hookrightarrow L^\infty((0,\ell);L^2(\omega;\mathbf{R}^3))$. In fact, these latter, along with assumptions $(m_1,m_2) \in Z(T;E_0)\times Z(T;H_0)$ (cf. (4.8)) imply that the linear operators $M_j u(t,x) = m_j(t,x_1)u(x)$ satisfy *H15*. □

5 Identification problems for Maxwell integrodifferential equations related to a cylinder

In this section we begin by analysing our integrodifferential identification problem **I. P.** stated in the introduction (cf. (1.23)–(1.29)) and related to the cylinder $\Omega = (0,\ell)\times\omega$, where ω is a C^2–bounded domain in $\mathbf{R}^2$.

We recall that problem **I. P.** is an *inverse problem:* as a rule, such a problem is (generally speaking) *ill-posed* (cf., e.g., [22], [10]).

Our basic task consists in determining sufficiently general conditions concerning the functionals Φ_j ($j=1,2$) and the data f, ψ_j ($j=1,2$), in order that our identification problem turns out to be *well-posed* in a *suitable functional setting.* By this we mean that it is well–posed in the sense of Hadamard (e.g. the solution exists, is unique and continuously depends on the data).

As usual in this paper we consider the following reformulation (of more symmetric form) of our explicit identification problem (1.23)–(1.29): *determine two pairs of functions* $u_j : [0,T]\times\Omega \to \mathbf{R}^3$ ($j=1,2$) *and* $b_j : [0,T]\times[0,\ell] \to \mathbf{R}$ ($j=1,2$) *satisfying the following equations:*

$$l_1(x_1)D_t u_1(t,x) + \mathcal{A}_1 u_2(t,x) + \lambda_1(x_1)u_1(t,x) \tag{5.1}$$

$$+\int_0^t b_1(t-s,x_1)D_t u_1(s,x)\,ds = f(t,x) \quad (t,x)\in[0,T]\times\Omega$$

$$l_2(x_1)D_t u_2(t,x) + \mathcal{A}_2 u_1(t,x) + \lambda_2(x_1)u_2(t,x)$$

$$(5.2) \qquad + \int_0^t b_2(t-s,x_1)D_t u_2(s,x)\,ds = 0 \quad (t,x) \in [0,T] \times \Omega$$

$$(5.3) \qquad u_1(0,x) = u_2(0,x) = 0 \qquad x \in \Omega$$

$$(5.4) \qquad n(x) \times u_1(t,x) = 0 \quad (t,x) \in [0,T] \times \Omega$$

$$(5.5) \qquad n(x) \cdot u_2(t,x) = 0 \quad (t,x) \in [0,T] \times \Omega$$

$$(5.6) \qquad \Phi_1(u_{1,i}(t,x_1,\cdot,\cdot)) = \psi_1(t,x_1) \qquad (t,x_1) \in [0,T] \times [0,\ell]$$

$$(5.7) \qquad \Phi_2(u_{2,j}(t,x_1,\cdot,\cdot)) = \psi_2(t,x_1) \qquad (t,x_1) \in [0,T] \times [0,\ell]$$

where $(i,j) \in \{(1,1),(2,3),(3,2)\}$, $\mathcal{A}_k = (-1)^k \mathrm{rot}$ $(k = 1,2)$, *while* l_k, λ_k *and* Φ_k $(k = 1,2)$ *are four given functions and two given linear functionals continuous in* $L^2(\omega;\mathbf{R})$, *respectively.*

The aim of this section consists in transforming our identification problem (5.1)–(5.7), essentially by a differentiation technique, into an integrodifferential problem which admits a *normal form* with respect to a new quadruplet of unknowns (v_1, v_2, m_1, m_2). For this purpose we introduce the new unknowns

$$(5.8) \qquad v_j = D_t^2 u_1 \quad j = 1,2 \iff u_j(t,x) = tf(0,x)\delta_{1,j} + \int_0^t v_j(s,x)\,ds \quad j = 1,2$$

$$(5.9) \qquad m_j = D_t b_j \quad j = 1,2 \iff b_j(t,x_1) = b_j(0,x_1) + \int_0^t m_j(s,x_1)\,ds \quad j = 1,2$$

Consider, then, the following basic formula

$$(5.10) \qquad D_t^2 \int_0^t b(s,x_1)u(t-s,x)\,ds = D_t \left\{ b(t,x_1)v(0,x) + \int_0^t b(t-s,x_1)D_t u(s,x)\,ds \right\}$$

$$= D_t b(t,x_1)u(0,x) + b(0,x_1)D_t u(0,x) + \int_0^t D_t b(t-s,x_1)D_t u(s,x)\,ds$$

It is easy to check (by differentiation and integration with respect to t) that, according to (5.8)–(5.10), problem (5.1)–(5.7) is equivalent to the following

$$(5.11) \qquad l_1(x_1)D_t v_1(t,x) + \mathcal{A}_1 v_2(t,x) + [\lambda_1(x_1) + b_1(0,x_1)]v_1(t,x)$$

$$+ \int_0^t m_1(t-s,x_1)v_1(s,x)\,ds = D_t^2 f(t,x) \quad (t,x) \in [0,T] \times \Omega$$

$$(5.12) \qquad l_2(x_1)D_t v_2(t,x) + \mathcal{A}_2 u_1(t,x) + [\lambda_2(x_1) + b_2(0,x_1)]v_2(t,x)$$

$$+ \int_0^t m_2(t-s,x_1)v_2(s,x)\,ds = 0 \quad (t,x) \in [0,T] \times \Omega$$

$$v_1(0,x) = l_1(x_1)^{-2}\{l_1(x_1)D_t f(0,x) - [\lambda_1(x_1) + b_1(0,\cdot)]f(0,x)\} := v_{0,1}(x) \qquad x \in \Omega \tag{5.13}$$

$$v_2(0,x) = -l_1(x_1)^{-1}\mathrm{rot}\,[f(0,\cdot)/l_1](x) := v_{0,2} \qquad x \in \Omega \tag{5.14}$$

$$n(x) \times v_1(t,x) = 0 \quad (t,x) \in (0,T) \times \Omega \tag{5.15}$$

$$n(x) \cdot u_2(t,x) = 0 \quad (t,x) \in (0,T) \times \Omega \tag{5.16}$$

$$\Phi_1(v_{1,i}(t,x_1,\cdot,\cdot)) = D_t^2\psi_1(t,x_1) \qquad (t,x_1) \in [0,T] \times [0,\ell] \tag{5.17}$$

$$\Phi_2(v_{2,j}(t,x_1,\cdot,\cdot)) = D_t^2\psi_2(t,x_1) \qquad (t,x_1) \in [0,T] \times [0,\ell] \tag{5.18}$$

Taking advantage of the results proved in sections 2 and 3 about the *well-posedness* of the *direct problem* **D. P.** (cf. (1.32)–(1.36)), we shall assume that l_j, λ_j, $(j = 1,2)$ satisfy assumptions *H1–H13* listed in sections 2 and 3, while, for the time being, we assume that ψ_j and Φ_j, $(j = 1,2)$ are a pair of functions and a pair of bounded functionals acting in $L^2(\omega;\mathbf{R})$, respectively.
As far as the unknowns v_1, v_2, m_1, m_2 are concerned, we shall assume, for the time being, that $(v_1,v_2) \in X(T) \times X(T;J)$ and $(m_1,m_2) \in \mathcal{W}^{2,1}(T) \times \mathcal{W}^{1,1}(T)$, where $X(T)$, $X(T;J)$ are defined by (4.7), (4.8) and

$$\mathcal{W}^{j,1}(T) = [W^{j,1}((0,T);L^2([0,\ell];\mathbf{R})) \cap W^{j-1,1}((0,T);H^1((0,\ell);\mathbf{R})] \quad j \in \mathbf{N} \tag{5.19}$$

Our first task consists in determining the "consistency" conditions, up to the fourth order, relative to problem (5.1)–(5.7). Such conditions are the following (cf. assumption (3.7)):

$$\psi_1(0,x_1) = \psi_2(0,x_1) = D_t\psi_2(0,x_1) = 0 \tag{5.20}$$

$$\Phi_1[f(0,x_1,\cdot)] = l_1(x_1)D_t\psi_1(0,x_1), \quad \Phi_2[\mathcal{A}_2(f(0,\cdot)/l_1)(x_1,\cdot)] = -l_1(x_1)D_t^2\psi_2(0,x_1) \tag{5.21}$$

$$D_t^k\psi_1(t,k\ell) = D_t^k\psi_2(t,k\ell) = 0 \ \ \textit{if } i = j = 1, \ k = 0,1,2,3,4 \tag{5.22}$$

$$D_t^k\psi_1(t,k\ell) = 0 \ \ \textit{if } i \in \{2,3\}, \ k = 0,1,2,3,4 \tag{5.23}$$

$$\begin{cases} b_1(0,x_1)\Phi_1[f(0,x_1,\cdot)] = l_1(x_1)\Phi_1[D_t f(0,x_1,\cdot)] + l_1(x_1)\lambda_1(x_1)D_t\psi_1(0,x_1) \\ - l_1(x_1)^2 D_t^2\psi_1(0,x_1) \\ b_2(0,x_1)\Phi_2\Big[\mathrm{rot}\,\Big(\dfrac{f(0,\cdot)}{l_1}\Big)\Big](x_1) = l_2(x_1)\Phi_2\Big[\dfrac{1}{l_2^2}\mathrm{rot}\,\Big(\dfrac{D_t f(0,\cdot)}{l_1} \\ - \dfrac{1}{l_1^2}(\lambda_1 + b_1(0,\cdot))f(0,\cdot)\Big)\Big](x_1) + l_2(x_1)\lambda_2(x_1)D_t\psi_2(0,x_1) + l_2(x_1)^2 D_t^3\psi_2(0,x_1) \\ m_1(0,x_1)\Phi_1[f(0,x_1,\cdot)] = l_1(x_1)D_t^2 f(0,x_1,\cdot) + l_1(x_1)\Phi_1\Big[\mathrm{rot}\,\Big(\dfrac{1}{l_2}\mathrm{rot}\,\Big(\dfrac{f(0,\cdot)}{l_1}\Big)\Big] \\ - l_1(x_1)[\lambda_1(x_1) + b_1(0,\cdot)]D_t^2\psi_1(0,x_1) - l_1(x_1)^2 D_t^3\psi_1(0,x_1) \\ m_2(0,x_1)\Phi_2\Big[\mathrm{rot}\,\Big(\dfrac{f(0,\cdot)}{l_1}\Big)\Big](x_1) - l_2(x_1)\Phi_2\Big\{\mathrm{rot}\,\Big[\dfrac{1}{l_1}\mathrm{rot}\,\Big(\dfrac{1}{l_2}\mathrm{rot}\,\Big(\dfrac{f(0,\cdot)}{l_1}\Big)\Big) \\ + \dfrac{1}{l_1^2}(\lambda_1 + b_1(0,\cdot))[D_t f(0,\cdot) - (\lambda_1 + b_1(0,\cdot))f(0,\cdot)] + \dfrac{1}{l_1}f(0,\cdot)D_t b_1(0,\cdot) \\ - D_t^2 f(0,\cdot)\Big]\Big\}(x_1) + l_2(x_1)[\lambda_2(x_1) + b_2(0,\cdot)]D_t^3\psi_2(0,x_1) + l_2(x_1)^2 D_t^4\psi_2(0,x_1) \\ D_t m_1(0,x_1)\Phi_1[f(0,x_1,\cdot)] = l_1(x_1)D_t^3 f(0,x_1,\cdot) + l_1(x_1)\Phi_1\Big\{\mathrm{rot}\,\Big[-\dfrac{1}{l_2}\mathrm{rot}\,\Big(\dfrac{1}{l_1^2} \\ \times [D_t f(0,\cdot) - (\lambda_1 + b_1(0,\cdot))f(0,\cdot)]\Big) + \dfrac{\lambda_2 + b_2(0,\cdot)}{l_2}\mathrm{rot}\,\Big(\dfrac{f(0,\cdot)}{l_1}\Big)\Big]\Big\}(x_1) \\ - l_1(x_1)(\lambda_1 + b_1(0,x_1))D_t^3\psi_1(0,x_1) - l_1(x_1)D_t b_1(0,x_1)D_t^2\psi_1(0,x_1) \\ - l_1(x_1)^2 D_t^4\psi_1(0,x_1) \end{cases} \tag{5.24}$$

System (5.24) can explicitly derived differentiating three times at $t = 0$ both members of the differential equations (5.1) and (5.2) and using the equations

$$D_t^j \psi_k(0, x_1) = \Phi_k[D_t^j u_k(0, x_1, \cdot)] \qquad k = 1, 2, \; j = 0, 1, 2, 3, 4, 5 \tag{5.25}$$

We also observe that equations (5.20)–(5.23) are actual consistency conditions, while system (5.24) yields the initial values of our unknowns b_1 and b_2 (recall that $m_j = D_t b_j$!).

In order to uniquely determine the initial values of b_1, b_2, m_1, m_2, $D_t m_1$ in functional classes suitable for solving our identification problem we make the assumption

H17 f belongs to $W^{4,1}((0,T); H) \cap W^{2,1}((0,T); F)$ and satisfies

$$|\Phi_1[f(0, x_1, \cdot)| \geq m, \quad |\Phi_2[\mathrm{rot}\,(f(0,\cdot)/l_1)(x_1, \cdot)]| \geq m \quad \forall x_1 \in [0, \ell] \tag{5.26}$$

for some positive m. The set of all f enjoying such a property is denoted by $\mathcal{F}(m)$;

H18 $\psi_1, \psi_2 \in C^5([0,T]; L^2((0,\ell); \mathbf{R})) \cap C^4([0,T]; H^1((0,\ell); \mathbf{R}))$ and satisfy (5.20)–(5.23). The set of all such ψ_1 and ψ_2 is denoted by $\Psi_1(m)$ and $\Psi_2(m)$, respectively;

H19 $l_1 \in W^{4,\infty}((0,T))$, $l_2 \in W^{3,\infty}((0,T))$;

H20 $\lambda_1, \lambda_2 \in W^{2,\infty}((0,T))$.

Under assumptions *H17–H20* we can show that for any $f \in \mathcal{F}(m)$ and any pair $(\psi_1, \psi_2) \in \Psi_1(m) \times \Psi_2(m)$ system (5.24) has a unique solution $(b_1(0,\cdot), b_2(0,\cdot), m_1(0,\cdot), m_2(0,\cdot), D_t m_1(0,\cdot)) \in H^2((0,\ell); \mathbf{R}) \times H^1((0,\ell); \mathbf{R})^2 \times L^2((0,\ell); \mathbf{R})^2$ which admits the representation

$$\begin{aligned}&(b_1(0,\cdot), b_2(0,\cdot), m_1(0,\cdot), m_2(0,\cdot), D_t m_1(0,\cdot))\\ &= (P_1(f,\psi_1,\psi_2), P_2(f,\psi_1,\psi_2), P_3(f,\psi_1,\psi_2), P_4(f,\psi_1,\psi_2), P_5(f,\psi_1,\psi_2))\end{aligned} \tag{5.27}$$

The nonlinear operators P_j ($j = 1,2,3,4,5$) are bounded and Lipschitz continuous from $\mathcal{F}(m) \times [C^4([0,T]; L^2((0,\ell); \mathbf{R})) \cap C^3([0,T]; L^2((0,\ell); \mathbf{R}))]^2$ to $H^1((0,\ell); \mathbf{R})$ (if $j = 1,2,3$) or $L^2((0,\ell); \mathbf{R})$ (if $j = 4,5$). More explicitly, there exist two positive functions C_j ($j = 0,1$), which are continuous and non decreasing in each of their arguments, such that

$$\begin{aligned}&\|P_1(f,\psi_1,\psi_2)\|_{2,0} + \sum_{j=2}^{3} \|P_j(f,\psi_1,\psi_2)\|_{1,0} + \sum_{j=4}^{5} \|P_j(f,\psi_1,\psi_2)\|_{0,0}\\ &\leq C_0(m, \|D_t^3 f(0)\|, \sum_{j=0}^{2} \|D_t^j f(0)\|_F, \sum_{k=1}^{2}\sum_{j=1}^{3} \|D_t^j \psi_k(0)\|_{1,0}, \sum_{k=1}^{2} \|D_t^4 \psi_k(0)\|_{0,0})\\ &\qquad \forall f \in \mathcal{F}(m), \forall \psi_1, \psi_2 \in C^4([0,T]; L^2((0,\ell); \mathbf{R})) \cap C^3([0,T]; H^1((0,\ell); \mathbf{R}))\end{aligned} \tag{5.28}$$

$$\begin{aligned}&\|P_1(f,\psi_1,\psi_2) - P_1(\tilde{f},\tilde{\psi}_1,\tilde{\psi}_2)\|_{1,0}\\ &+ \sum_{j=2}^{3} \|P_j(f,\psi_1,\psi_2) - P_j(\tilde{f},\tilde{\psi}_1,\tilde{\psi}_2)\|_{1,0} + \sum_{j=4}^{5} \|P_j(f,\psi_1,\psi_2) - P_j(\tilde{f},\tilde{\psi}_1,\tilde{\psi}_2)\|_{0,0}\\ &\leq C_1(m, \|D_t^3 f(0)\|, \|D_t^3 \tilde{f}(0)\|, \sum_{j=0}^{2} \|D_t^j f(0)\|_F, \sum_{j=0}^{2} \|D_t^j \tilde{f}(0)\|_F, \sum_{k=1}^{2}\sum_{j=1}^{3} \|D_t^j \psi_k(0)\|_{1,0},\end{aligned} \tag{5.29}$$

$$\sum_{k=1}^{2}\sum_{j=1}^{3}\|D_t^j\tilde{\psi}_k(0)\|_{1,0},+\sum_{k=1}^{2}\|D_t^4\psi_k(0)\|_{0,0},+\sum_{k=1}^{2}\|D_t^4\tilde{\psi}_k(0)\|_{0,0}\Big)(\|D_t^3 f(0)-D_t^3\tilde{f}(0)\|,$$

$$+\sum_{k=1}^{3}\|D_t^j f(0)-D_t^j\tilde{f}(0)\|_F)+\sum_{k=1}^{2}\sum_{j=1}^{3}\|D_t^j\psi_k(0)-D_t^j\tilde{\psi}_k(0)\|_{1,0}+\sum_{k=1}^{2}\|D_t^4\psi_k(0)-D_t^4\tilde{\psi}_k(0)\|_{0,0})$$

$$\forall f,\tilde{f}\in\mathcal{F}(m),\forall\psi_1,\tilde{\psi}_1,\psi_2,\tilde{\psi}_2\in C^4([0,T];L^2((0,\ell);\mathbf{R}))\cap C^3([0,T];H^1((0,\ell);\mathbf{R}))$$

In order to derive the integrodifferential problem for (v_1, v_2, m_1, m_2) in a normal form we need the following assumption

H21 there exist a real constant a and two linear bounded functionals $\tilde{\Phi}_k$ defined on $L^2(\omega;\mathbf{R}^3)$ $(k = 1, 2)$ satisfying

$$\Phi_1[\mathcal{A}_1 u_2(x_1,\cdot)] = aD_1\Phi_2[u_2(x_1,\cdot)] + \tilde{\Phi}_1[u_2(x_1,\cdot)] \quad \forall u_2\in V,\ \textit{for a.e. } x_1\in(0,\ell) \tag{5.30}$$

$$\Phi_2[\mathcal{A}_2 u_1(x_1,\cdot)] = aD_1\Phi_1[u_1(x_1,\cdot)] + \tilde{\Phi}_2[u_1(x_1,\cdot)] \quad \forall u_1\in V,\ \textit{for a.e. } x_1\in(0,\ell) \tag{5.31}$$

Remark 5.1. If we make the same choice as in (1.22) with

$$\eta_1,\eta_2\in H_0^1(\omega;\mathbf{R})) \tag{5.32}$$

and, in addition, we assume

$$\eta_1=\eta_2 \quad \textit{if } (i,j)\in\{(2,3),(3,2)\} \tag{5.33}$$

in the case $(i, j) = (1.1)$, recalling that $\mathcal{A}_k = (-1)^k\mathrm{rot}$, we easily obtain $a = 0$ and the following equations, where $k = 1, 2$:

$$\tilde{\Phi}_k[u(x_1,\cdot)] = (-1)^k\int_\omega [u_2(x_1,x_2,x_3)D_3\eta_k(x_2,x_3) - u_3(x_1,x_2,x_3)D_2\eta_k(x_2,x_3)\,dx_2dx_3 \tag{5.34}$$

while in the case $(i, j) \in \{(2.3), (3, 2)$ we get $a = 1$ and

$$\tilde{\Phi}_k[u(x_1,\cdot)] = (-1)^i\int_\omega u_1(x_1,x_2,x_3)D_{4-k}\eta(x_2,x_3)\,dx_2dx_3 \quad k=1,2 \tag{5.35}$$

We can now derive the desired integrodifferential problem for (v_1, v_2, m_1, m_2) in a normal form. For this purpose we apply functionals Φ_1 and Φ_2 to both members in (5.11) and (5.12), respectively, and use assumptions *H21*. From (5.17), (5.18) we easily obtain the following equations, where $(t, x_1) \in [0, T] \times [0, \ell]$:

$$\begin{aligned}&l_1(x_1)D_t^3\psi_1(t,x_1) + aD_1D_t^2\psi_2(t,x_1) + \tilde{\Phi}_1[v_2(t,x_1,\cdot)] + [\lambda_1(x_1)+b_1(0,x_1)]D_t^2\psi_1(t,x_1)\\&+\int_0^t m_1(s,x_1)D_t^2\psi_1(t-s,x_1)\,ds = \Phi_1[D_t^2 f(t,x_1,\cdot)] - m_1(t,x_1)l_1(x_1)^{-1}\Phi_1[f(0,x_1,\cdot)]\end{aligned} \tag{5.36}$$

$$\begin{aligned}&l_2(x_1)D_t^3\psi_2(t,x_1) + aD_1D_t^2\psi_1(t,x_1) + \tilde{\Phi}_2[v_2(t,x_1,\cdot)] + [\lambda_2(x_1)+b_2(0,x_1)]D_t^2\psi_2(t,x_1)\\&+\int_0^t m_2(s,x_1)D_t^2\psi_2(t-s,x_1)\,ds = 0\end{aligned} \tag{5.37}$$

We now observe that problem (5.11)–(5.18) is equivalent to problem (5.11)–(5.16), (5.36), (5.37). In fact, applying operators Φ_1 and Φ_2 to both members in (5.11) and (5.12), respectively, and using *H18*, we get the following system, where $(t,x_1) \in [0,T] \times [0,\ell]$:

$$(5.38)\quad l_1(x_1)\Phi_1[D_t v_1(t,x_1,\cdot)] + aD_1\Phi_2[v_2(t,x_1,\cdot)] + \tilde{\Phi}_1[v_2(t,x_1,\cdot)] + [\lambda_1(x_1)+b_1(0,x_1)]\Phi_1[v_1(t,x_1\cdot)]$$

$$+\int_0^t m_1(s,x_1)\Phi_1[v_1(t-s,x_1,\cdot)]\,ds = \Phi_1[D_t^2 f(t,x_1,\cdot)] - m_1(t,x_1)l_1(x_1)^{-1}\Phi_1[f(0,x_1,\cdot)]$$

$$(5.39)\quad l_2(x_1)\Phi_2[D_t v_2(t,x_1,\cdot)] + aD_1\Phi_1[v_1(t,x_1,\cdot)] + \tilde{\Phi}_2[v_1(t,x_1,\cdot)] + [\lambda_2(x_1)+b_2(0,x_1)]\Phi_2[v_2(t,x_1\cdot)]$$

$$+\int_0^t m_2(s,x_1)\Phi_2[v_2(t-s,x_1,\cdot)]\,ds = 0$$

Subtracting term by term the pairs of equations (5.36), (5.38) and (5.37), (5.39), we obtain the following homogeneous problem for the pair $(w_1,w_2) = (\Phi_1[v_1] - D_t^2\psi_1, \Phi_1[v_2] - D_t^2\psi_2)$, where $(t,x_1) \in [0,T] \times [0,\ell]$:

$$(5.40)\qquad l_1(x_1)D_t w_1(t,x_1) + aD_1 w_2(t,x_1) + [\lambda_1(x_1) + b_1(0,x_1)]w_1(t,x_1)$$

$$+\int_0^t m_1(s,x_1)w_1(t-s,x_1)\,ds = 0$$

$$(5.41)\qquad l_2(x_1)D_t w_2(t,x_1) + aD_1 w_1(t,x_1) + [\lambda_2(x_1) + b_2(0,x_1)]w_2(t,x_1)$$

$$+\int_0^t m_2(s,x_1)w_2(t-s,x_1)\,ds = 0$$

$$(5.42)\qquad w_1(0,x_1) = 0, \quad w_2(0,x_1) = 0$$

$$(5.43)\qquad w_1(t,k\ell) = w_2(t,k\ell) = 0 \qquad k = 0,1,\ \textit{if}\ (i,j) = (1,1)$$

$$(5.44)\qquad w_1(t,k\ell) = 0 \qquad k = 0,1,\ \textit{if}\ (i,j) \in \{(2,3),(3.2)\}$$

We now scalarly multiply both members of equations (5.40) and (5.41) by $w_(t,\cdot)$ and $w_2(t,\cdot)$, respectively. Then we sum memberwise such equations and take advantage of the following identity, which is implied by conditions (5.43) or (5.44):

$$(5.45)\qquad (D_1 w_2(t,\cdot), w_1(t,\cdot)) = -(w_2(t,\cdot), D_1 w_1(t,\cdot)) \qquad \forall t \in [0,\ell]$$

Consequently, we obtain the equation

$$(5.46)\qquad D_t\left\{\sum_{k=1}^{2}(l_k w_k(t,\cdot), w_k(t,\cdot))\right\} + \sum_{k=1}^{2}([\lambda_k + b_k(0,\cdot)]w_k(t,\cdot), w_k(t,\cdot))$$

$$+\int_0^t \sum_{k=1}^{2}((m_k(t-s,\cdot)w_k(s,\cdot), w_k(t,\cdot))\,ds = 0 \quad \forall t \in [0,T]$$

From (5.46), (5.42) and (3.7) we easily derive the inequalities

$$\gamma^{-1}\sum_{k=1}^{2}\|w_k(t,\cdot)\|_0^2 \le \max_{1\le k\le 2}(\|\lambda_k\|_{C([0,T])} + (\|b_k(0,\cdot)\|_{C([0,T])})\int_0^t \sum_{k=1}^{2}\|w_k(s,\cdot)\|_0^2\,ds \tag{5.47}$$

$$+\frac{1}{2}\int_0^t ds\int_0^s \left\{\sum_{k=1}^{2}\|w_k(\sigma,\cdot)\|_0^2 + \max_{1\le k\le 2}\|m_k\|^2_{C([0,T];L^2((0;\ell)))}\sum_{k=1}^{2}\|w_k(\sigma,\cdot)\|_0^2\right\}d\sigma$$

$$\le \max_{1\le j\le 2}(\|\lambda_k\|_{C([0,T])}+(\|b_k(0,\cdot)\|_{C([0,T])}+t+t\|m_k\|^2_{C([0,T];L^2((0;\ell)))})\int_0^t\sum_{k=1}^{2}\|w_k(s,\cdot)\|_0^2\,ds \quad \forall t\in[0,T]$$

Finally, from Gronwall's inequality we deduce that $w_1(t) = w_2(t) = 0\ \forall t \in [0,T]$: hence v_1 and v_2 satisfy (5.18) and (5.19).
We now proceed to transform the system of Volterra convolution integral equations (5.38), (5.39), which is of the first kind with respect to m_2, into a system of Volterra convolution integral equations of the *second* kind by a differentiation with respect to time. According to the consistency conditions (5.27) and definition (5.20), we derive the following system of Volterra integral equations, which is easily seen to be equivalent to (5.36), (5.37):

$$l_1(x_1)^{-1}\Phi_1[f(0,x_1,\cdot)]m_1(t,x_1) + \int_0^t m_1(s,x_1)D_t^2\psi_1(t-s,x_1)\,ds + aD_1D_t^2\psi_2(t,x_1) \tag{5.48}$$

$$+\tilde{\Phi}_1[v_2(t,x_1,\cdot)] = \Phi_1[D_t^2 f(t,x_1,\cdot)] - l_1(x_1)D_t^3\psi_1(t,x_1)$$

$$-[\lambda_1(x_1) + P_1(f,\psi_1,\psi_2)(x_1)]D_t^2\psi_1(t,x_1) \qquad (t,x_1)\in[0,T]\times[0,\ell]$$

$$-l_2(x_1)^{-1}(x_1)\Phi_2[\mathcal{A}_2(f(0,\cdot)/l_1)(x_1,\cdot)]m_2(t,x_1) + \int_0^t m_2(s,x_1)D_t^3\psi_2(t-s,x_1)\,ds \tag{5.49}$$

$$+aD_1D_t^3\psi_1(t,x_1) + \tilde{\Phi}_2[D_t v_1(t,x_1,\cdot)] = -l_1(x_1)D_t^4\psi_2(t,x_1)$$

$$-[\lambda_2(x_1) + P_2(f,\psi_1,\psi_2)(x_1)]D_t^3\psi_2(t,x_1) \qquad (t,x_1)\in[0,T]\times[0,\ell]$$

We have thus shown that our identification problem is equivalent to the following:

$$l_1(x_1)D_t v_1(t,x) + \mathcal{A}_1 v_2(t,x) + [\lambda_1(x_1) + b_1(0,x_1)]v_1(t,x) \tag{5.50}$$

$$+\int_0^t m_1(t-s,x_1)v_1(s,x)\,ds = D_t^2 f(t,x) \qquad (t,x)\in[0,T]\times\Omega$$

$$l_2(x_1)D_t v_2(t,x) + \mathcal{A}_2 u_1(t,x) + [\lambda_2(x_1) + b_2(0,x_1)]v_2(t,x) \tag{5.51}$$

$$+\int_0^t m_2(t-s,x_1)v_2(s,x)\,ds = 0 \qquad (t,x)\in[0,T]\times\Omega$$

$$v_1(0,x) = v_{0,1}(x) = 0, \quad v_2(0,x) = v_{0,2}(x) = 0 \qquad x\in\Omega \tag{5.52}$$

$$n(x)\times v_1(t,x) = 0 \quad (t,x)\in[0,T]\times\Omega \tag{5.53}$$

$$n(x)\cdot v_2(t,x) = 0 \quad (t,x)\in[0,T]\times\Omega \tag{5.54}$$

$$m_1(t,x_1) = K_1(f)(x_1)\int_0^t m_1(s,x_1)D_t^2\psi_1(t-s,x_1)\,ds + K_1(f)\tilde{\Phi}_1[v_2(t,x_1,\cdot)]$$
$$+K_1(f)(x_1)\{-\Phi_1[D_t^2 f(t,x_1,\cdot)] + l_1(x_1)D_t^3\psi_1(t,x_1) + aD_1D_t^2\psi_2(t,x_1)$$
$$+[\lambda_1(x_1) + P_1(f,\psi_1,\psi_2)(x_1)]D_t^2\psi_1(t,x_1)\} \qquad (t,x_1)\in[0,T]\times[0,\ell] \tag{5.55}$$

$$m_2(t,x_1) = K_2(f)(x_1)\int_0^t m_2(s,x_1)D_t^3\psi_2(t-s,x_1)\,ds + K_2(f)\tilde{\Phi}_2[D_t v_1(t,x_1,\cdot)]$$
$$+K_2(f)(x_1)\{l_1(x_1)D_t^4\psi_2(t,x_1) + aD_1D_t^3\psi_1(t,x_1)$$
$$+[\lambda_2(x_1) + P_2(f,\psi_1,\psi_2)(x_1)]D_t^2\psi_1(t,x_1)\} \qquad (t,x_1)\in[0,T]\times[0,\ell] \tag{5.56}$$

where we have set

$$K_1(f) = -l_1(x_1)\{\Phi_1[f(0,x_1,\cdot)]\}^{-1}, \quad K_2(f) = l_1(x_1)^{-1}\{\Phi_2[\mathcal{A}_2(f(0,\cdot)/l_1)(x_1,\cdot)]\}^{-1} \tag{5.57}$$

and $v_{0,j}$ $(j=1,2)$ are defined by (5.13) and (5.14).

6 Solving the identification problem (5.1)–(5.7)

In order to solve our identification problem we transform the equivalent problem (5.50)–(5.56) into an equivalent fixed–point system for the pair (m_1,m_2). For this purpose we introduce the three pairs of Hilbert spaces

$$\mathbf{X} = (H;V), \quad \mathbf{Y} = (H,F), \quad \mathbf{Z} = (Z_0,Z_1) \tag{6.1}$$

where

$$H = L^2(\Omega;\mathbf{R}^3), \quad F = H(\Omega;\mathrm{div}), \quad V = H^1(\Omega;\mathbf{R}^3), \quad Z_j = H^j((0,\ell);\mathbf{R}) \quad j\in\mathbf{N} \tag{6.2}$$

Then, with $\mathbf{Q} = (Q_0,Q_1)\in\{\mathbf{X},\mathbf{Y},\mathbf{Z}\}$ we associate the Banach spaces

$$C^j([0,T];\mathbf{Q}) = C^j([0,T];Q_0)\cap C^{j-1}([0,T];Q_1) \quad j\in\mathbf{N} \tag{6.3}$$

$$W^{j,p}((0,T);\mathbf{Q}) = W^{j,p}((0,T);Q_0)\cap W^{j-1,p}((0,T);Q_1) \quad j\in\mathbf{N},\ p\in[1,+\infty) \tag{6.4}$$

We can now write down the final form of our identification problem (5.50)–(5.56). For this purpose we recall that our assumptions guarantee the existence in $X(T)\times X(T;J)$ (cf. theorem 4.1) of the solution to the direct problem (5.50)–(5.54), which we denote by $(N_1(m_1,m_2,f), N_2(m_1,m_2,f))$. We note that operators N_1 and N_2 are nonlinear with respect to (m_1,m_2) and linear with respect to f.

Consequently, problem (5.50)–(5.56) is equivalent to the following nonlinear fixed–point system

$$m_1 = K_1(f)m_1 * D_t^2\psi_1 + K_1(f)\tilde{\Phi}_1[N_2(m_1,m_2,f)] + \mathcal{V}_1(f,\psi_1,\psi_2)$$
$$:= \mathcal{N}_1(m_1,m_2,f,\psi_1,\psi_2) \tag{6.5}$$

$$m_2 = K_2(f)m_2 * D_t^3\psi_2 + K_2(f)\tilde{\Phi}_2[D_tN_1(m_1,m_2,f)] + \mathcal{V}_2(f,\psi_1,\psi_2)$$
$$:= \mathcal{N}_2(m_1,m_2,f,\psi_1,\psi_2) \tag{6.6}$$

$$m_1(0,\cdot) = P_3(f,\psi_1,\psi_2), \quad m_2(0,\cdot) = P_4(f,\psi_1,\psi_2), \quad D_tm_1(0,\cdot) = P_3(f,\psi_1,\psi_2) \tag{6.7}$$

where the operators $\mathcal{V}_1$, $\mathcal{V}_2$ are defined for all $t \in [0,T]$ by

$$\begin{aligned}(6.8)\qquad \mathcal{V}_1(f,\psi_1,\psi_2)(t,x_1) &= K_1(f)\{-\Phi_1[D_t^2 f(t,x_1,\cdot)] + l_1(x_1)D_t^3\psi_1(t,x_1)\\ &+aD_1D_t^2\psi_2(t,x_1) + [\lambda_1(x_1) + P_1(f,\psi_1,\psi_2)(x_1)]D_t^2\psi_1(t,x_1)\}\end{aligned}$$

$$\begin{aligned}(6.9)\qquad \mathcal{V}_2(f,\psi_1,\psi_2)(t,x_1) &= K_2(f)\{l_1(x_1)D_t^4\psi_2(t,x_1)\\ &+aD_1D_t^3\psi_2(t,x_1) + [\lambda_2(x_1) + P_2(f,\psi_1,\psi_2)(x_1)]D_t^3\psi_1(t,x_1)\}\end{aligned}$$

We now associate with any *fixed* positive real number m the (complete metric) space $\mathcal{G}_{i,j}(m)$ of our admissible data defined by

$$\begin{aligned}(6.10)\qquad \mathcal{G}_{i,j}(m) = \Big\{&(f,\psi_1,\psi_2) \in \mathcal{F}_1(T) \times C^5([0,T];\mathbf{Z}) \times C^5([0,T];\mathbf{Z}) :\\ &D_t^h\psi_1(t,k\ell) = D_t^h\psi_2(t,k\ell) = 0,\ \forall t \in [0,T],\ if\ (i,j) = (1,1),\ h = 0,1,2,3,4,\ k = 0,1\\ &D_t^h\psi_1(t,k\ell) = 0\ \ if\ (i,j) \in \{(2,3),(3,2)\},\ \forall t \in [0,T],\ h = 0,1,2,3,4,\ k = 0,1\\ &\psi_1(0,x_1) = \psi_2(0,x_1) = D_t\psi_2(0,x_1) = 0,\ |D_t\psi_1(0,x_1)| \geq m,\ |D_t^2\psi_1(0,x_1)| \geq m\quad \forall x_1 \in [0,\ell]\\ &l_1(x_1)D_t\psi_1(0,x_1) = \int_\omega \eta_1(x_2,x_3)f_i(0,x_1,x_2,x_3)\,dx_2dx_3\quad \forall x_1 \in [0,\ell],\quad D_t^2\psi_2(0,x_1)\\ &= -\int_\omega \eta_2(x_2,x_3)[l_1(x_1)\mathrm{rot}\, f(0,x_1,x_2,x_3) - l_1'(x_1)e_1 \times f(0,x_1,x_2,x_3)]_j\,dx_2dx_3\quad \forall x_1 \in [0,\ell]\Big\}\end{aligned}$$

In fact, since in our case $(g, v_0, w_0) = (0, f(0), 0)$ (cf. (5.43)–(5.44)), the space $\mathcal{E}(T)$ defined by (4.9) reduces to (cf. *H17*)

$$\begin{aligned}(6.11)\qquad \mathcal{F}_1(T) = \{&f \in W^{4,1}((0,T);\mathbf{Y}) : \mathrm{rot}^j f(0,\cdot) \in V,\ j = 0,1,2,\\ &n \cdot \mathrm{rot}\, f(0,\cdot) = 0,\quad n \times f(0,\cdot) = 0, n \times D_t f(0,\cdot) = 0\ \ on\ \partial\Omega\}\end{aligned}$$

Remark 6.1. From (5.57), *H17–H20*, and (6.11) we conclude that $K_j(f) \in Z_1\ \forall f \in \mathcal{F}_1(T)$.

Finally, we introduce, as the space of unknowns, the following complete metric space:

$$\begin{aligned}(6.12)\qquad S(T) = \{&(m_1,m_2) \in W^{2,1}((0,T);\mathbf{Z}) \cap W^{1,1}((0,T);\mathbf{Z}) : m_1(0,\cdot) = P_3(f,\psi_1,\psi_2),\\ &m_2(0,\cdot) = P_4(f,\psi_1,\psi_2),\ D_t m_1(0,\cdot) = P_5(f,\psi_1,\psi_2)\}\end{aligned}$$

We observe that $S(T)$ is a closed metric subspace of $W^{2,1}((0,T);\mathbf{Z}) \times W^{1,1}((0,T);\mathbf{Z})$ when endowed with the induced metrics.

We can now state our existence, uniqueness and stability result concerning our identification problem (1.23)–(1.29). For this purpose we recall that $(i,j) \in \{(1,1),(2,3),(3,2)\}$ and assume

H22 $\eta_1, \eta_2 \in H_0^1(\omega;\mathbf{R})$, $\eta_1 = \eta_2$ if $(i,j) \in \{(2,3),(3,2)\}$.

THEOREM 6.1 *Let assumptions H19, H20, H22 be satisfied and let $(f,\psi_1,\psi_2) \in \mathcal{G}_{i,j}(m)$ for some fixed positive m and some pair $(i,j) \in \{(1,1),(2,3),(3,2)\}$. Then the identification problem (1.23)–(1.29) admits a unique strong solution (E,H,m_1,m_2) such that*

$$(6.13)\qquad E, H \in C^4([0,T];\mathbf{X}),\quad (m_1,m_2) \in W^{2,1}((0,T);\mathbf{Z}) \times W^{1,1}((0,T);\mathbf{Z})$$

Moreover, such a solution continuously depends on the data (f, ψ_1, ψ_2) with respect to the metrics of $\mathcal{G}_{i,j}(m)$.

In order to solve problem (6.5)–(6.7) we need to equip the Banach space $W^{m,r}((0,T);\mathbf{Q})$ $(m \in \mathbf{N},\ r \in [1,+\infty))$, related to a pair $\mathbf{Q} = (Q_0, Q_1)$ of Banach spaces with property (4.29), with the equivalent norm defined by

$$(6.14) \qquad \|u\|_{m,r,\lambda,\mathbf{Q}} = \sum_{j=0}^{m} \left[\int_0^T \left(\|u^{(m)}(t)\|_{Q_j}^r + \sum_{j=0}^{m-1} \|u^{(j)}(t)\|_{Q_j}^r \right) e^{-rt\lambda}\, dt \right]^{1/r} \qquad (\lambda \geq 0)$$

When $Q_0 = Q_1 = \mathbf{R}$, instead of $\|u\|_{m,r,\lambda,\mathbf{R}^m}$ we simply write $\|u\|_{m,r,\lambda}$. When the space $W^{m,r}((0,T);\mathbf{Q})$ is endowed with norm (6.14), we use the notation $W_\lambda^{m,r}((0,T);\mathbf{Q})$.

Then, we denote by $B_j(r;\lambda)$ $(j = 0,1)$ the closed ball with radius r and center at 0 in $W_\lambda^{j,1}((0,T);\mathbf{Z})$.

We recall that for any Banach space Z we have the embedding

$$(6.15) \qquad \|g\|_{0,1,\lambda,Z} \leq \left(\frac{1}{p'}\right)^{1/p'} \lambda^{-1/p'} \|g\|_{0,p,0,Z} \qquad \forall p \in (1,+\infty]$$

Moreover, Young's inequality for convolutions in a Banach algebra A in the present case reads as

$$(6.16) \qquad \|g_1 * g_2\|_{0,p,\lambda,A} \leq \|g_1\|_{0,1,\lambda,A} \|g_2\|_{0,p,\lambda,A} \quad \forall p \in [1,+\infty)$$

Before proving our existence and uniqueness theorem, we state some lemmas involving convolutions and operators Θ and Ξ defined by (4.75) and (4.82). For the sake of simplicity, for the time being, we point out the dependence of such operators on m_1 and m_2, only.

LEMMA 6.1. *Let $\psi \in W^{j-1,p}((0,T);Z_1)$ $(p \in (1,+\infty])$ and $m \in W^{j,1}((0,T);\mathbf{Z})$ $(j = 1,2)$. Then $m * \psi \in W^{j,p}((0,T);\mathbf{Z})$ and satisfies the following estimates:*

$$(6.17) \qquad \|D_t^j(m * \psi)\|_{0,1,\lambda,Z_{1-j}} \leq (1/p')^{1/p'} \lambda^{-1/p'}$$
$$\times \{\|D_t^j m\|_{0,1,\lambda,Z_{1-j}} + \delta_{0,j} \|m(0,\cdot)\|_{Z_1}\} \|\psi\|_{0,p,\lambda,Z_1} \quad j = 1$$

$$(6.18) \qquad \|D_t^2(m * \psi)\|_{0,1,\lambda,Z_0} + \sum_{j=0}^{1} \|D_t^j(m * \psi)\|_{0,1,\lambda,Z_1} \leq (1/p')^{1/p'} \lambda^{-1/p'}$$

$$\times \Big\{ \sum_{j=1}^{2} \{\|D_t^j m\|_{0,1,\lambda,Z_0} \|\psi\|_{0,p,\lambda,Z_1} + \|m(0,\cdot)\|_{Z_1} \|\psi\|_{0,p,\lambda,\mathbf{Z}} + \|D_t m(0,\cdot)\|_{Z_1} \|\psi\|_{0,p,\lambda,Z_0} \Big\} \quad j = 2$$

where $\delta_{i,j}$ denotes the Kronecker delta.

LEMMA 6.2. *For any pair $(m_1, m_2) \in B_2(r,\lambda) \times B_1(r,\lambda)$ and any $\lambda > \beta$ the following estimate holds true:*

$$(6.19) \qquad \|\Theta(m_1, m_2)\|_{1,\lambda} \leq Q(f,\psi_1,\psi_2)(\lambda - \beta)^{-1}(1 + r)$$

where

$$(6.20) \qquad Q(f,\psi_1,\psi_2) = C_8 C_{12}(m^{-1}) \Big[\Big(\sum_{j=0}^{2} \|\mathrm{rot}^j f(0,\cdot)\|_V + \|D_t^2 \psi_1(0,\cdot)\|_V \Big)^2$$

$$+ \sum_{0\le j+k\le 2}^{1} \|\mathrm{rot}^j D_t^{2+k} f(0,\cdot)\|_V\Big] + \|f\|_{4,1,\lambda,Y}\Big]$$

C_{12} denoting a positive, continuous and nondecreasing function.

LEMMA 6.3. *For any pair $(m_{1,2}) \in B_2(r,\lambda) \times B_1(r,\lambda)$ and any $\lambda > \beta + C_{8+i}r$ $(i = 1,2)$ the following estimates holds true:*

$$\|k_i(m_1,m_2)\|_{0,1,\lambda} \le 2C_{8+i}r(\lambda - \beta - 2C_{8+i}r)^{-1} \qquad i = 1,2 \tag{6.21}$$

LEMMA 6.4. *For any $m_1, m_2 \in B_2(r,\lambda)$, $\tilde{m}_1, \tilde{m}_2 \in B_1(r,\lambda)$ and any $\lambda > \beta + C_8 r$ the following estimate holds true:*

$$\|\Xi(m_1,m_2,\tilde{m}_1,\tilde{m}_2)\|_{0,1,\lambda} \le C_{13}(\lambda-\beta)^{-1}Q(f,\psi_1,\psi_2) \tag{6.22}$$

$$\times\Big[\sum_{j=1}^{2} \|m_j - \tilde{m}_j\|_{1,1,\lambda,Z} + 1 + 2r + C_8(1+r)(\lambda - \beta - 2C_9 r)^{-1}\Big]$$

where C_{13} is a positive, continuous and nondecreasing function depending only on the norms of the functions l_j and λ_j $(j = 1,2)$ as pointed out in H19, H20.

Proof of lemma 6.1. It is an immediate consequence of the formulas

$$D_t(m * \psi) = D_t m * \psi + m(0,\cdot)\psi \tag{6.23}$$

$$D_t^2(m * \psi) = D_t^2 m * \psi + m(0,\cdot)D_t\psi + D_t m(0,\cdot)\psi \tag{6.24}$$

and estimates (6.15) and (6.16). □

Proof of lemma 6.2. First we note that from definition (4.75), with $(f, g, v_0.w_0)$ replaced by $(D_t^2 f, 0, v_{0,1}, v_{0,2})$ (cf. (5.23), (5.13), (5.14)), and Young's inequality (6.16) with $p = 1$ we deduce

$$\|\Theta(m_1,m_2)\|_{0,1,\lambda} \le C_8(\lambda-\beta)^{-1}\Big[\Theta_0(D_t^2 f(0), 0, v_{0,1}, v_{0,2}) \tag{6.25}$$

$$+r\sum_{j=1}^{2}\Big(\|v_{0,j} + \Theta_j(D_t^2 f(0), 0, v_{0,1}, v_{0,2})\Big) + \|\Theta_3(f,g)\|_{1,\lambda}\Big]$$

Then we observe that (6.19) is an immediate consequence of (6.25) and the following estimates, which follow from definitions (5.13), (5.14), (4.12) and estimate (5.28):

$$\sum_{j=1}^{2} \|v_{0,j}\| + \sum_{j=0}^{2} \|\Theta_j(D_t^2 f(0), 0, v_{0,1}, v_{0,2})\|_V \tag{6.26}$$

$$\le C_{12}(m^{-1})\Big[\Big(\sum_{j=0}^{2} \|\mathrm{rot}^j f(0,\cdot)\|_V + \|D_t^2\psi_1(0,\cdot)\|_V\Big)^2 + \sum_{0\le j+k\le 2}^{1} \|\mathrm{rot}^j D_t^{2+k} f(0,\cdot)\|_V\Big]$$

$$\|\Theta_3(D_t^2 f(0), 0)\|_{1,\lambda} - \|f\|_{4,1,\lambda,Y} \tag{6.27}$$

Here C_{12} denotes a positive, continuous and nondecreasing function.

Proof of lemma 6.3. First we observe that the following identity holds for any function $\phi \in L^1_\lambda([0,T];\mathbf{R}_+)$ and any $\lambda \in [0,+\infty)$:

$$k := \sum_{n=1}^{+\infty} [(e^{\beta t} * \phi)*]^{n-1}[(e^{\beta t} * \phi)] = \sum_{n=1}^{+\infty} [e^{\beta t}*]^{n-1}e^{\beta t}] * [(\phi*)^{n-1}\phi] \tag{6.28}$$

$$= \sum_{n=1}^{+\infty} [e^{\beta t}(1*)^{n-1}1] * [(\phi*)^{n-1}\phi] = \sum_{n=1}^{+\infty} \left[e^{\beta t}\frac{t^{n-1}}{(n-1)!}\right] * [(\phi*)^{n-1}\phi]$$

From (6.28) we easily deduce

$$e^{-\lambda t}k = \sum_{n=1}^{+\infty} \left[e^{(\beta-\lambda)t}\frac{t^{n-1}}{(n-1)!}\right] * [(e^{-\lambda t}\phi*)^{n-1}(e^{-\lambda t}\phi)] \tag{6.29}$$

From (6.29) and (6.16) we obtain the following inequalities

$$\|e^{-\lambda t}k\|_{0,1,\lambda} \le \sum_{n=1}^{+\infty} \|\phi\|^n_{0,1,\lambda} \int_0^T \left[e^{(\beta-\lambda)t}\frac{t^{n-1}}{(n-1)!}\right] dt \tag{6.30}$$

$$= \|\phi\|_{0,1,\lambda} \int_0^T e^{(\beta-\lambda)t} \sum_{n=1}^{+\infty} \|\phi\|^{n-1}_{0,1,\lambda}\frac{t^{n-1}}{(n-1)!}\, dt = \|\phi\|_{0,1,\lambda} \int_0^T e^{(\beta-\lambda+\|\phi\|_{0,1,\lambda})t}\, dt$$

$$= \|\phi\|_{0,1,\lambda}(\lambda-\beta-\|\phi\|_{0,1,\lambda})^{-1}\{1-\exp[(\lambda-\beta-\|\phi\|_{0,1,\lambda})T]\}$$

If we now assume $\|\phi\|_{0,1,\lambda} \le C_{8+i}r$ and $\lambda \ge \beta + C_{8+i}r$, from (6.30) we get the basic estimate

$$\|k\|_{0,1,\lambda} \le \|\phi\|_{0,1,\lambda}(\lambda-\beta-\|\phi\|_{0,1,\lambda})^{-1} \tag{6.31}$$

Whence and from definitions (4.76) and (4.81) we immediately derive (6.19). □

Proof of lemma 6.4. First we consider definition (4.82) with $g = 0$, $v_0 = v_{0,j}$, $w_0 = v_{0,2}$, $v_0' = \{D_t^2 f(0,\cdot) - \mathcal{A}_1 v_{0,2} - [\lambda_1 + P_1(f,\psi_1,\psi_2)]v_{0,1}\}/l_1$, $w_0' = -\{\mathcal{A}_2 v_{0,1} + [\lambda_2 + P_2(f,\psi_1,\psi_2)]v_{0,2}\}/l_2$ and $\tilde{v}_0$, $\tilde{w}_0$, $\tilde{v}_0'$, $\tilde{w}_0'$ defined likewise with (f,ψ_1,ψ_2) replaced by $(\tilde{f},\tilde{\psi}_1,\tilde{\psi}_2)$.
Whence and from Young's inequality (6.16) with $p = 1$ we deduce the estimates

$$\|\Xi(m_1,m_2) - \Xi(\tilde{m}_1,\tilde{m}_2)\|_{0,1,\lambda} \le C_{11}(\lambda-\beta)^{-1}\Big[\Theta_0(D_t^2 f(0) - D_t^2\tilde{f}_t(0), 0, v_0-\tilde{v}_0, w_0-\tilde{w}_0) \tag{6.32}$$

$$+\sum_{j=1}^{2}\Big(\|D_t m_j - D_t\tilde{m}_j\|_{0,1,\lambda,Z_0}\|v_{0,j}\| + r\|v_{0,j}-\tilde{v}_{0,j}\|\Big)$$

$$+r\sum_{j=1}^{2}\Theta_j(D_t^2 f(0) - D^2\tilde{f}_t(0), 0, v_0-\tilde{v}_0, w_0-\tilde{w}_0) + \sum_{j=1}^{2}\|m_j-\tilde{m}_j\|_{0,1,\lambda,Z_1}\Theta_j(D_t^2 f(0), 0, v_0, w_0)$$

$$+C_8\|\Theta(m_1,m_2)\|_{0,1,\lambda}(1+\|k_1(m_1,m_2)\|_{0,1,\lambda}) + \|\Theta(m_1,m_2)\|_{0,1,\lambda}\Big]$$

Finally, we note that (6.22) is implied by (6.32), definition (6.20) and estimates (6.26).
□

Proof of theorem 6.1. First we introduce the following complete metric space $\mathcal{B}(r,\lambda)$ depending on the pair of positive real numbers (λ, r) (to be determined in the sequel):

$$\mathcal{B}_1(r,\lambda) = \{m_1 \in S(T) : \|m_1\|_{2,1,\lambda,\mathbf{Z}} + \|m_2\|_{1,1,\lambda,\mathbf{Z}} \le r\} \tag{6.33}$$

According to lemmas 6.1, 6.2 and theorem 4.2 we easily deduce that the nonlinear operators N_1 and N_2, related to the solution of problem (4.18)–(4.20), satisfy the following estimates, where we drop out the explicit dependence of N_1 and N_2 on the data:

$$(6.34)\qquad \|N_1(m_1,m_2)\|_{2,1,\lambda,\mathbf{Z}} + \|N_2(m_1,m_2)\|_{1,1,\lambda,\mathbf{Z}} \le \|\Theta(m_1,m_2)\|_{0,1,\lambda}(1+\|k_1(m_1,m_2)\|_{0,1,\lambda})$$
$$\le C_8(\lambda-\beta-2C_9r)^{-1}(1+r)Q(f,\psi_1,\psi_2) \qquad \lambda>\beta+2C_9r$$

As a consequence, according to assumption *H17–21* and lemmas 6.1, 6.2, 6.3, 6.4 the nonlinear operators $\mathcal{N}_1$ and $\mathcal{N}_2$, defined by the right hand sides in (6.5) and (6.6), satisfy the estimate

$$(6.35)\qquad \|\mathcal{N}_1(m_1,m_2)\|_{2,1,\lambda,\mathbf{Z}} + \|\mathcal{N}_2(m_1,m_2)\|_{1,1,\lambda,\mathbf{Z}} \le r$$

if we choose the pair (r,λ) to be a solution to the system of inequalities

$$(6.36)\qquad \lambda>\beta+2C_9r$$

$$(6.37)\qquad \alpha_1(r,\lambda,f,\psi_1,\psi_2) := \|\mathcal{V}_1(f,\psi_1,\psi_2)\|_{2,1,0,Z_1} + \|\mathcal{V}_2(f,\psi_1,\psi_2)\|_{1,1,0,Z_1} + C_8(\lambda-\beta-2C_9r)^{-1}$$
$$\times(1+r)Q(f,\psi_1,\psi_2)\sum_{j=1}^{2}\|\tilde{\Phi}_j\|_{\mathcal{L}(L^2(\omega;\mathbf{R});\mathbf{R})} + \left(\frac{1}{p'}\right)^{1/p'}\lambda^{-1/p'}r$$
$$\times\Big\{\|K_1(f)\|_{Z_1}[\|D_t^2\psi_1\|_{0,p,0,Z_1} + \|D_t^3\psi_1\|_{0,p,0,Z_1} + \|P_1(f,\psi_1,\psi_2)\|_{Z_2}$$
$$+\sum_{j=2}^{3}\|P_j(f,\psi_1,\psi_2)\|_{Z_1} + \|K_2(f)\|_{Z_1}[\|D_t^3\psi_2\|_{0,p,0,Z_1}\Big\} \le r$$

We note that estimate (5.27) and assumptions *H17–H20, H22* easily imply that $\mathcal{V}_k(f,\psi_1,\psi_2)$ belongs to $W^{3-k,1}((0,T);\mathbf{Z})$ $(k=1,2)$ for any triplet $(f,\psi_1,\psi_2)\in\Gamma_{i,j}(m)$.
Finally, we observe that from (6.5), (6.6) and the definitions of $\mathcal{V}_1$, $\mathcal{V}_2$ we obtain the equations

$$(6.38)\qquad \mathcal{N}_j(m_1,m_2)(0) = P_{2+j}(f,\psi_1,\psi_2)\quad j=1,2,\quad D_t\mathcal{N}_1(m_1,m_2)(0) = P_5(f,\psi_1,\psi_2)$$

We have thus shown that the vector operator $\mathcal{N}$ defined by

$$(6.39)\qquad \mathcal{N} = (\mathcal{N}_1(m_1,m_2),\mathcal{N}_2(m_1,m_2))$$

maps $\mathcal{B}(r,\lambda)$ into itself for any pair (r,λ) solution to (6.36), (6.37).

We now note that from lemmas 6.3, 6.4 with $(\tilde{f},\tilde{\psi}_1,\tilde{\psi}_2)=(f,\psi_1,\psi_2)$ and estimate (4.24), we easily deduce that N_1 and N_2 satisfy the estimate

$$(6.40)\qquad \|N_1(m_1,m_2)-N_1(\tilde{m}_1,\tilde{m}_2)\|_{2,1,\lambda,\mathbf{Z}} + \|N_2(m_1,m_2)-N_2(\tilde{m}_1,\tilde{m}_2)\|_{2,1,\lambda,\mathbf{Z}}$$
$$\le \gamma(r,\lambda,f,\psi_1,\psi_2)[\|m_1-\tilde{m}_1\|_{2,1,\lambda,\mathbf{Z}} + \|m_2-\tilde{m}_2\|_{1,1,\lambda,\mathbf{Z}}] \qquad \text{if } \lambda>\beta+2C_{10}r$$

where

$$\gamma(r,\lambda,f,\psi_1,\psi_2) = C_8C_{12}(\lambda-\beta-2C_{10}r)^{-1}Q(f,\psi_1,\psi_2)$$
$$(6.41)\qquad \times\Big[\sum_{j=1}^{2}\|m_j-\tilde{m}_j\|_{1,1,\lambda,\mathbf{Z}} + 1 + 2r + C_8(1+r)(\lambda-\beta-2C_9r)^{-1}\Big]$$

From (6.40), (6.41) and lemma 6.1 we conclude that operator $\mathcal{N}$ is a contraction mapping in $\mathcal{B}(r,\lambda)$ if the pair (r,λ) satisfies, in addition to (6.36), (6.37), also the inequality

$$\alpha_2(r,\lambda,f,\psi_1,\psi_2) := \left(\frac{1}{p'}\right)^{1/p'} \lambda^{-1/p'} \max\Big(\|K_1(f)\|_{Z_1}[\|D_t^2\psi_1\|_{0,p,0,Z_1}, \tag{6.42}$$

$$+\|K_2(f)\|_{Z_1}[\|D_t^3\psi_2\|_{0,p,0,Z_1}\Big) + \gamma(r,\lambda,f,\psi_1,\psi_2)\sum_{j=1}^{2} \|K_j(f)\|_{Z_1}\|\tilde{\Phi}_j\|_{\mathcal{L}(L^2(\omega;\mathbf{R});\mathbf{R})} < 1$$

Then we observe that the set of solutions to the system of inequalities

$$\begin{cases} \alpha_1(r,+\infty,f,\psi_1,\psi_2) \le \dfrac{1}{2}r \\ \alpha_2(r,+\infty,f,\psi_1,\psi_2) \le \dfrac{1}{2} \end{cases} \tag{6.43}$$

contains an interval of the form $(r_0,+\infty)$ for some positive r_0 depending on the norms of f, ψ_1, ψ_2 pointed out in this section.
Finally, for any *fixed* $r \in (r_0,+\infty)$ we choose λ so as to satisfy the system

$$\begin{cases} \lambda > \beta + 2r\max(C_9, C_{10}) \\ \alpha_1(r,+\infty,f,\psi_1,\psi_2) - \alpha_1(r,\lambda,f,\psi_1,\psi_2) \le \dfrac{1}{2}r \\ \alpha_2(r,+\infty,f,\psi_1,\psi_2) - \alpha_2(r,\lambda,f,\psi_1,\psi_2) < \dfrac{1}{2} \end{cases} \tag{6.44}$$

Consequently, we can conclude that problem (6.5)–(6.7) admits a unique solution (m_1,m_2) in $S(T)$.

We omit the proof of the continuous dependence of the solution upon the data (f,ψ_1,ψ_2) owing to the lack of space. We limit ourselves to observing that such a result is now a standard consequence of lemmas 6.1–6.4 and the techniques developed in this proof.

We conclude this proof by observing that the assertion in theorem 6.1 is an immediate consequence of the equivalence of problems (5.1)–(5.7) and (6.9)–(6.11). □

References

[1] D. Bainov and P. Simeonov: *Integral inequalities and applications*, Kluwer Academic Publishers (1992).

[2] F. Bloom: *Ill-posed problems for integrodifferential equations in mechanics and electromagnetic theory*, S.I.A.M. Studies in Applied Mathematics vol 18, (1981).

[3] C. Cavaterra and A. Lorenzi: *An identification problem for the Maxwell equations in a non homogeneous dispersive medium*, to appear in J. Int. Diff. Eq.s.

[4] B. D. Coleman and E. H. Hill: *On the thermodynamics of electromagnetic fields in materials with memory*, Arch. Rat. Mech. Anal. 41, (1971), pp. 132-162.

[5] G. Duvaut and J. L. Lions: *Inequalities in Mechanics and Physics*, Springer–Verlag (1976).

[6] J. D. Jackson: *Classical Electrodynamics*, John Wiley and Sons, New York (1975) .

[7] S. I. Kabanikhin and A. Lorenzi: *An identification problem related to the integrodifferential Maxwell's equations*, Inv. Prob. 7 (1991), pp. 863–886.

[8] G. Kristensson and S. Rikte: *Scattering of transient electromagnetic waves in bi-isotropic media*, technical report LUTEDX/(TEAT 7015) (1991),pp. 1–17.

[9] L. D. Landau and E. M. Lifshitz: *Electrodynamics of continuous media*, Pergamon Press (1960).

[10] M. M. Lavrent'ev, V. G. Romanov–S. P. Shishatskii: *Ill-posed problems of Mathematical Physics and Analysis*, Transl. Math. Monographs, vol. 64, Amer. Math. Soc. (1986).

[11] B. Lazzari and E. Vuk: *Sul comportamento asintotico dell'energia elettromagnetica in un materiale dielettrico lineare*, Boll. U. M. I. (7) 4–B (1990), pp. 155–177.

[12] J. L. Lions: *Controlabilité exacte, perturbations et stabilisation de systemes distribués*, vol 1, Masson (1988).

[13] A. Lorenzi: *Determination of a time-dependent coefficient in a quasi-linear parabolic equation*, Ric. Mat. Napoli 32 (1984), pp. 263-284.

[14] A. Lorenzi and E. Paparoni : *Direct and inverse problems in the theory of materials with memory*, Rend. Sem. Mat. Univ. Padova 87 (1992), pp. 105-138.

[15] J. Nečas: *Les méthodes directes en théorie des équations elliptiques*, Masson, Paris (1967).

[16] A. I. Prilepko, D. G. Orlovsky and I. A. Vasin: *Inverse problems in mathematical physics*, in *Ill posed problems in natural sciences*, A. Tikhonov ed., pp. 1–15, VSP/TVP, Moscow (1992).

[17] V. G. Romanov: *Inverse problems of mathematical physics*, VNU Science Press, Utrecht (1987).

[18] V. G. Romanov and S. I. Kabanikhin: *Inverse problems of geoelectrics*, Nauka, Moscow (1991) (in Russian).

[19] D. Sheen: *A generalized Green's theorem*, Appl. Math. Lett. 5 (1992), pp. 95–99.

[20] Z. Sun and G. Uhlmann: *An inverse boundary value problem for Maxwell's equations*, Arch. Rational Mech. Anal. 119 (1992), pp. 71–93,

[21] R. Temam: *Navier-Stokes equations*, North-Holland (1977).

[22] A. N. Tikhonov and V. Ya. Arsenin: *Méthodes de résolution de problèmes mal posés*, Mir (1976).

[23] L. v. Wolfersdorf: *On an electromagnetic inverse problem for dispersive media*, Quart. Appl. Math. 49 (1991), pp. 237–246.

[24] K. Yosida: *Functional Analysis*, Springer–Verlag (1969).

Inverse Problem of Ground Water Filtration: Identifiability, Discretization and Regularization

Gennadi Vainikko
Department of Mathematics
University of Tartu
EE2400 Tartu, Estonia

Abstract
Projection discretization schemes with Tikhonov regularization for the inverse problem of ground water filtration and other linear ill-posed problems with noncompact operators and noisy data are proposed. The noisy operators may converge only pointwise. The identifiability of the filtration coefficient among the functions of the class $L^2(\Omega)$ is discussed.

Introduction

It is well known that the steady state filtration of ground water in a porous media can be described by the equation

$$-\mathrm{div}(a(x)\nabla u(x)) = f(x), \quad x \in \Omega \subset I\!R^n$$

accompanied by the Neumann boundary condition. A physical background, together with a short derivation of the equation can be found in Sections 1.1 – 1.2. If u, f, and the boundary flux are given, one can try to determine the filtration coefficient $a(x)$ from the same equation (the inverse problem). The method of characteristics is here most popular enabling a numerical solution as well as a theoretical investigation of the identifiability (see Richter [4]). A disadvantage of the method is that it needs the second derivatives of u in whole Ω. Usually u is measured only on a irregular grid. Therefore different other methods have been proposed to relax the information demands on u, see e.g. Yeh [13], Kunisch [3], Kohn and Lowe [2] and the literature in those works.

In this paper we develop projection discretizations of the inverse problem together with Tikhonov regularization. We start from the weak formulation of the problem and give an equivalent operator equation formulation (Sections 1.3 – 1.5). Mathematically, we obtain a linear ill-posed problem with a noncompact operator. In some physically important case the noisy operators and their adjoints converge only pointwise, nevertheless the convergence of the methods takes place (Section 3). Moreover, for some special scheme (for the method of minimal error), the discretization parameter and the regularization parameter may tend to zero independently from one another. The method of minimal error accompanied by a more general class of regularization methods was already analyzed in [8]; see also [9]. Here we restrict ourselves to the Tikhonov regularization since this allows us to present rather elementary convergence analysis and relax the assumptions. The convergence analysis in the case of general projection methods is new in this paper. We call attention also to some (possibly new) matrix form of the general regularized projection method.

Only first derivatives of u are needed to apply the proposed methods. An advantage of our approach is that the boundary condition is needed only on a part of $\partial\Omega$ guaranteeing the identifiability. In Section 2, the results on the identifiability of $a(x)$ in the class $L^2(\Omega)$ or even $L^1(\Omega)$ is breafly discussed following Vainikko and Kunisch [11].

1 Mathematical model of ground water filtration

1.1 Quantities characterising the filtration.

Filtration of ground water (or oil or other fluids) means the soaking through a porous medium (soil) in a situation where the fluid fully fills the pores. The *porosity* $\mu = \mu(x)$ of the medium is defined by $\mu = V_p/V$ where V is the volume of a medium element and V_p is the volume of pores in it. Typically, the porosity is of order 0.5, e.g. for sand and gravel μ varies between 0.3 and 0.45, for clay between 0.4 and 0.55. The *filtration velocity* $\underline{v} = \underline{v}(x)$ is defined by $\underline{v} = \mu\underline{w}$ where $\underline{w}(x)$ is the mean velocity of water at the point x; the mean is assumed to be taken on a small medium element but not too small one so that the influence of the architecture of the pores will be eliminated. The *piezometric head* $u(x)$ is the level of the water in a pipe or well with water proof walls and open bottom at the point x of porous medium; the level is measured from a common reference plain. According to the Darcy law,

$$\underline{v}(x) = -a(x)\nabla u(x)$$

where $a = a(x) > 0$ is called the *filtration* (or transmissivity, or conduction) *coefficient.* For different kinds of soil, the filtration coefficient varies in a large scale, e.g.

	a [cm/sec]
gravel	0.1 ... 0.05
sand	1.0 ... 0.005
limestone	0.01 ... 0.003
clay	0.0005 ... 0.000000001

1.2 Filtration equation.

Consider a small region V in the porous medium with the boundary S and unit outer normal $\nu = \nu(x)$ to S at $x \in S$. The integral $\int_S \underline{v}(x) \cdot \nu(x) dS$ gives the difference between the water volumes soaking through S out from V and into V. Assuming the water to be incompressable, this amount must be compensated by sources and sinks of water in V:

$$\int_S \underline{v}(x) \cdot \nu(x) dS = \int_V f(x) dx$$

where $f(x)$ is the density of the sinkings (out and in). Using the Gauss-Ostrogradski formula, the equality takes the form

$$\int_V \operatorname{div}\underline{v}(x) dx = \int_V f(x) dx.$$

Since this holds for any region V, we have

$$\operatorname{div}\underline{v}(x) = f(x),$$

and due to the Darcy law, we obtain the *filtration equation*

$$(1) \qquad -\operatorname{div}(a(x)\nabla u(x)) = f(x), \quad x \in \Omega,$$

where $\Omega \subset I\!R^3$ is a region in the porous medium under consideration. To determine u from (1) uniquely, one has to put an appropriate boundary condition, usually the Neumann condition

$$(2) \qquad a(x)\nabla u(x) \cdot \nu(x) = g(x), \quad x \in \partial\Omega.$$

Now $\nu = \nu(x)$ means the unit outer normal to $\partial\Omega$. Note that $g = a\nabla u \cdot \nu = -\underline{v} \cdot \nu$ is the filtration velocity through $\partial\Omega$.

1.3 Inverse problem

For mining works, the water will be pumped out from a subregion of Ω causing changes of ground water level $u(x)$ in whole Ω. One can find $u(x)$ from the *direct problem* (1), (2) with f corresponding to (steady state) pumping the water out of a subregion. A serious difficulty on this way arises since the filtration coefficient $a(x)$ is usually known very coarsely. There is a possibility to determine $a(x)$ from (1), (2) using previous measurements of $u(x)$ which correspond to "natural" $f(x)$ (without pumpings). Thus we reach to the *inverse problem*: given f, g, u, find a. Usually, $a(x)$ can be uniquelly determined even if the boundary condition is given only on an appropriate part Γ of $\partial\Omega$ (see Section 2). Thus we formulate the inverse problem as follows: given a bounded region $\Omega \subset I\!R^n$ with a piecewise smooth boundary $\partial\Omega$, a relatively open set $\Gamma \subset \partial\Omega$ with a piecewise smooth (relative) boundary on $\partial\Omega$, functions $f \in L^2(\Omega)$, $g \in L^2(\Gamma)$ and $u \in W^{1,\infty}(\Omega)$, find $a(x) > 0$, $x \in \Omega$, such that

$$(3) \qquad \begin{aligned} -\operatorname{div}(a(x)\nabla u(x)) &= f(x), && x \in \Omega, \\ a(x)\nabla u(x) \cdot \nu(x) &= g(x), && x \in \Gamma \subset \partial\Omega \end{aligned}$$

(the derivatives in the equation are understood in the sense of distributions). The *weak formulation* of the inverse problem (3) reads as follows: find $a \in L^2(\Omega)$ such that

$$(4) \qquad \int_\Omega a(x)\nabla u(x) \cdot \nabla w(x) dx = \int_\Omega f(x)w(x)dx + \int_\Gamma g(x)w(x)dS \quad \forall w \in H^1(\Omega, \Gamma)$$

where

$$H^1(\Omega, \Gamma) = \{w \in H^1(\Omega) : w(x) = 0 \text{ for } x \in \partial\Omega \setminus \Gamma\} \subseteq H^1(\Omega).$$

One obtains (4) multiplying the equation by $w(x)$, integrating over Ω, integrating by parts in the left side and using the boundary condition. Note that $\Gamma \subset \partial\Omega$ may be void — in this case the boundary condition is omitted in (3) and the surfice integral is omitted in (4).

1.4 Operator equation formulation of the inverse problem.

Introduce the function $\psi \in H^1(\Omega, \Gamma)$ as the solution of the problem

$$(5) \qquad \begin{aligned} -\Delta\psi(x) = f(x), &\quad x \in \Omega, \\ \nabla\psi(x) \cdot \nu(x) = g(x), &\quad x \in \Gamma, \\ \psi(x) = 0, &\quad x \in \partial\Omega \setminus \Gamma. \end{aligned}$$

In the case $\Gamma \neq \partial\Omega$ (5) is uniquely solvable. In the case $\Gamma = \partial\Omega$ we have the Neumann boundary condition on whole $\partial\Omega$, and (5) is solvable provided that

$$\int_\Omega f(x)dx + \int_{\partial\Omega} g(x)dS = 0.$$

This condition is necessary also for the solvability of the inverse problem (3) if $\Gamma = \partial\Omega$.

Due (5), for any $w \in H^1(\Omega, \Gamma)$ we have

$$\int_\Omega \nabla\psi(x) \cdot \nabla w(x)dx = \int_\Omega f(x)w(x)dx + \int_\Gamma g(x)w(x)dS.$$

Thus, (4) can be reformulated as follows:

$$\int_\Omega [a(x)\nabla u(x) - \nabla\psi(x)] \cdot \nabla w(x)dx = 0 \quad \forall w \in H^1(\Omega, \Gamma).$$

Introduce the space of gradient vector functions

$$G = G(\Omega, \Gamma) = \{\nabla w : w \in H^1(\Omega, \Gamma)\} \subset (L^2(\Omega))^n.$$

Note that G is a closed subspace of $(L^2(\Omega))^n$; let Q_G denote the corresponding orthogonal projection. The reformulation of (4) means that

$$Q_G(a\nabla u - \nabla\psi) = 0$$

or, since $\nabla\psi \in G$,

$$Q_G(a\nabla u) = \nabla\psi. \tag{6}$$

This the *operator equation formulation* of (4).

Let us denote by $T = T_u \in \mathcal{L}(L^2(\Omega), G(\Omega, \Gamma))$ the operator occuring in equation (6):

$$Ta = T_u a = Q_G(a\nabla u), \quad a \in L^2(\Omega). \tag{7}$$

It depends on u as a functional parameter. Thus we have the equation

$$Ta = \nabla\psi \tag{8}$$

to determine $a \in L^2(\Omega)$.

1.5 Ill-posedness of the inverse problem.

Let us present the formula for the adjoint operator $T^* \in \mathcal{L}(G(\Omega, \Gamma), L^2(\Omega))$. For any $a \in L^2(\Omega)$, $\nabla w \in G(\Omega, \Gamma)$ we have

$$(Ta, \nabla w)_G = (Q_G(a\nabla u), \nabla w)_{L^2(\Omega)^n} = (a\nabla u, Q_G\nabla w)_{L^2(\Omega)^n}$$

$$= (a\nabla u, \nabla w)_{L^2(\Omega)^n} = (a, \nabla u \cdot \nabla w)_{L^2(\Omega)},$$

hence

$$T^*\nabla w = \nabla u \cdot \nabla w, \quad \nabla w \in G(\Omega, \Gamma). \tag{9}$$

It is easy to check that if $u \in C^1(\overline{\Omega})$ and $\nabla u(x^0) = 0$ for a $x^0 \in \overline{\Omega}$ then T^* has a non-closed range $\mathcal{R}(T^*) \subset L^2(\Omega)$. Moreover, for $n \geq 2$ it occurs that $\mathcal{R}(T^*)$ is non-closed even if $|\nabla u(x)| \geq c_0 > 0$ everywhere in $\overline{\Omega}$ (here the case $n \geq 2$ differs from the case $n = 1$). This is connected with the fact that solving the equation $\nabla u(x) \cdot \nabla w(x) = a(x)$ with respect to w one has to integrate $a(x)$ only in one direction given by the vector field $\nabla u(x)$.

It is also clear from (9) that the operator $T^* \in \mathcal{L}(G(\Omega, \Gamma), L^2(\Omega))$ is noncompact. Consequently, $T \in \mathcal{L}(L^2(\Omega), G(\Omega, \Gamma))$ is noncompact and has a non-closed range $\mathcal{R}(T) \subset G(\Omega, \Gamma)$. The last property of T causes the ill-posedness of problem (8).

A pecuarility of the inverse problem (8) is that an application of T to a $a \in L^2(\Omega)$ is much more expensive operation than an application of T^* to a $\nabla w \in G(\Omega, \Gamma)$, cf. (9). Namely, to find $Ta = \nabla w$ one has to determine $w \in H^1(\Omega, \Gamma)$ from the bondery value problem

$$\begin{gathered} \Delta w = \operatorname{div}(a(x)\nabla u(x)), \quad x \in \Omega, \\ \nabla w(x) \cdot \nu(x) = a\nabla u(x) \cdot \nu(x), \quad x \in \Gamma, \\ w(x) = 0, \quad x \in \partial\Omega \setminus \Gamma. \end{gathered}$$

Therefore discretizations of (8) should be organized so that only T^* and not T is involved.

1.6 Approximations of T.

The data u, f, g are available with some errors. Here we discuss the influence of errors in u, the piezometric head. In practice this function is given (with some measurement errors) on a non-regular grid and must be somehow extended to Ω. Thus, instead of the exact $u \in W^{1,\infty}(\Omega)$ and $T = T_u$ (see (7)) we have at our disposal some $u_\eta \in W^{1,\infty}(\Omega)$ and the corresponding $T_\eta = T_{u_\eta}$:

$$Ta = Q_G(a\nabla u), \quad T_\eta a = Q_G(a\nabla u_\eta), \quad a \in L^2(\Omega).$$

It is essential to try to construct the approximation u_η so that

$$\sup_{x\in\Omega} |\nabla u_\eta(x) - \nabla u(x)| \le \eta \tag{10}$$

with a small $\eta \ge 0$. Then we obviously have also

$$\|T_\eta - T\|_{\mathcal{L}(L^2(\Omega), G(\Omega,\Gamma))} \le \eta. \tag{11}$$

In some cases (10) becomes unrealistic. Namely, if Ω contains different kinds of soil with unknown interfaces between them, then the "physical" $a(x)$ has jumps on these surfaces causing jumps of ∇u. From the grid values of $u(x)$ one cannot locate the interfaces exactly, and $\sup |\nabla u_\eta(x) - \nabla u(x)|$ becomes of the order of jumps of ∇u. On the other hand, estimates of the type

$$\sup_{x\in\Omega} |\nabla u_\eta(x)| \le c, \quad \|\nabla u_\eta - \nabla u\|_{L^2(\Omega)^n} \le \eta \tag{12}$$

with some small $\eta > 0$ are realistic. A consequence of the second inequality is that

$$\|(T_\eta - T)a\|_{L^2(\Omega)^n} \le \eta \|a\|_{L^\infty(\Omega)}. \tag{13}$$

It is also easy to see that (12) implies the pointwise convergence of T_η to T and T_η^* to T^* as $\eta \to 0$:

$$\|(T_\eta - T)a\|_{L^2(\Omega)^n} \to 0 \quad \text{as} \quad \eta \to 0 \quad \forall a \in L^2(\Omega), \tag{14}$$

(15) $$\|(T_\eta^* - T^*)\nabla w\|_{L^2(\Omega)} \to 0 \quad \text{as} \quad \eta \to 0 \quad \forall\, \nabla w \in G(\Omega, \Gamma).$$

Thus we are in a situation where only a pointwise approximation of T and T^* is available.

1.7 Approximation of $\nabla\psi$.

Errors of f and g cause errors in $\nabla\psi$, the right hand term of equation (8). Recall that $\psi \in H^1(\Omega, \Gamma)$ is determined as the solution of boundary value problem (5).

Assume that, instead of exact data $f \in L^2(\Omega)$ and $g \in L^2(\Gamma)$ we have at our disposal some approximations $f_{\delta_1} \in L^2(\Omega)$ and $g_{\delta_2} \in L^2(\Gamma)$ such that

(16) $$\|f_{\delta_1} - f\|_{L^2(\Omega)} \le \delta_1, \qquad \|g_{\delta_2} - g\|_{L^2(\Gamma)} \le \delta_2.$$

We assert that for the corresponding ψ and ψ_{δ_1,δ_2} we have

(17) $$\|\nabla\psi_{\delta_1,\delta_2} - \nabla\psi\|_{L^2(\Omega)^n} \le c_1\delta_1 + c_2\delta_2 \equiv \delta;$$

the values of constants c_1 and c_2 will be discussed too. Indeed, $\psi_{\delta_1,\delta_2} - \psi$ is the solution of (5) with the right hand terms $f_{\delta_1} - f$ and $g_{\delta_2} - g$. Therefore we only need to establish an estimate

$$\|\nabla\psi\|_{L^2(\Omega)^n} \le c_1\|f\|_{L^2(\Omega)} + c_2\|g\|_{L^2(\Gamma)}$$

for the solution of (5). Taking the scalar product of both sides of the equation in (5) with ψ, integrating by parts and using the boundary conditions of (5) we find

(18) $$\int_\Omega |\nabla\psi(x)|^2 dx = \int_\Omega f(x)\psi(x)dx + \int_\Gamma g(x)\psi(x)dS.$$

Hence

$$\|\nabla\psi\|^2_{L^2(\Omega)^n} \le \|f\|_{L^2(\Omega)}\|\psi\|_{L^2(\Omega)} + \|g\|_{L^2(\Gamma)}\|\psi\|_{L^2(\Gamma)}.$$

Making use of the Friedrichs inequality

(19) $$\|\psi\|_{L^2(\Omega)} \le c_1\|\nabla\psi\|_{L^2(\Omega)^n}, \quad \psi \in H^1(\Omega, \Gamma)$$

and the trace inequality

(20) $$\|\psi\|_{L^2(\Gamma)} \le c_2\|\nabla\psi\|_{L^2(\Omega)^n}, \quad \psi \in H^1(\Omega, \Gamma)$$

we obtain (17).

The smallest values of c_1 and c_2 in (19) and (20) are connected with some eigenvalue problems. Namely, one can prove that

$$c_1 = (1/\lambda_1)^{1/2}, \quad c_2 = (1/\mu_1)^{1/2}$$

where λ_1 is the smallest eigenvalue of the problem

$$\begin{aligned} -\Delta w(x) &= \lambda w(x), \quad x \in \Omega, \\ \nabla w(x) \cdot \nu(x) &= 0, \quad x \in \Gamma, \\ w(x) &= 0, \quad x \in \partial\Omega \setminus \Gamma, \end{aligned}$$

and μ_1 is the smallest eigenvalue of the problem

$$
\begin{aligned}
-\Delta w(x) &= 0, \quad x \in \Omega, \\
\nabla w(x) \cdot \nu(x) &= \mu w(x), \quad x \in \Gamma, \\
w(x) &= 0, \quad x \in \partial\Omega \setminus \Gamma.
\end{aligned}
$$

With other constans c_1 and c_2, (17) can be obtained also from conditions

$$\|f_{\delta_1} - f\|_{H^{-1}(\Omega,\Gamma)} \leq \delta_1, \qquad \|g_{\delta_2} - g\|_{H^{-1/2}(\Gamma)} \leq \delta_2$$

where $\|f\|_{H^{-1}(\Omega,\Gamma)} = \sup_{u \in H^1(\Omega,\Gamma), \|u\|_{H^1(\Omega)} \leq 1} \int_\Omega f(x)u(x)dx$.

2 Identifiability

2.1 L^1-identifiability.

The discretization and regularization methods which will be considered in Section 3 converge to the solution $a(x)$ of problem (4) of minimal $L^2(\Omega)$ norm. A question about the uniqueness (identifiability) of $a(x)$ among the functions of the class $L^2(\Omega)$ acutely arises. For sufficiently smooth data, the identifiability of $a(x)$ from (3) in the class of smooth functions has been sufficiently fully analysed by Richter [4], Chicone and Gerlach [1], and Kunish [3]. Vainikko and Kunish [11] extended the results to the identifiability within the class of $L^1(\Omega)$-functions and relaxed the smoothness conditions on data. In this section we reproduce the main results of [11]. For proofs, see [11].

We slightly generalize our inverse problem (cf. (4)): having data $u \in W^{1,\infty}(\Omega)$, $f \in L^1(\Omega)$, $g \in L^1(\Omega)$ at our disposal, we look for a function $a \in L^1(\Omega)$ such that

$$\int_\Omega a\nabla u \cdot \nabla w dx = \int_\Omega f w dx + \int_\Gamma g w dS \quad \text{for all} \quad w \in W^{1,\infty}(\Omega, \Gamma) \tag{21}$$

where $W^{1,\infty}(\Omega,\Gamma) = \{w \in W^{1,\infty}(\Omega) : w(x) = 0 \text{ for } x \in \partial\Omega \setminus \Gamma\}$. Let us recall that $\partial\Omega$, the boundary of an open bounded region $\Omega \subset \mathbb{R}^n$, is assumed to be piecewise smooth and $\Gamma \subset \partial\Omega$ is relatively open subset of $\partial\Omega$ with a piecewise smooth (relative) boundary on $\partial\Omega$.

We say that the filtration coefficient a is *L^1-identifiable from problem (21) on a subregion* $\Omega' \subseteq \Omega$ if, for any solutions $a_1, a_2 \in L^1(\Omega)$ to problem (21), $a_1(x) = a_2(x)$ for a.e. $x \in \Omega'$. Our goal is to describe subregions $\Omega' \subseteq \Omega$ where a is L^1-identifiable. It is clear that L^1-identifiability of a from (21) on Ω' implies L^2-identifiability of a from (4) on the same set Ω'.

It is clear also that a is L^1-identifiable from (21) on $\Omega' \subseteq \Omega$ if and only if any solution $a \in L^1(\Omega)$ to the homogeneous problem

$$\int_\Omega a\nabla u \cdot \nabla w dx = 0 \quad \text{for all} \quad w \in W^{1,\infty}(\Omega, \Gamma) \tag{22}$$

vanishes almost everwhere on Ω'. Thus, for L^1-identifiability, assumptions on $u \in W^{1,\infty}(\Omega)$ are deciding. Instead of $f \in L^1(\Omega)$, $g \in L^1(\Gamma)$ we could assume that f, g define linear continuous functionals on $W^{1,\infty}(\Omega,\Gamma)$. First we consider the case where ∇u is continuous; later we shall treat the case where ∇u may have jumps.

2.2 Flow curves.

More precisely, we assume here that

$$u \in W^{1,\infty}(\Omega) \cap W^{2,\infty}(\Omega_\epsilon) \text{ for all } \epsilon > 0 \tag{23}$$

where Ω_ϵ consists of all points $x \in \Omega$ such that the distance from x to a nearest nonsmoothness point of $\partial\Omega$ exceeds ϵ; if $\partial\Omega$ is smooth, then (23) means that $u \in W^{2,\infty}(\Omega)$. Introduce the flow curve $x = \varphi(t, y)$ through a point $y \in \Omega$ as the maximal solution (the solution on the maximal time interval) to the Cauchy problem

$$dx/dt = -\nabla u(x), \quad x(0) = y. \tag{24}$$

Physically, the ground water flows along those curves but the speed depends on ∇u and the filtration coefficient (Darcy's law).

Due to (23), ∇u is bounded and locally Lipschitz continuous on Ω, with a possible unbounded growth of the Lipschitz coefficient only as x tends to a non-smoothness point of $\partial\Omega$. Therefore, problem (24) is uniquely solvable and $\varphi(t, y)$ is defined on a finite or infinite time interval (t_y^-, t_y^+); if t_y^- or t_y^+ is finite, then $\varphi(t, y)$ tends to a point on $\partial\Omega$ as $t \downarrow t_y^-$ or $t \uparrow t_y^+$, respectively. If $\nabla u(y) \neq 0$, then $\nabla u(\varphi(t, y)) \neq 0$ for all $t \in (t_y^-, t_y^+)$ and $u(\varphi(t, y))$ is strictly decreasing:

$$du(\varphi(t, y))/dt = \nabla u(\varphi(t, y)) \cdot d\varphi(t, y)/dt = -|\nabla u(\varphi(t, y))|^2 < 0 \quad (t \in (t_y^-, t_y^+)).$$

A corollary is that problem (24) allows no periodic solutions.

Introduce further the following subsets of Ω:

$$\Omega_C = \{y \in \Omega : \nabla u(y) = 0\},$$

$$\Omega^+ = \{y \in \Omega : \nabla u(y) \neq 0, t_y^+ = +\infty\}, \quad \Omega^- = \{y \in \Omega : \nabla u(y) \neq 0, t_y^- = -\infty\},$$

$$\Omega_\Gamma^+ = \left\{ y \in \Omega \,\middle|\, \begin{array}{l} \nabla u(y) \neq 0,\ t_y^+ < +\infty,\ \varphi(t, y) \text{ transversely (non-tangen-} \\ \text{tially) reaches a smoothness point of } \Gamma \subseteq \partial\Omega \text{ as } t \uparrow t_y^+ \end{array} \right\},$$

$$\Omega_\Gamma^- = \left\{ y \in \Omega \,\middle|\, \begin{array}{l} \nabla u(y) \neq 0,\ t_y^- < -\infty,\ \varphi(t, y) \text{ transversely (non-tangen-} \\ \text{tially) reaches a smoothness point of } \Gamma \subseteq \partial\Omega \text{ as } t \downarrow t_y^- \end{array} \right\}.$$

Since $\Gamma \subset \partial\Omega$ is relatively open, Ω_Γ^+ and Ω_Γ^- are open subsets of Ω. The interior of Ω^+ and Ω^- will be denoted by $\text{int}\Omega^+$ and $\text{int}\Omega^-$, respectively.

2.3 Main results and comments.

If Ω_C, the set of critical points of u, has a positive Lebesgue measure, then the function $a \in L^\infty(\Omega)$ defined by $a(x) = 1$ if $\nabla u(x) = 0$ and $a(x) = 0$ elsewhere in Ω satisfies the homogeneous problem (22) but does not vanish a.e. on Ω_C. Thus, a cannot be L^1-identifiable from (21) on Ω_C if meas $\Omega_C > 0$.

THEOREM 2.1. *Under condition (23), the filtration coefficient a is L^1-identifiable from problem (21) on the sets* $\text{int}\Omega^+$, $\text{int}\Omega^-$ *and* Ω_Γ^+, Ω_Γ^-; *on* $\text{int}\Omega_\Gamma^+$ *and* $\text{int}\Omega^-$, *L^1-identifiability holds even if* $\Gamma = \emptyset$.

Remark 2.1. Theorem 2.1 fails if assumption (23) is replaced by $u \in W^{1,\infty}(\Omega) \cap W^{2,p}(\Omega)$, $p < \infty$ (see a counter-example in [11]). Thus, for L^1-identifiability, the conditions of Theorem 2.1 are rather close to the necessary ones. For the L^2-identifiability of a on $\text{int}\Omega^{\pm}$ and $\Omega_{\Gamma}^{\pm}$ the conditions of Theorem 2.1 are only sufficient and seem to be far from the necessity – examining the examples one can conjecture that here (23) may be replaced by $u \in C^1(\overline{\Omega}) \cap H^2(\Omega)$. Unfortunately, our proof method does not work in this case since the flow curves may be non-uniquely determined from (24) if (23) fails.

Remark 2.2. Under condition $u \in W^{2,\infty}(\Omega)$, it is easy to see that $\text{meas}\partial\Omega_{\Gamma}^{\pm} = 0$, therefore the L^1-identifiability result of Theorem 2.1 can be extended from Ω_{Γ}^{+} and Ω_{Γ}^{-} to their closures. On the other hand, for some (rather exotic) functions $u \in W^{2,\infty}(\Omega)$ and even $u \in C^2(\overline{\Omega})$, the sets $\partial(\text{int}\Omega^{\pm})$ may be of positive measure, and the result of Theorem 2.1 about the L^1- identifiability of a on $\text{int}\Omega^{+}$ and $\text{int}\Omega^{-}$ cannot be extended to the closures of those sets (see a counter-example in [11]). Here a difference between the results about L^1-identifiability and C-identifiability appears.

Remark 2.3. If $u \in W^{2,\infty}(\Omega)$ and $u(x) = 0$ for $x \in \partial\Omega \setminus \Gamma$, then

$$\overline{\Omega^{+}} \cup \overline{\Omega^{-}} \cup \overline{\Omega_{\Gamma}^{+}} \cup \overline{\Omega_{\Gamma}^{-}} \supset \Omega \setminus \Omega_C.$$

2.4 Case of piecewise smooth u.

Now consider the case where u remains continuous on Ω but ∇u may have discontinuities on piecewise smooth surface M_i $(i = 1, \dots, m)$ in Ω. Physically, M_i are the surfaces between different types of soil. Denote $M = \cup_{i=1}^{m} M_i$. We assume that

$$(25) \qquad u \in W^{1,\infty}(\Omega) \cap W^{2,\infty}(\Omega_{\epsilon,M}) \quad \text{for all} \quad \epsilon > 0$$

where $\Omega_{\epsilon,M}$ consists of all points $x \in \Omega \setminus M$ such that the distance from x ta a nearest non-smoothness point of $\partial\Omega$ and M_i $(i = 1, \dots, m)$ as well to a nearest intersection point of a pair of surfaces $\partial\Omega$, M_i $(i = 1, \dots, m)$ exceeds ϵ. Further, we introduce the following *consistency condition*:

$$(26) \qquad \left\{ \begin{array}{l} \text{There is a strictly positive piecewise smooth function } a_{test}, \\ \text{with possible jumps on } M, \text{ such that } \operatorname{div}(a_{test}\nabla u) \in L^1(\Omega) \end{array} \right.$$

where the derivatives are understood in the sense of distributions. Usually, the "physical" solution of the inverse problem (3) meets this requirement.

LEMMA 2.1. *Let $x^0 \in M_i$ be in the smooth part of M_i, and let $\nu_i(x^0)$ denote a unit normal to M_i at this point. Then, under conditions (25) and (26),*

$$(27) \qquad \lim_{\substack{(x-x^0)\cdot\nu_i(x^0)>0 \\ x\to x^0}} a_{test}(x)\nabla u(x) \cdot \nu_i(x^0) = \lim_{\substack{(x-x^0)\cdot\nu_i(x^0)<0 \\ x\to x^0}} a_{test}(x)\nabla u(x) \cdot \nu_i(x^0).$$

Note that Lemma 2.1 holds without the positiveness assumption of a_{test}. The positiveness of a_{test} is needed when the flow curves are considered. The following assertion is a direct corollary from (25) – (27): If a flow curve $x = \varphi(t, x)$, in a finite time moment, transversely reaches a smoothness point of M_i, then this flow curve passes M_i transversely to the other side of M_i and continues there. We can define sets Ω^{+}, Ω^{-}, Ω_{Γ}^{+}, Ω_{Γ}^{-} as in Subsection 2.2 adding a requirement about the transversal cuttings of M, e.g.,

$$\Omega^+ = \left\{ y \in \Omega \;\middle|\; \begin{array}{l} \nabla u(y) \neq 0,\ t_y^+ = +\infty \text{ and, for } 0 \leq t < \infty,\ \varphi(t,y) \text{ cuts } M \text{ not} \\ \text{more than finite times whereby every cutting is transversal} \\ \text{and takes place at a smoothness point of an } M_i, \quad 1 \leq i \leq m \end{array} \right\},$$

$$\Omega_\Gamma^- = \left\{ y \in \Omega \;\middle|\; \begin{array}{l} \nabla u(y) \neq 0,\ t_y^- > -\infty,\ \varphi(t,y) \text{ transversely reaches a smooth-} \\ \text{ness point of } \Gamma \subseteq \partial\Omega \text{ as } t \downarrow t_y^- \text{ and, for } t_y^- < t \leq 0 \text{ cuts } M \\ \text{not more than finite times whereby every cutting is transversal} \\ \text{and takes place at a smoothness point of an } M_i, \quad 1 \leq i \leq m \end{array} \right\}.$$

THEOREM 2.2. *Under conditions (25) and (26), the transmissivity coefficient a is L^1-identifiable from problem (21) on the sets* $\text{int}\Omega^+$, $\text{int}\Omega^-$ *and* Ω_Γ^+, Ω_Γ^- *specified above; on* $\text{int}\Omega^+$ *and* Ω^- *the L^1-identifiability holds even if* $\Gamma = \emptyset$.

3 Discretization and regularization

3.1 Generalization of the inverse problem.

There is no reason to confine ourselves to the concrete problem (8) when the projection discretizations and Tikhonov or other regularizations are treated. We shall consider a general linear ill-posed problem

$$Ta = \mathbf{f} \tag{28}$$

where $T \in \mathcal{L}(E, F)$ is an arbitrary linear bounded operator between (possibly complex) Hilbert spaces E and F. To problem (8) we return putting $E = L^2(\Omega)$, $F = G(\Omega, \Gamma)$, and defining $T \in \mathcal{L}(E, F)$ by (7) and the right hand term $\mathbf{f} = \nabla\psi$ through the solution ψ of problem (5). We shall use only the properties of $T \in \mathcal{L}(E, F)$ established in Section 1 for the inverse problem (8).

3.2 What to discretize?

The Tikhonov regularization of (28) provides the equation

$$\alpha a + T^*Ta = T^*\mathbf{f}, \quad \alpha > 0 \tag{29}$$

or equivalently,

$$a = T^*\mathbf{g}, \quad \alpha\mathbf{g} + TT^*\mathbf{g} = \mathbf{f}, \qquad \alpha > 0. \tag{30}$$

There are two possible ways to perform a discretization:

(i) discretize (28) and then regularize the discretized equation;

(ii) discretize the regularized equation (29) or (30).

We mainly follow the way (i). Our purpose is to design methods where the limit processes in discretization and regularization parameters, $h \to 0$ and $\alpha \to 0$, may be performed independently.

3.3 Projection discretization

of equation (28) consists in the approximation by the equation

$$Q_hTP_ha = Q_h\mathbf{f} \tag{31}$$

where $P_h \in \mathcal{L}(E, E)$ and $Q_h \in \mathcal{L}(F, F)$ are orthogonal projection operators corresponding to some finite dimensional subspaces $E_h \subset E$ and $F_h \subset F$ respectively. The Tikhonov method replaces (31) by

$$\alpha a + P_h T^* Q_h T P_h a = P_h T^* Q_h \mathbf{f}, \tag{32}$$

or equivalently,

$$a = P_h T^* Q_h \mathbf{g}, \quad \alpha \mathbf{g} + Q_h T P_h T^* Q_h \mathbf{g} = Q_h \mathbf{f}. \tag{33}$$

The matrix schemes of (32) and (33) depend on the bases of E_h and F_h. Let a basis $\varphi_1, \ldots, \varphi_l$ of E_h and a basis $\psi_1, \ldots, \psi_m$ of F_h be available (actually, $\varphi_j = \varphi_{j,h}$, $j = 1, \ldots, l = l_h$ and $\psi_i = \psi_{i,h}$, $i = 1, \ldots, m = m_h$, depend on the discretization parameter h but we omit the index h). Introduce the following matrices and vectors:

$$G_\varphi = \begin{pmatrix} (\varphi_1, \varphi_1) & (\varphi_2, \varphi_1) & \cdots & (\varphi_l, \varphi_1) \\ (\varphi_1, \varphi_2) & (\varphi_2, \varphi_2) & \cdots & (\varphi_l, \varphi_2) \\ \cdots & \cdots & \cdots & \cdots \\ (\varphi_1, \varphi_l) & (\varphi_2, \varphi_l) & \cdots & (\varphi_l, \varphi_l) \end{pmatrix},$$

$$G_\psi = \begin{pmatrix} (\psi_1, \psi_1) & (\psi_2, \psi_1) & \cdots & (\psi_m, \psi_1) \\ (\psi_1, \psi_2) & (\psi_2, \psi_2) & \cdots & (\psi_m, \psi_2) \\ \cdots & \cdots & \cdots & \cdots \\ (\psi_1, \psi_m) & (\psi_2, \psi_m) & \cdots & (\psi_m, \psi_m) \end{pmatrix},$$

$$B = \begin{pmatrix} (\varphi_1, T^*\psi_1) & (\varphi_2, T^*\psi_1) & \cdots & (\varphi_l, T^*\psi_1) \\ (\varphi_1, T^*\psi_2) & (\varphi_2, T^*\psi_2) & \cdots & (\varphi_l, T^*\psi_2) \\ \cdots & \cdots & \cdots & \cdots \\ (\varphi_1, T^*\psi_m) & (\varphi_2, T^*\psi_m) & \cdots & (\varphi_l, T^*\psi_m) \end{pmatrix},$$

$$\underline{c} = \begin{pmatrix} c_1 \\ c_2 \\ \vdots \\ c_l \end{pmatrix}, \quad \underline{d} = \begin{pmatrix} d_1 \\ d_2 \\ \vdots \\ d_m \end{pmatrix}, \quad \underline{f} = \begin{pmatrix} (\mathbf{f}, \psi_1) \\ (\mathbf{f}, \psi_2) \\ \vdots \\ (\mathbf{f}, \psi_m) \end{pmatrix}.$$

The Gram matrixes G_φ and G_ψ are hermitian, positive definite and invertible. The matrix scheme of (32) reads as follows:

$$a_{h,\alpha} = \sum_{j=1}^{l} c_j \varphi_j, \quad \alpha G_\varphi \underline{c} + B^* G_\psi^{-1} B \underline{c} = B^* G_\psi^{-1} \underline{f}. \tag{34}$$

Through (33) we obtain an equivalent form of (34):

$$a_{h,\alpha} = \sum_{j=1}^{l} c_j \varphi_j, \quad \underline{c} = G_\varphi^{-1} B^* \underline{d}, \quad \alpha G_\psi \underline{d} + B G_\varphi^{-1} B^* \underline{d} = \underline{f}. \tag{35}$$

Here B^* is the transposed complex conjugate matrix to B.

There is a third matrix scheme of (32) not involving the inverses of Gram matrices:

$$a_{h,\alpha} = \sum_{j=1}^{l} c_j \varphi_j, \quad \begin{cases} \alpha^{1/2} G_\varphi \underline{c} + B^* \underline{d} = 0 \\ B \underline{c} - \alpha^{1/2} G_\psi \underline{d} = \underline{f}. \end{cases} \tag{36}$$

Here $\underline{d}$ occurs as an auxiliary unknown. Eliminating $\underline{d}$ we obtain (34).

Let us denote by $\lambda_1 = \lambda_{1,h}$ and $\mu_1 = \mu_{1,h}$ the smallest eigenvalues of Gram matrices G_φ and G_ψ, respectively. The stability properties of systems (34), (35) and (36) are characterized by the norms of the corresponding inverse matrices. It is easy to prove that the spectral norms of the inverses can be estimated as follows:

$$\|(\alpha G_\varphi + B^* G_\psi^{-1} B)^{-1}\| \leq \lambda_1^{-1}\alpha^{-1},$$

$$\|(\alpha G_\psi + B G_\psi^{-1} B^*)^{-1}\| \leq \mu_1^{-1}\alpha^{-1},$$

$$\left\| \begin{pmatrix} \alpha^{1/2} G_\varphi & B^* \\ B & -\alpha^{1/2} G_\psi \end{pmatrix}^{-1} \right\| \leq \max\{\lambda^{-1}, \mu^{-1}\}\alpha^{-1/2}.$$

For some special B, these inequalities turn into equalities. We see that system (36) is essentially more stable than (34) or (35). Unfortunately, system (36) is not (positively or negatively) definite. Nevertheless, it seems that the matrix form (36) is worth to be examined in more details. This form seems to be not discussed in literature.

3.4 Convergence in the case of noisy data.

Let, instead of $\mathbf{f} \in \mathcal{R}(T) \subset F$ and $T \in \mathcal{L}(E, F)$, their approximations $\mathbf{f}_\delta \in F$ and $T_\eta \in \mathcal{L}(E, F)$ be given.

THEOREM 3.1. *Assume that* $\mathbf{f} \in \mathcal{R}(T)$, *and let* $a_0 \in \overline{\mathcal{R}(T^*)}$ *be the normal solution (i.e. the solution of minimal* E *norm) to equation (28). Let*

$$\|Q_h(\mathbf{f}_\delta - \mathbf{f})\|_F \leq \delta, \tag{37}$$

$$\|Q_h(T_\eta - T)a_0\|_F \leq c_0\eta, \quad c_0 = \text{const}, \tag{38}$$

$$\|T_\eta\| \leq c', \quad T_\eta^* g \to T^* g \quad \text{as} \quad \eta \to 0 \quad \forall g \in \overline{\mathcal{R}(T)} \subseteq F. \tag{39}$$

Further, assume that

$$\|a - P_h a\|_E \to 0 \quad \text{as} \quad h \to 0 \quad \forall a \in \overline{\mathcal{R}(T^*)} \subseteq E, \tag{40}$$

$$\|\mathbf{g} - Q_h \mathbf{g}\|_F \to 0 \quad \text{as} \quad h \to 0 \quad \forall \mathbf{g} \in \overline{\mathcal{R}(T)} \subseteq F, \tag{41}$$

Choose $\alpha = \alpha(h, \delta, \eta) > 0$ *in the regularized projection method (cf. (32))*

$$\alpha a + P_h T_\eta^* Q_h T_\eta P_h a = P_h T_\eta^* Q_h \mathbf{f}_\delta \tag{42}$$

so that

$$\alpha(h, \delta, \eta) \to 0, \quad \frac{\delta^2 + \eta^2 + \xi_{h,\eta}^2}{\alpha(h, \delta, \eta)} \to 0 \quad \text{as} \quad h \to 0, \delta \to 0, \eta \to 0 \tag{43}$$

where (see (39) and (40))

$$\xi_{h,\eta} \geq \|Q_h T_\eta (I - P_h) a_0\| \to 0 \quad \text{as} \quad h \to 0, \eta \to 0.$$

Then

$$\|a_{h,\alpha(h,\delta,\eta)} - a_0\|_E \to 0 \quad \text{as} \quad h \to 0,\ \delta \to 0,\ \eta \to 0$$

where $a_{h,\alpha}$ *is the (unique) solution of (42).*

Proof. Introduce the Tikhonov functional

$$\Phi_{h,\alpha}(a) = \|Q_h T_\eta P_h a - Q_h \mathbf{f}_\delta\|^2 + \alpha\|a\|^2, \quad a \in E, \alpha > 0.$$

It is minimized by $a_{h,\alpha}$, the solution of (42). In particular, $\Phi_{h,\alpha}(a_{h,\alpha}) \le \Phi_{h,\alpha}(a_0)$, i.e.

$$\|Q_h T_\eta P_h a_{h,\alpha} - Q_h \mathbf{f}_\delta\|^2 + \alpha\|a_{h,\alpha}\|^2$$

$$\le \|Q_h T_\eta P_h a_0 - Q_h \mathbf{f}_\delta\|^2 + \alpha\|a_0\|^2, \quad \alpha > 0. \tag{44}$$

Due to (37) and (38),

$$\|Q_h T_\eta P_h a_0 - Q_h \mathbf{f}_\delta\| \le \delta + c_0\eta + \xi_{h,\eta}. \tag{45}$$

From (43) – (45) we conclude that

$$\|Q_h T_\eta P_h a_{h,\alpha(h,\delta,\eta)} - Q_h \mathbf{f}_\delta\| \to 0 \quad \text{as} \quad h \to 0, \delta \to 0, \eta \to 0, \tag{46}$$

$$\limsup_{h\to 0,\delta\to 0,\eta\to 0} \|a_{h,\alpha(h,\delta,\eta)}\| \le \|a_0\|. \tag{47}$$

We first prove the weak convergence $a_{h,\alpha(h,\delta,\eta)} \rightharpoonup a_0$ as $h \to 0$, $\delta \to 0$ $\eta \to 0$. On the ground of (47), $a_{h,\alpha(h,\delta,\eta)}$ is bounded and consequently weakly compact as $h \to 0$, $\delta \to 0$, $\eta \to 0$; we just have to prove that a_0 is the unique weak limit point. Take any sequences $h_n \to 0$, $\delta_n \to 0$, $\eta_n \to 0$ such that a_{h_n,α_n} with $\alpha_n = \alpha(h_n, \delta_n, \eta_n) \to 0$ weakly converges to an element $a' \in E$ as $n \to \infty$. Then (see (47))

$$\|a'\| \le \liminf \|a_{h_n,\alpha_n}\| \le \|a_0\|. \tag{48}$$

Further, due to (39) – (41), for any $\mathbf{g} \in \overline{\mathcal{R}(T)}$ we have

$$(Q_{h_n} T_{\eta_n} P_{h_n} a_{h_n,\alpha_n}, \mathbf{g}) = (a_{h_n,\alpha_n}, P_{h_n} T^*_{h_n} Q_{h_n}\mathbf{g}) \to (a', T^*\mathbf{g}) = (Ta', \mathbf{g}).$$

Noticing also that $Q_h\mathbf{f}_\delta = Q_h T a_0 + Q_h(\mathbf{f}_\delta - \mathbf{f}) \to Ta_0$ (see (37) and (41)), we obtain

$$(Q_{h_n} T_{\eta_n} P_{h_n} a_{h_n,\alpha_n} - Q_{h_n}\mathbf{f}_{\delta_n}, \mathbf{g}) \to (Ta' - Ta_0, \mathbf{g}), \quad \forall \mathbf{g} \in \overline{\mathcal{R}(T)}.$$

On the ground of (46) the limit is zero:

$$(Ta' - Ta_0, \mathbf{g}) = 0, \quad \forall \mathbf{g} \in \overline{\mathcal{R}(T)}.$$

Taking $\mathbf{g} = T(a' - a_0)$ we obtain $\|Ta' - Ta_0\|^2 = 0$, $Ta' = Ta_0 = \mathbf{f}$. Now (48) together with the uniqueness of the normal solution implies $a' = a_0$. This proves the weak convergence $u_{h,\alpha(h,\delta,\eta)} \rightharpoonup a_0$. Moreover, making use of (47) we obtain

$$\|a_0\| \le \liminf_{h,\delta,\eta\to 0} \|a_{h,\alpha(h,\delta,\eta)}\| \le \limsup_{h,\delta,\eta\to 0} \|a_{h,\alpha(h,\delta,\eta)}\| \le \|a_0\|,$$

i.e. $\|a_{h,\alpha(h,\delta,\eta)}\| \to \|a_0\|$ as $h \to 0$, $\delta \to 0$, $\eta \to 0$, and the H-propety of Hilbert spaces quarantees that the convergence of $a_{h,\alpha(h,\delta,\eta)}$ to a_0 is actually strong. □

To choose the regularization parameter α by the residual principle, recall some properties of the residual function $\rho(\alpha) = \|Q_h T_\eta P_h a_{h,\alpha} - Q_h \mathbf{f}_\delta\|$:

$\rho(\alpha)$ is smooth and increasing in α for $\alpha > 0$,

$$\lim_{\alpha\to 0} \rho(\alpha) = \inf_{a\in H} \|Q_h T_\eta P_h a - Q_h \mathbf{f}_\delta\|,$$

$$\lim_{\alpha\to\infty} \rho(\alpha) = \|Q_h \mathbf{f}_\delta\|.$$

(see e.g. [5] or [12]). If $\mathbf{f} \in \mathcal{R}(T)$ and (37), (38) are satisfied then (see (45))

$$\lim_{\alpha\to 0} \rho(\alpha) \le \delta + c_0\eta + \xi_{h,\eta}.$$

THEOREM 3.2. *Assume the conditions of Theorem 3.1 with $f \neq 0$ but choose $\alpha = \alpha(h, \delta, \eta) > 0$ so that*

$$(49) \qquad \delta + c_0\eta + \xi_{h,\eta} \le \|Q_h(T_\eta a_{h,\alpha} - \mathbf{f}_\delta)\| \le b(\delta + c_0\eta + \xi_{h,\eta}), \quad b = \text{const} \ge 1.$$

Then $\|a_{h,\alpha(h,\delta,\eta)} - a_0\|_E \to 0$ as $h \to 0$, $\delta \to 0$, $\eta \to 0$.

Proof. Note that $P_h a_{h,\alpha} = a_{h,\alpha}$. For $\alpha = \alpha(h, \delta, \eta)$ satisfing (49) we obtain from (44) and (45)

$$(\delta + c_0\eta + \psi_{h,\eta})^2 + \alpha(h, \delta, \eta)\|a_{h,\alpha(h,\delta,\eta)}\|^2$$

$$\le \|Q_h T_\eta P_h a_0 - Q_h \mathbf{f}_\delta\|^2 + \alpha(h, \delta, \eta)\|a_0\|^2$$

$$\le (\delta + c_0\eta + \psi_{h,\eta})^2 + \alpha(h, \delta, \eta)\|a_0\|^2,$$

therefore

$$\|a_{h,\alpha(h,\delta,\eta)}\| \le \|a_0\|.$$

We see that (47) is fulfilled again; (46) is a direct consequence of (49). Now the arguments of the proof of Theorem 3.1 can be repeated. □

Remark 3.1. If $\|\mathbf{f}_\delta - \mathbf{f}\| \le \delta$, $\|T_\eta - T\| \le \eta$ then conditions (37) – (39) are fulfilled with $c_0 = \|a_0\|_E$.

Remark 3.2. We always may put $\xi_{h,\eta} = c'\|a_0 - P_h a_0\|$ with c' from (39). If $T \in \mathcal{L}(E, F)$ is compact and $\|T_\eta - T\| \le \eta$ then we may put $\xi_{h,\eta} = \eta\|a_0\| + \|(I - P_h)T^*\|\|a_0 - P_h a_0\|$; in (43), $\xi_{h,\eta}$ may be replaced by $\|(I - P_h)T^*\|\|a_0 - P_h a_0\|$. Note that $\|(I - P_h)T^*\| \to 0$ due to (40) and the compactness of $T^* \in \mathcal{L}(F, E)$.

3.5 Method of least squares

The method of least squares is a most popular projection method of the type (31): one chooses freely $E_h \subset E$ and puts $F_h = TE_h$. Due to this relation between E_h and F_h, one has $Q_h T P_h = T P_h$, and (31) specifies to the form

$$(50) \qquad T P_h a = Q_h \mathbf{f} \quad (F_h = T E_h).$$

Together with the Tikhonov regularization we obtain

$$\alpha a + P_h T^* T P_h a = P_h T^* \mathbf{f} \tag{51}$$

which may be considered as the Galerkin discretization of the equation (29). The matrix form of (51) reads as follows:

$$a_{h,\alpha} = \sum_{j=1}^{l} c_j \varphi_j, \quad (\alpha G_\varphi + G_\psi)\underline{c} = \underline{f}$$

where $\varphi_1, \ldots, \varphi_l$ is a basis of E_h; the Gram matrix G_ψ now corresponds to $\psi_i = T\varphi_i$, $i = 1, \ldots, l$, and $\underline{f}$ has the components (f, ψ_i), $i = 1, \ldots, l$. There is no possibility to avoid the computation of $T\varphi_i$, $i = 1, \ldots, l$, therefore the method of least squares is unsuitable for the inverse problem of ground water filtration (see Section 1.5).

Without any regularization, method (50) with exact data usually diverges: the natural conditions

$$\mathcal{N}(T) := \{a \in E : Ta = 0\} = \{0\}, \ \mathbf{f} \in \mathcal{R}(T), \ P_h a \to a \text{ as } h \to 0 \quad \forall a \in E$$

do not imply the convergence $a_{h,0} := T^{-1} Q_h \mathbf{f} \to T^{-1}\mathbf{f} =: a_0$ as $h \to h$. Convergence results for the regularized version (51) of the method in the case of noisy data can be obtained from Theorems 3.1 and 3.2 as a special case putting formally $Q_h = I$ (compare (42) and (51)).

3.6 Method of minimal error

The method of minimal error, or the dual method of least squares, is another special case of the general projection method (31): this time one freely chooses $F_h \subset F$ and puts $E_h = T^* F_h$. Due to this relation, $P_h T^* Q_h = T^* Q_h$, or $Q_h T P_h = Q_h T$, and (31) specifies to the form

$$Q_h T a = Q_h \mathbf{f}. \tag{52}$$

Together with the Tikhonov regularization we obtain

$$a = T^* Q_h \mathbf{g}, \quad \alpha \mathbf{g} + Q_h T T^* Q_h \mathbf{g} = Q_h \mathbf{f} \tag{53}$$

which may be considered as the Galerkin discretization of the equation (30).

Let $\psi_1, \ldots, \psi_m$ be a basis of F_h, and denote $\varphi_i = T^* \psi_i$, $i = 1, \ldots, m$. The matrix form of (53) reads as follows:

$$a_{h,\alpha} = \sum_{j=1}^{m} c_j T^* \psi_j, \quad (\alpha G_\psi + G_\varphi)\underline{c} = \underline{f}. \tag{54}$$

Here only T^* and not T itself is involved. Therefore, the method of minimal error is rather suitable for the inverse problem of the ground water filtration.

With exact data, method (52) converges without any regularization (cf.[10]):

(i) if $\mathbf{f} \in \mathcal{R}(T)$, and a_0 is any solution to (28) then $a_{h,0} = P_h a_0 \in T^* F_h$ is a solution to (52) and this solution is unique in $T^* F_h$;

(ii) if $\mathbf{f} \in \mathcal{R}(T)$ and $Q_h \mathbf{g} \to \mathbf{g}$ as $h \to 0$ for any $g \in F$ then $a_{h,0} \to a_0$ as $h \to 0$ where $a_{h,0}$ is the unique solution of (52) in $T^* F_h$ and a_0 is the normal solution of (28).

Often the approximation properties of the subspaces $E_h = T^* F_h$ are weak, and one has to solve (52) in high dimensions. In higher dimensions, the corresponding system (54) with

$\alpha = 0$ becomes more and more instable, and a regularization is useful even in the case of exact data.

Let us turn to the case of noisy data. In the general regularized projection method, the discretization parameter h and the regularization parameter α must be coordinated by the relation $\xi^2_{h,\eta}/\alpha(h,\delta,\eta) \to 0$ as $h \to 0$, $\delta \to 0$, $\eta \to 0$ (see condition (43)). It occurs that in the regularized method of minimal error, no coordination between α and h is needed — they may tend to zero independently.

THEOREM 3.3. *Let* $\mathbf{f} \in \mathcal{R}(T)$, *and let* a_0 *be the normal solution of (28). Let conditions (37) - (39) and (41) be satisfied. Choose* $\alpha = \alpha(\delta,\eta) > 0$ *in the regularized method of minimal error*

$$a = T^*_\eta Q_h \mathbf{g}, \quad \alpha \mathbf{g} + Q_h T_\eta T^*_\eta Q_h \mathbf{g} = Q_h \mathbf{f}_\delta \tag{55}$$

so that

$$\alpha(\delta,\eta) \to 0, \quad (\delta^2+\eta^2)/\alpha(\delta,\eta) \to 0 \quad as \quad \delta \to 0, \eta \to 0. \tag{56}$$

Then for the solution $a_{h,\alpha}$ *of (55) we have*

$$\|a_{h,\alpha(\delta,\eta)} - a_0\|_E \to 0 \quad \text{as} \quad h \to 0,\ \delta \to 0,\ \eta \to 0.$$

Proof. The solution $a_{h,\alpha}$ of (55) satisfies also the equation $\alpha a + T^*_\eta Q_\eta T_\eta a = T^*_\eta Q_\eta \mathbf{f}_\delta$ and minimizes the Tikhonov functional $\Phi_{h,\alpha}(u) = \|Q_h T_\eta a - Q_h \mathbf{f}_\delta\|^2 + \alpha\|a\|^2$, $a \in E$. The inequality $\Phi_{h,\alpha}(a_{h,\alpha}) \le \Phi_{h,\alpha}(a_0)$ provides

$$\|Q_h T_\eta a_{h,\alpha} - Q_h \mathbf{f}_\delta\|^2 + \alpha\|a_{h,\alpha}\|^2 \le \|Q_h T_\eta a_0 - Q_h \mathbf{f}_\delta\|^2 + \alpha\|a_0\|^2.$$

Due to (37) and (38), $\|Q_h T_\eta a_0 - Q_h \mathbf{f}_\delta\| \le \delta + c_0\eta$, therefore for $\alpha = \alpha(\delta,\eta)$ satisfying (56) we obtain (cf. (46), (47))

$$\|Q_h T_\eta a_{h,\alpha(\delta,\eta)} - Q_h \mathbf{f}_\delta\| \to 0 \quad \text{as} \quad \delta \to 0, \eta \to 0,$$

$$\limsup_{h\to 0,\delta\to 0,\eta\to 0} \|a_{h,\alpha(h,\delta)}\| \le \|a_0\|$$

Now the arguments of the proof of Theorem 3.1 may be repeated. □

By similar arguments the following results concerning the residual principle of the choice of α can be proved.

THEOREM 3.4. *Assume the conditions of Theorem 3.3 with* $\mathbf{f} \ne 0$ *but choose* $\alpha = \alpha(h,\delta,\eta)$ *in (55) so that*

$$\delta + c_0\eta \le \|Q_h(T_\eta a_{h,\alpha} - \mathbf{f}_\delta)\| \le b(\delta + c_0\eta), \quad b = \text{const} \ge 1.$$

Then $\|a_{h,\alpha(\delta,\eta)} - a_0\|_E \to 0$ *as* $h \to 0$, $\delta \to 0$, $\eta \to 0$.

THEOREM 3.5. *Assume that* $0 \ne \mathbf{f} \in \mathcal{R}(T)$, $\|Q_h(\mathbf{f}_\delta - \mathbf{f})\| \le \delta$, $\|Q_h(T_\eta - T)\| \le \eta$ *and* $Q_h\mathbf{g} \to \mathbf{g}$ *as* $h \to 0$ *for any* $\mathbf{g} \in \overline{\mathcal{R}(T)} \subseteq F$. *Choose* $\alpha = \alpha(h,\delta,\eta)$ *in (55) so that*

$$\delta + \|u_{h,\alpha}\|\eta \le \|Q_h(T_\eta u_{h,\alpha} - \mathbf{f}_\delta)\| \le b(\delta + \|u_{h,\alpha}\|\eta), \quad b \ge 1.$$

Then $\|a_{h,\alpha(h,\delta,\eta)} - a_0\|_E \to 0$ *as* $h \to 0$, $\delta \to 0$, $\eta \to 0$.

Note that $\|Q_h(T_\eta a_{h,\alpha} - \mathbf{f}_\delta)\|^2 = \alpha^2(G_\psi \underline{c}, \underline{c})_{\mathbb{R}^n}$ where $\underline{c}$ is the solution of (54) with noisy data.

3.7 Application to the inverse problem.

Let us comment on discretization and regularization of the inverse problem of ground water filtration in the form of equation (8) with $E = L^2(\Omega)$, $F = G(\Omega, \Gamma)$, $T \in \mathcal{L}(E; F)$ defined in (7), and $\mathbf{f} = \nabla\psi$ where ψ is the solution of problem (5). It is natural to use finite elements approximations of $L^2(\Omega)$ and $G(\Omega, \Gamma)$. Let $S_h \subset H^1(\Omega, \Gamma)$ be some finite element subspaces. Unfortunately, natural bases $\{\psi_i\}_{i=1}^m$ of $S_h \subset H^1(\Omega, \Gamma)$ and the corresponding bases $\{\nabla\psi_i\}_{i=1}^m$ of $\nabla S \subset G(\Omega, \Gamma)$ are non-orthogonal, therefore $G_\psi = \left((\nabla\psi_j, \nabla\psi_i)_{L^2(\Omega)^n}\right)_{i,j=1}^m$ contains more than one non-zero diagonal, and the numerical scheme (34) of the regularised general projection method involving an explicit inversion of G_ψ seems to be unreasonable. Instead, the equivalent scheme (35) with diagonal matrix G_φ, i.e. piecewise constant approximations of $E = L^2(\Omega)$ may be recommended.

As we know from Theorems 3.3 – 3.5, the regularized method of minimal error has the advantage that the discretization and regularization parameters need not to be coordinated to one another. The numerical schemes (54) can be specified as follows (see (9)):

$$a_{h,\alpha} = \sum_{j=1}^m c_j \nabla u \cdot \nabla\psi_j,$$

$$\sum_{j=1}^m \left[\alpha \int_\Omega \nabla\psi_j \cdot \nabla\psi_i dx + \int_\Omega (\nabla u \cdot \nabla\psi_j)(\nabla u \cdot \nabla\psi_i) dx\right] c_j =$$

$$\int_\Omega \nabla\psi \cdot \nabla\psi_i dx, \quad i = 1, \dots, m. \tag{57}$$

Thereby, integrating by parts and using (5) we notice that

$$\int_\Omega \nabla\psi \cdot \nabla\psi_i dx = \int_\Omega f\psi_i dx + \int_\Gamma g\psi_i dS,$$

i.e. we actually do not need to solve the boundary value problem (5) at all. Most simple numerical schemes correspond to Courant elements in the two dimensional case ($n = 2$, filtration in a water conducting layer). Approximating u, f and g by their Courant interpolants, all integrals occuring in system (57) can be taken analytically, and one obtains a seven point difference scheme to determine $c_{j_1 j_2}$ (instead of the index $j \in \mathbb{N}$, here two indices $(j_1, j_2) \in \mathbb{N} \times \mathbb{N}$ are used to label the grid points and finite elements). For more details see E. Vainikko and G. Vainikko [7], E. Vainikko [6]. This way E. Vainikko [6] performed real life computations to find the filtration coefficient of the lowest water conducting layer of Estonian undergrounds. System (57) was solved by the method of conjugate gradients which is rather suitable due to the positive definiteness and a strong sparsity of the matrix.

References

[1] Chicone C. and Gerlach J. *Identifiability of distributed parameters*, In: Inverse and Ill-Posed Problems, Engl H.W. and Groetsch C.W. (Ed.), Academic Press, Boston etc., (1987) pp. 513-521.

[2] Kohn R.V. and Lowe B.C. *A variational method for parameter estimation*, RAIRO Math. Mod. and Numer. Anal. 22, (1988) pp. 119-158.

[3] Kunisch K. *A review of some recent results on the output least squares formulation of parameter estimation problem*, Automatica 24, (1988) pp. 531-539.

[4] Richter G.R. *An inverse problem for steady state diffusion equation*, SIAM J. Appl. Math. 41, (1981) pp. 210-221.

[5] Tikhonov A.N. and Arsenin V.Y. *Solution of Ill-Posed Problems*, Winston & Sons, Washington,(1977).

[6] Vainikko E. *Courant elements for the inverse problem in the filtration coefficient identification*, Lecture Notes in Pure and Appl. Math. Ser. 164, (1994) pp.451-459.

[7] Vainikko E. and Vainikko G. *Some numerical schemes for the identification of the filtration coefficient*, Acta et comment. univ. Tartuensis 937, (1992) pp. 90-102

[8] Vainikko G. *On the discretization and regularization of ill-posed problems with noncompact operators*, Numer. Funct. Anal. and Optimiz. 13, (1992) pp. 381-396.

[9] Vainikko G. *Identification of filtration coefficient*, Lecture Notes in Physics 422, (1993) pp. 251-256.

[10] Vainikko G. and Hämarik U. *Projection methods and self-regularization in ill-posed problems*, Sov. Math. 29, (1985) pp. 1-20.

[11] Vainikko G. and Kunisch K. *Identifiability of the transmission coefficient in an elliptic boundary value problem*, Z.Anal. Anwend., 12, (1993) pp. 327-341.

[12] Vainikko G. and Veretennikov A. *Iteration Procedures in Ill-Posed Problems*, Nauka, Moscow (1986) (in Russian).

[13] Yeh W.W.-G. *Review of parameter identification procedures in groundwater hydrology: the inverse problem*, Water Resources Res. 22, (1986) pp.95-108.

INTERPRETATIONS AND ESTIMATES FOR ILL-POSED PROBLEMS

THOMAS I. SEIDMAN *

Abstract. Following a discussion of the relation of these problems to applications, intended to clarify the considerations which must be handled in order to obtain genuinely useful results, we consider techniques for determining optimal approximations and consequent optimal error bounds for certain classes of ill-posed problems with appropriate *a priori* information.

Key words. ill-posed, filtering, error estimates. partial differential equations

AMS subject classifications. 65J20, 35R25

1. Introduction. The canonical ill-posed problem requires us to 'evaluate' an unevaluable map $\mathbf{S}$: we must 'solve', in some sense, an equation

$$\mathbf{F}(\bar{x}) = \bar{y} \tag{1.1}$$

in a context in which we have a continuous *direct map* $\mathbf{F}$

$$\begin{array}{ccc} y & \xleftarrow{\quad\mathbf{F}\quad} & x \\ \mathcal{D} & \xrightarrow{\quad\mathbf{S}\quad} & \mathcal{P} \end{array} \tag{1.2}$$

from a *parameter space* $\mathcal{P}$ to a *data space* $\mathcal{D}$ for which the inverse $\mathbf{S} = \mathbf{F}^{-1}$ is not continuous. Our nominal 'problem' (1.1) views $\bar{x}$ as an unknown 'parameter' which we are to determine (estimate) from available data. Necessarily, since the data is to be obtained by 'measurement', there is some uncertainty: one is not given $\bar{y}$ in (1.1) but knows only that $\bar{y} \in \mathcal{Y} \subset \mathcal{D}$. The corresponding information for the solution will then be expressed by

$$\bar{x} \in \mathcal{U} = \mathcal{U}(\mathcal{Y}) := \{x \in \mathcal{P} : \mathbf{F}(x) \in \mathcal{Y}\} \tag{1.3}$$

so we may refer to $\mathcal{U}(\mathcal{Y}) = \mathbf{F}^{-1}(\mathcal{Y})$ as the *'region of uncertainty'*. We emphasize that, with the information given, every $x \in \mathcal{U}(\mathcal{Y})$ is an equally valid candidate to be $\bar{x}$.

For well-posed problems, the usual practice is to represent $\mathcal{Y} = \mathcal{Y}_k$ — suggesting the possibility of considering a potential sequence of increasingly accurate measurements — by some nominal value y_k and then to take $x_k := \mathbf{S}(y_k)$ as a representative of $\mathcal{U}_k = \mathbf{S}(\mathcal{Y}_k)$. The measurement accuracy is typically indicated by $\delta = \delta_k$ so our available information is that

$$\bar{y} \in \mathcal{Y}_k \subset \{y \in \mathcal{D} : \|y - y_k\| \le \delta_k\} \tag{1.4}$$

for some suitable choice of norm on $\mathcal{D}$; for simplicity one generally takes $\subset$ to be a defining equality. The uncertainty in the result is then correspondingly characterized by

$$\bar{x} \in \mathcal{U}_k \subset \{x \in \mathcal{P} : \|x - x_k\| \le \varepsilon_k = \varepsilon(\delta_k)\} \tag{1.5}$$

* Department of Mathematics and Statistics, University of Maryland Baltimore County, Baltimore, MD 21228, USA e-mail: (seidman@math.umbc.edu).

with well-posedness defined by the property that

$$\varepsilon_k \to 0 \ \text{ as } \delta_k \to 0; \tag{1.6}$$

indeed, for well-posed problems one typically has a linear rate of convergence

$$\varepsilon(\delta) \le K\delta \tag{1.7}$$

for some (local) Lipschitz constant $K = K(\bar{x})$.

For ill-posed problems one cannot use $x_k := \mathbf{F}^{-1}(y_k)$ as a plausible approximation for $\bar{x}$ as above since (a) this may not be defined as y_k need not be in the range of $\mathbf{F}$, (b) there are severe difficulties, which must be addressed, in constructing any computational approximation $\hat{\mathbf{S}}$ to $\mathbf{F}^{-1}$, and (c) (arbitrarily) small perturbations of y can give (arbitrarily) large perturbations of x so (1.5) typically requires $\varepsilon_k \equiv \infty$ even for arbitrarily small δ_k, not only precluding (1.6) but rendering (1.5) entirely uninformative. As a substitute for this, we seek some other notion of 'solution' which will be feasible in the context.

The intention here is to follow [9] to some extent in discussing the questions, "What is the problem? What is a solution?" — with some prejudice towards my own work in examples. The thrust of [9] was to note that

- despite the heuristic arguments of [4] in bestowing upon our concerns the pejorative title *ill-posed* and the technical arguments of [13] which demonstrate them *infinitely complex*, we nevertheless find ourselves successful in contributing usefully to applications and
- this can be understood (only) in terms of a consideration of ill-posed problems in an applied context rather than in isolation.

It is this last which involves the questions above: our notion of *success* depends on an interpretation of *'solution'* which, in turn, depends on a contextual interpretation of the *problem*. In general, the emphasis of this paper will be at least as much on pointing out 'things to worry about' as on specific algorithms or estimates. We do, however, use the perspective developed in the next section to introduce, in the following section, the paradigms, constructions, and estimates of the *'Method of Optimal Filtering'*.

2. Paradigms of interpretation. The first key question for interpretation for (1.1) is: *What do we mean by 'small'/'large' perturbations?* — i.e., What determines the choice of spaces $\mathcal{P}, \mathcal{D}$ and the topologies imposed on them for discussing the invertibility of $\mathbf{F}$?

Our first observation is that these choices, necessarily part of the specification of the problem, represent an approximation to a question of economics and technology: the ultimate goal of our analysis is the minimization (over a range of technologically feasible possibilities) of

$$\begin{aligned} \text{total cost} &= \text{cost of measurements (to some accuracy)} \\ &\quad + \text{cost of error/uncertainty in the estimate} \\ &\quad [+ \text{cost of computation, etc.}] \\ \mathcal{J} &= \varphi(\mathcal{Y}) + \psi(\mathcal{U})\ [+\gamma(\hat{\mathbf{S}})]. \end{aligned} \tag{2.1}$$

[In this last formula, $\hat{\mathbf{S}}$ denotes some computational algorithm for approximating $\mathbf{F}^{-1}$ and this cost may or may not be included.] Before any further analysis of (2.1), we expand the diagram (1.2) as

$$\begin{array}{ccc} \mathcal{D} & \xleftarrow{\quad\mathbf{F}\quad} & \mathcal{P} \\ \Big\downarrow \mathbf{M}_k & & \Big\uparrow \mathbf{I}_k \\ \mathcal{D}_k & \xrightarrow{\quad\hat{\mathbf{S}}_k\quad} & \mathcal{P}_k \end{array} \tag{2.2}$$

to emphasize the facts that in the real world any actual implementation $\mathbf{M}_k$ of measurement *must* lead to a finite set of numbers (with finite precision) so $\mathcal{D}_k = \mathbb{R}^{n(k)}$ and, correspondingly, that computation must also be finitary so we must work with a finite dimensional approximating representation in $\mathcal{P}_k = \mathbb{R}^{m(k)}$ (e.g., using splines, etc., related to $\mathcal{P}$ by $\mathbf{I}_k$). As noted earlier, the index k suggests that we will later embed this schema in a sequence of such diagrams, e.g., with increasingly accurate measurements and increasingly refined computations as $m(k), n(k) \to \infty$.

Our second observation is that ill-posed problems of practical significance come with an applied context so our efforts are teleologically directed: we will assume now that there is some specified application in which a 'result' z depends on the unknown parameter $\bar{x}$ through a map

$$\mathcal{P} \xrightarrow{\quad\mathbf{A}\quad} \mathcal{R} : x \mapsto z \tag{2.3}$$

so that 'what we really want' is the result $\bar{z} = \mathbf{A}(\bar{x})$. From this point of view, **the solution of the ill-posed problem is only an intermediate construct intervening between the available data and the intended application.** In isolation, this problem (2.3) would be a 'standard' well-posed problem and we would eventally also introduce computational approximations $\mathbf{A}_k$ to $\mathbf{A}$.

Of course, we have an obvious difficulty — the 'value' of $\bar{x} \in \mathcal{P}$ is not 'given' *a priori* but is to be obtained from the data via the ill-posed problem (1.1). We concatenate (2.3) with (2.2) to obtain

$$\begin{array}{ccccc} \mathcal{D} & \xleftarrow{\quad\mathbf{F}\quad} & \mathcal{P} & \xrightarrow{\quad\mathbf{A}\quad} & \mathcal{R} \\ \Big\downarrow \mathbf{M}_k & & & & \Big\uparrow \\ \hat{\mathbf{G}}_k : \ \mathcal{D}_k & \xrightarrow{\quad\hat{\mathbf{S}}_k\quad} & \mathcal{P}_k & \xrightarrow{\quad\mathbf{A}_k\quad} & \mathcal{R}_k \end{array} \tag{2.4}$$

in recognition of our awareness that $\bar{x}$ is only an 'instrumental variable'.

Our true concern, then, is with the approximation of $\bar{z} = \mathbf{G}(\bar{y})$ with $\mathbf{G} := \mathbf{A} \circ \mathbf{F}^{-1}$ by $\hat{\mathbf{G}}_k(y_k)$ with $y_k := \mathbf{M}_k(\bar{y})$. In this formulation, the ill-posed $\mathbf{F}^{-1}$ appears only as a factor of $\mathbf{G}$. The factorization here and also in taking $\hat{\mathbf{G}}_k = \mathcal{A}_k \circ \mathbf{S}_k$ is a convenience for our techniques of analysis and algorithm development; the ill-posedness of (1.1) is an artifact of this but may not be intrinsic to the real problem since it is entirely possible for $\mathbf{G}$ to be continuous even though *one* of its factors $\mathbf{F}^{-1}$ is not. To the extent that, at each stage, our computational procedures maintain this factorization, we do actually work with the ill-posed problem but the representation and topology for the *intermediate* variable $x \in \mathcal{P}$ are now seen to be somewhat arbitrary — subject to our convenient choice with respect to (2.3). [E.g., if $\mathbf{A}$ is linear, one convenient choice might be to use $x \mapsto \|\mathbf{A}x\|_{\mathcal{R}}$ as a norm for $\mathcal{P}$.]

EXAMPLE 1: The simplest example in which ill-posedness 'disappears' in our present framework of considering the composed problem is that in which $\mathcal{R} = \mathbb{R}$ in (2.2), particularly when $\mathbf{A}$ is linear; compare [1].

Let us suppose that the (direct) map $\mathbf{F} : \mathcal{P} \to \mathcal{D}$ is also linear and write $\mathbf{A} = \boldsymbol{\alpha} \in \mathcal{P}^*$. Since we have $\mathbf{G} := \mathbf{A} \circ \mathbf{F}^{-1} = [\mathbf{F}^{-1}]^* \boldsymbol{\alpha} : \mathcal{D} \to \mathcal{R} = \mathbb{R}$ by definition, we now can solve the equation

$$\mathbf{F}^* \boldsymbol{\gamma} = \boldsymbol{\alpha} \tag{2.5}$$

to obtain $\boldsymbol{\gamma} \in \mathcal{D}^*$ — provided, of course, one has $\boldsymbol{\alpha} \in ran(\mathbf{F}^*)$. Since $\mathbf{F}^{-1}$ is unbounded (by assumption), so is $[\mathbf{F}^{-1}]^* = [\mathbf{F}^*]^{-1}$; the problem (2.5) is 'just as ill-posed' as the original ill-posed problem (1.1) *but*: while y is only 'given' approximately (by measurement), the functional $\boldsymbol{\alpha}$ is given analytically *a priori* in defining the application so we may use *analytic* methods to solve the new 'ill-posed problem' (2.5). Once one has precomputed $\boldsymbol{\gamma}$ in this way, the problem is no longer ill-posed: the genuinely significant map: $y \mapsto z$ is now easily realized as: $y \mapsto \langle \boldsymbol{\gamma}, y \rangle$ and *this* computation is well-posed.

One can also treat, in almost the same way, the situation in which $\mathcal{R}$ is finite dimensional by working with $\mathbf{A}$ 'component by component' and we again precompute $\mathbf{G} := \mathbf{A}\mathbf{F}^{-1}$ and proceed as above. This also gives some insight into the situation in which $\mathbf{F}$ has a nontrivial nullspace: if this happens to be contained in the nullspace of $\mathbf{A}$ (which is necessarily nontrivial if $\mathcal{R}$ is finite dimensional), then $\mathbf{G}$ is well-defined and we can either ignore the intermediate construct $\bar{x}$ entirely or else work with a representative of its equivalence class, factoring out the nullspace in some fashion (which may be fairly arbitrary if there are no genuine grounds for selection). Somewhat similar considerations apply to nonlinear problems although the analysis may then be rather more complicated. The issue, however is the same: to precompute (analytically) an effectively usable realization of the functional $\boldsymbol{\gamma} = \boldsymbol{\alpha} \circ \mathbf{F}^{-1}$ under hypotheses ensuring that this is continuous.

It is interesting that a variant of this approach [6], [8] can be used directly for an infinite dimensional ill-posed problem by noting the equivalence of (1.1) to

$$\boldsymbol{\gamma}_\nu(\bar{x}) = \langle \boldsymbol{\alpha}_\nu, \bar{y} \rangle \qquad \nu = 1, 2, \ldots \tag{2.6}$$

for a suitable sequence $\{\boldsymbol{\alpha}_\nu\}$ in $\mathcal{D}^*$ — although this no longer gives a well-posed problem except in approximating by truncation: (2.6) for $\nu = 1, \ldots, N$ with minimization of the solution norm determines x_N with convergence $x_N \to \bar{x}$, more precisely, to the minimum norm solution of (1.1) if that is unique. ∎

EXAMPLE 2: Now suppose $\mathbf{A} : x \mapsto z : \mathcal{P} \to \mathcal{R}$ represents finding the solution z of a PDE with coefficients involving a parameter $x \in \mathcal{P}$ (system identification). Then the result may well be quite insensitive to certain (structured) perturbations of $x \in \mathcal{P}$. In this context it is very possible that $\mathbf{F}$ is roughly the same as $\mathbf{A}$, say, with a diffent choice of initial or boundary data for the PDE. This can be expected to make inversion of $\mathbf{F}$ ill-posed as we well know — **but** *the structure of the uncertainty is* *precisely such that this uncertainty probably does not really matter much*! The general principle, here, is that if some 'aspects' of the unknown parameter x do not (significantly) affect the output y for purposes of identification, then those same aspects will not (significantly) affect the result z desired for the application — the compactness of $\mathbf{A}$ 'cancels' the compactness of $\mathbf{F}$ so one may reasonably hope that $\mathbf{G}$ is continuous. As in Example 1, this may also 'handle' a lack of injectivity in $\mathbf{F}$. Compare, also, the discussion in [2]. ■

At this point we note another highly significant effect of working in the context of an application. We have implicitly been viewing the parameter space $\mathcal{P}$ as a Banach space with only the implicit restriction from (1.1) that $\bar{x}$ is in the domain of $\mathbf{F}$. We now note that one often has additional auxiliary information about $\bar{x}$ — e.g., from prior measurements or from the physical context of the model. The effective use of such information (data fusion) is often computationally difficult but essential to getting reliable results. For the present, we represent this information by adjoining to (1.1) the *a priori* constraint

$$\bar{x} \in \mathcal{K} \tag{2.7}$$

for some specified set $\mathcal{K} \subset \mathcal{P}$. [It should be emphasized that, unlike occasional practice in the community, (2.7) is only legitimate in an applied context to the extent that it represents genuine knowledge about the unknown parameter (or, as above, imposition of a consciously arbitrary selection to handle nonuniqueness) and is not merely wishful thinking for the convenience of proving theorems of 'justification'.]

It is fortunate that in many applications the available information provides us with a *compact* set $\mathcal{K}$. As a first consequence, we note that it is this compactness which makes it possible for us to utilize the simplifying assumption (1.4) in working with an infinite dimensional problem with data given, as in (2.2), by a finite number of measurements of finite accuracy. As a second consequence, we also note that the ill-posedness of the problem may again disappear in some sense: a standard theorem of topology asserts that if the continuous map $\mathbf{F}$ is injective and is considered on a compact set $\mathcal{K}$, then $\mathbf{F}^{-1}$ is necessarily also continuous: if one now replaces $\mathbf{F} : \mathcal{P} \to \mathcal{D}$ by $\mathbf{F} : \mathcal{K} \to \mathcal{D}$, one will not only get ε_k finite in (1.5) but will have (1.6). It is this theorem and related considerations which suggests the importance for ill-posed problems of injectivity proofs *per se*, although we note that the significance of this injectivity is for the restriction of $\mathbf{F}$ to $\mathcal{K}$, possibly after the imposition of a 'selection'.

Assuming we have chosen an appropriately application-related norm for $\mathcal{P}$, we may now change notation and rewrite (2.1) as

$$\mathcal{J} = \varphi(\delta) + \psi(\varepsilon(\delta)) \; [+\gamma(\hat{\mathbf{S}})]. \tag{2.8}$$

Here, $\varphi(\delta)$ is the cost of making measurement accurate enough to ensure, utilizing the constraint (2.7), that

$$\bar{y} := \mathbf{F}(\bar{x}) \in \mathcal{Y}_\delta := \{y \in \mathcal{D} : \|y - \hat{y}\| \leq \delta\} \tag{2.9}$$

where $\hat{y} = y_k$ is the nominal data value; $\varepsilon(\delta)$ is the radius of $\mathcal{U}(\mathcal{Y}_\delta) \cap \mathcal{K}$ so

$$\bar{x} \in \mathcal{U}(\mathcal{Y}_\delta) \cap \mathcal{K} \subset \{x \in \mathcal{P} : \|x - \hat{x}\| \leq \varepsilon(\delta)\} \tag{2.10}$$

where $\hat{x}$ is a 'center' for $\mathcal{U}(\mathcal{Y}_\delta) \cap \mathcal{K}$.

The original ill-posedness of the problem reasserts itself in that one now has no hope of (1.7), but can expect some extremely slow rate of convergence $\varepsilon(\delta) \to 0$ as $\delta \to 0$. This is further exacerbated on noting that, since we are using the same compactness of $\mathcal{K}$ and so of $\mathbf{F}(\mathcal{K})$ to control the relation between $n(k)$ in (2.2) and δ_k (even with idealized measurements of infinite accuracy), we may expect the cost $\varphi(\mathcal{Y}_\delta)$ in (2.1) to increase quite rapidly as one might hope to let $\delta = \delta_k$ become small. This may, of course, mean that the practical application is simply infeasible: even neglecting computational costs, one may have $\psi(\varepsilon(\delta))$ unacceptably large unless δ becomes small enough that one has $\varphi(\delta)$ unacceptably large.

One interpretation of practical importance is that of '*a posteriori* analysis', in which the measurements have already been made so $\mathcal{Y}$ is fixed. A 'solution' now means some effective determination of the *set* $\mathcal{U} = \mathcal{U}(\mathcal{Y})$. The analyst's first task is to use the *a priori* specification of $\mathcal{K}$, an understanding of $\mathbf{M}_k$, and the measured data to represent $\hat{y}$ in some convenient form with a bound δ on its uncertainty. The second task is then to construct some algorithm denoted by $\hat{\mathbf{S}}$, enabling us to compute an approximation, and a bound ε such that **reliably** one has

$$\bar{x} \in \mathcal{U}(\mathcal{Y}) \subset \{x \in \mathcal{P} : \|x - \hat{x}\| \leq \varepsilon\} \qquad [\hat{x} := \hat{\mathbf{S}}(\hat{y})]. \tag{2.11}$$

In this case the cost will be $\mathcal{J} = \psi(\varepsilon) + \gamma(\hat{\mathbf{S}})$. Note that the data for the construction of $\hat{\mathbf{S}}$ actually consists of $\mathcal{K}$ and δ as well as $\hat{y}$. While much is often made of the 'regularized' continuity properties of $\hat{y} \mapsto \hat{x}$ as compared with $\mathbf{S}$, we assert that the *only* significance of $\hat{\mathbf{S}}$ comes through (2.11) and, indeed, that one should be careful not to let such regularization obscure the genuine uncertainty involved in replacing $\bar{x}$ by $\hat{x}$, which is intrinsic to the problem.

In contrast, one may consider the interpretation as '*a priori* analysis'. While eventually this must be interpreted as above, one now has an opportunity to analyze the problem *before* the measurements have been already made and, within the limitations of technological feasibility, to treat (2.8) as an optimization problem, with (2.10) replaced by (2.11) in defining $\varepsilon = \varepsilon(\delta)$. It is this mode of analysis which leads to the considerations of our next section and which is closely related to to the suggestion above of embedding the problem in a sequence of potential computations of increasing accuracy.

Without further comment, we note — as an obvious parallel to adaptive control theory and the sequential decision theory of statistical analysis — an intermediate mode of analysis in which one might incrementally optimize (2.8) for the proposal of new measurements based on a sensitivity analysis, etc., in the light of the new $\mathcal{K}$ determined by the computations thus far.

The formulation already implicit in (1.3) — viewing $\mathcal{Y}$ as an undifferentiated fixed set — implies a deterministic context with specified error bounds for measurements and the treatment of *a priori* auxiliary information as a constraint (2.7) is consistent with that. In this paper we confine ourselves to such a deterministic setting. A brief comment is in order here, however, as to the related nature of interpretation in contexts involving statistical considerations for these uncertainties. We first note that any probability distribution for the noise determines a corresponding probability distribution for the uncertainty in the solution by

$$Prob\{\bar{x} \in \mathcal{S}\} = Prob\{\bar{y} \in \mathbf{F}(\mathcal{S})\} \tag{2.12}$$

and that this may be combined with an uncertainty distribution in the *a priori* information by a relatively standard Bayesian computation. A plausible choice for $\hat{\mathbf{S}}$ for the *a posteriori* mode of interpretation might now be to use a 'maximum likelihood estimator' (i.e., selecting $\hat{x}$ to maximize the relative probability of observing the given data); we will, however, view our goal in the statistical context to be much as above for the deterministic setting, except that the total cost of (2.1) would now be replaced by an 'expected total cost' with respect to the Bayesian distribution or one might seek an assurance that the *risk* (= *'probability that the total cost might exceed some acceptable level'*) is acceptably small. To the extent that the relevant distributions can be reasonably expressed simply in terms of distributions for the norms of uncertainties, it may be possible to use the deterministic estimates below for the relevant statistical computations.

3. Estimates and asymptotics. For well-posed problems the usual analysis leads to the result that:

One can find the desired result to any degree of specified precision — provided one is prepared to work hard enough.

Here, what we mean by 'working hard enough' is: making enough measurements with enough accuracy and then doing enough computation. The history of analysis of ill-posed problems has consisted primarily of demonstrations that this remains true for ill-posed problems even without considering the composed framework introduced in (2.4). As already noted, the principal difference is that the ill-posedness shows up in an explosive growth of the accuracy requirements for the measurements as one seeks more and more precision in the result. We observe that *ill-posedness resides primarily in the asymptotics* so that for some specific level of desired precision it is quite possible that the problem may still be entirely reasonable.

It is traditional to be concerned about the convergence of an approximation scheme, i.e., that

$$x_k := \hat{\mathbf{S}}_k(y_k) \to \bar{x} \;\text{ as } y_k \to \bar{y}\ (\delta_k \to 0), \tag{3.1}$$

and in many cases one can prove this. It is not at all clear, however, that this has much relevance to actual computation, with or without a 'rate of convergence' estimate as in (1.6). In practice one never actually approaches the asymptotic limit, so such results are significant primarily as a guide — e.g., providing an error estimate at any stage so as to select appropriate computational parameters, such as a stopping criterion. Note, too, that one is only *guessing* the magnitudes of any 'constants' appearing in an asymptotic expression unless it is reasonable to view these as available *a priori* from the model — there is a big difference between knowing that some constant exists and knowing (even a rough estimate of) its value. [Ironically, one may also have an easily computable estimate yet have no reasonable way to determine whether one is in its domain of validity.] This is a consideration which applies equally well to instances of the same tradition for standard (well-posed) problems and techniques of *a posteriori* error analysis have been developed for those concerned by this. These techniques are typically based on the asymptotics of the residuals and are inapplicable to ill-posed problems. We have not observed, however, any development of corresponding modes of analysis which *would* be applicable in addressing this difficulty.

Our emphasis, then, is on the estimation of $\varepsilon = \varepsilon(\delta)$ — for fixed δ and with computable constants. This addresses both the context of *a posteriori* analysis and also permitting the optimization involved in an *a priori* analysis, while *a fortiori* providing the asymptotics as $\delta \to 0$. Note that the ε we seek is a bound on the 'worst case error':

$$\varepsilon = \varepsilon(\delta) := \inf_{\hat{\mathbf{S}}} \max_{\hat{y}} \sup_{x} \{\|x - \hat{\mathbf{S}}(\hat{y})\| : x \in \mathcal{K},\ \|\mathbf{F}(x) - \hat{y}\| \le \delta\} \tag{3.2}$$

with the implicit assumptions that the $\mathcal{D}$-norm used to measure data perturbations reasonably corresponds to the realistic uncertainties of measurement and that, similarly, the $\mathcal{P}$-norm reasonably corresponds to the related uncertainties in the result when $\hat{x}$ is to be used for $\bar{x}$ in the application under consideration. While one might possibly obtain a sharper bound dependent on $\hat{y}$ (e.g., by allowing for a possible preprocessing step of projecting $\hat{y}$ to $\mathbf{F}(\mathcal{K})$), we will content ourselves with estimating $\varepsilon(\delta) = \varepsilon(\delta; \mathcal{K})$ so (3.2) includes, on the right, a 'max$_{\hat{y}}$' (which may be taken formally over all $\hat{y} \in \mathcal{D}$, but actually reduces to considering only $\hat{y} \in \mathbf{F}(\mathcal{K}) \subset \mathcal{D}$). We must consider the 'sup$_x$' in (3.2) since this is necessary to ensure a bound on $\|\bar{x} - \hat{x}\|$: as already noted, with the information given, every $x \in \{x \in \mathcal{K} : \|\mathbf{F}(x) - \hat{y}\| \le \delta\}$ is an equally valid candidate to be $\bar{x}$. While a true 'min$_{\hat{x}}$' is attained for each $\mathcal{Y}_\delta$ when $\hat{x}$ is the unknown center of $\mathcal{U}(\mathcal{Y}_\delta)$, the minimization here is actually taken as an inf over feasibly implementable quasi-inverse maps

$$\hat{\mathbf{S}} = \hat{\mathbf{S}}(\cdot; \mathcal{K}, \delta) : \hat{y} \mapsto \hat{x} = \hat{\mathbf{S}}(\hat{y}). \tag{3.3}$$

In general this is an 'inf' rather than a 'min', but we will be able to determine an optimal $\hat{\mathbf{S}}(\cdot; \mathcal{K}, \delta)$ in certain special cases for which an explicit construction is possible; cf., e.g., [12], [11] and below.

A fairly typical form for the constraint (2.7) is that

$$\mathcal{K} = \mathcal{K}_M := \{\mathbf{E}\xi : \|\xi - \xi_*\| \le M\} \tag{3.4}$$

for a suitable (linear) map $\mathbf{E} : \tilde{\mathcal{X}} \to \mathcal{P}$ and some $\xi_* \in \tilde{\mathcal{X}}$ (which, for simplicity, we will take to be $\xi_* = 0$); for example, $\mathbf{E}$ might be the embedding map of $\tilde{\mathcal{X}} = C^2(\bar{\Omega})$ into $\mathcal{P} = L^2(\Omega)$, representing a regularity condition on the unknown parameter $\bar{x}$ with an explicit bound[1] M in the stronger norm. With $\bar{x} = \mathbf{E}\bar{\xi}$ and assuming $\hat{\mathbf{S}}$ is Lipschitzian so

$$\|\hat{\mathbf{S}}(y_1) - \hat{\mathbf{S}}(y_2)\| \le \beta \|y_1 - y_2\| \tag{3.5}$$

for some β, we have, using (1.1), the estimate

$$\|\bar{x} - \hat{\mathbf{S}}(\hat{y})\| = \|[\mathbf{1} - \hat{\mathbf{S}} \circ \mathbf{F}](\mathbf{E}\bar{\xi}) - [\hat{\mathbf{S}}(\bar{y}) - \hat{\mathbf{S}}(\hat{y})]\|. \tag{3.6}$$

Then, setting

$$\mathbf{\Gamma}(\beta) := \inf_{\hat{\mathbf{S}}} \{\| \left[\mathbf{1} - \hat{\mathbf{S}} \circ \mathbf{F}\right] \circ \mathbf{E}\|_{Lip} : \|\hat{\mathbf{S}}\|_{Lip} \le \beta\} \tag{3.7}$$

[where $\|\cdot\|_{Lip}$ denotes the minimal Lipschitz constant of a map], we have

$$\varepsilon(\delta) \le \mathbf{\Gamma}(\beta) M + \beta\delta \tag{3.8}$$

[1] We remark that, although the assumptions are comparable, the approach presented here, which we term the 'Method of Optimal Filtering', is quite different from the weighted 'Least Squares' optimization used in [5]. As we shall see, this approach gives an estimate which is genuinely optimal in a specified sense.

for any β. Clearly we can minimize the right hand side of (3.8) over β and, setting

$$\tilde{\mathbf{\Gamma}}(\tilde{\beta}) := \begin{cases} -\mathbf{\Gamma}(-\beta) & \text{if } \tilde{\beta} \leq 0 \\ \infty & \text{else,} \end{cases}$$

this gives the fundamental estimate of *'Optimal Filtering'*

$$\varepsilon(\delta; \mathcal{K}_M) \leq M \min_{\beta \geq 0} \{\mathbf{\Gamma}(\beta) + \beta\delta/M\} = -M\tilde{\mathbf{\Gamma}}^*(\delta/M) \tag{3.9}$$

where $\tilde{\mathbf{\Gamma}}^*$ denotes the Legendre-Fenchel conjugate of $\tilde{\mathbf{\Gamma}}$. In many cases this estimate is 'asymptotically sharp', in that (with M fixed) there is a sequence $\delta_k \to 0$ and a sequence of pairs $\bar{x}_k, \bar{x}'_k \in \mathcal{K}_M$ such that

$$\|\mathbf{F}(\bar{x}) - \hat{y}_k\|, \|\mathbf{F}(\bar{x}) - \hat{y}_k\| \leq \delta_k \quad \|\bar{x}_k - \bar{x}'_k\| = M\tilde{\mathbf{\Gamma}}^*(\delta_k/M).$$

Our next problem, of course, is to show how to compute $\mathbf{\Gamma}(\cdot)$ (or, better, $\tilde{\mathbf{\Gamma}}(\cdot)$) and the optimizing $\hat{\mathbf{S}}$. The principal result of [10] is that a fairly explicit construction is possible in the Hilbert space setting (in which case we may identify the spaces $\mathcal{X}$, $\mathcal{P}$, $\mathcal{D}$), provided $\mathbf{E}$ and the map $\mathbf{F} \circ \mathbf{E}$ are commuting normal linear operators. This is a considerable restriction, of course, but still admits a variety of interesting and significant examples. For simplicity, we describe the construction assuming discrete spectra so there is an orthonormal basis $\{w_k\}$ with

$$\mathbf{FE}w_k = \lambda_k w_k \qquad \mathbf{E}w_k = \epsilon_k w_k$$

and we take $\hat{\mathbf{S}}$ in the form $\hat{\mathbf{S}} : w_k \mapsto \rho_k w_k$ so

$$\hat{\mathbf{S}}\hat{y} = \sum_k \rho_k c_k w_k \quad \text{with } c_k := \langle \hat{y}, w_k \rangle. \tag{3.10}$$

From (3.7) we wish to minimize

$$\|\mathbf{E} - \hat{\mathbf{S}}\mathbf{FE}\| = \max_k \{|\epsilon_k - \rho_k \lambda_k|\} \tag{3.11}$$

by our choice of the coefficients $\{\rho_k\}$, subject to $|\rho_k| \leq \beta$. Treating each k independently and noting that λ_k/ϵ_k is precisely the corresponding eigenvalue φ_k of $\mathbf{F}$, this minimization is clearly realized by taking

$$\rho_k = \begin{cases} 1/\varphi_k & \text{if } \beta|\varphi_k| \geq 1 \\ \beta \arg\{1/\varphi_k\} & \text{else} \end{cases} \tag{3.12}$$

so $|\epsilon_k - \rho_k \lambda_k| = (|\epsilon_k| - \beta|\lambda_k|)_+$. At this point we introduce the spectral function of $|\mathbf{E}|$ relative to $|\mathbf{FE}|$:

$$s(\lambda) := \max\{|\epsilon_k| : |\lambda_k| = \lambda\} \tag{3.13}$$

with the interpretation that this 'max' is $-\infty$ if λ is not in the spectrum of $\mathbf{FE}$. We now have

$$\mathbf{\Gamma}(\beta) = \max_k \{|\epsilon_k| - \beta|\lambda_k|\} = \sup_\lambda \{s(\lambda) - \beta\lambda\}. \tag{3.14}$$

With slight manipulation, this gives both

$$\beta = \beta(\tilde{\nu}) = \operatorname{argmin}\{\mathbf{\Gamma}(\beta) + \beta\tilde{\nu}\} = \operatorname{argmin}\{\beta\tilde{\nu} - [-s]^*(-\tilde{\nu})\} \tag{3.15}$$

with $\tilde{\nu} := \delta/M$ for use in (3.12) and the new version of our fundamental estimate:

$$\tilde{\mathbf{\Gamma}}(\tilde{\nu}) = [-s]^*(\tilde{\nu}) \qquad \varepsilon(\delta;\mathcal{K}_M) = -M[-s]^{**}(\delta/M) \tag{3.16}$$

where * again indicates the Legendre-Fenchel conjugate, whence

$$\varepsilon(\delta;\mathcal{K}_M) = Ms(\delta/M) \qquad \text{if } s(\cdot) \text{ is concave.} \tag{3.17}$$

It is important to note that we may often have a fairly simple expression $\tilde{s}(\cdot)$, giving $s(\lambda) = \tilde{s}(\lambda)$ when $\lambda = \lambda_k$, and we may then use (3.15), (3.16), (3.17) with s replaced by $\tilde{s}$ and retain asymptotic correctness.

At this point it is interesting to compare the estimates obtained here with those obtainable by some other standard schemes. Following [3], we note that the assumption there (p. 18) that $Kg \in \mathcal{R}(\tilde{K}^{\nu})$ is, in our present notation, just our requirement that $\bar{x} \in \mathcal{K}_M$ as in (3.4) with $\mathbf{E} = |\mathbf{F}|^{2\nu}$ and $\xi_* = 0$, although with M unspecified. One then has $\tilde{s}(\lambda) = \lambda^{1-\vartheta}$ so (3.17) gives

$$\varepsilon(\delta;\mathcal{K}_M) = M^{\vartheta}\delta^{1-\vartheta} \text{ with } \vartheta = \frac{1}{2\nu+1} \tag{3.18}$$

if one uses (3.10), (3.12) with the correct choice of β. Note that the assumption on $\mathcal{K}_M$ makes (3.18) and the choice of β independent of the particular problem (i.e., of $\mathbf{F}$). In comparison, Groetsch notes accuracy $\mathcal{O}\left(\delta^{1-1/2\nu}\right)$ for the Landweber-Fridman iteration, which is reasonably comparable for large $\tilde{\nu}$ and notes exactly the same asymptotic *order* as in (3.18) for Tikhonov regularization when $\nu \le 1$ but shows that this never gives better accuracy than $\mathcal{O}\left(\delta^{2/3}\right)$ for larger ν.

Actually, it is quite easy to obtain such results by the mode of analysis we have used above. For Tikhonov regularization, for example, one is effectively using (3.10) with $\rho = \rho_T := \varphi/[\varphi^2 + \alpha]$ so maximization of $(1 - \rho_T\varphi)\epsilon$ (again with $\epsilon = \varphi^{2\nu}$ for $0 < \nu < 1$) easily gives $\mathbf{\Gamma}_T(\alpha) = \left[\nu^{\nu}(1-\nu)^{1-\nu}\right]\alpha^{\nu}$ and $\beta(\alpha) = \|\hat{\mathbf{S}}_T\| = 1/2\sqrt{\alpha}$ whence optimization of $M\mathbf{\Gamma}_T(\alpha) + \delta\beta_T(\alpha)$ over α gives the value: $\alpha = \alpha_{opt} = [C\delta/M]^{2\vartheta}$ (ϑ as in (3.18)) with $C = 1/4\nu^{\nu+1}(1-\nu)^{1-\nu}$. If we consider the approximation method based on truncated singular value decomposition, we are again essentially using (3.10), now with $\rho_{SV} := \{1/\varphi \text{ if } \varphi \ge \bar{\varphi};\ 0 \text{ else}\}$ so $\beta = 1/\bar{\varphi}$, etc., whence we get $\bar{\varphi}_{opt} = [(\delta/M)/2\nu]^{\vartheta}$ by elementary calculus and then $\varepsilon_{SV} = CM^{\vartheta}\delta^{1-\vartheta}$ for comparison with (3.18) with $C = \nu^{\vartheta}(1 + 1/2\nu)$ for $\nu > 0$.

EXAMPLE 3: We sketch the application of these ideas to the Backwards Heat Equation (BHE). Thus, we consider, say,

$$u_t = \Delta u \text{ on } \mathcal{Q} := (0,T) \times \Omega \text{ with } u\Big|_{\partial\Omega} = 0 \tag{3.19}$$

and take $\bar{x} = \bar{x}_{\tau} := u\Big|_{t=\tau}$ (for some $0 \le \tau < T$) as the unknown 'parameter' and $\bar{y} := u\Big|_{t=T}$ as data. Our treatment here is based on the fact that there is a semigroup associated with the forward problem:

$$\mathbf{H}(s) : u\Big|_{t=\sigma} \longmapsto u\Big|_{t=\sigma+s} \tag{3.20}$$

and that this semigroup can be expressed, following Fourier, in terms of the orthonormal spectral expansion of the generator $(-\Delta + \mathrm{BC})$:

$$-\Delta w_k = \mu_k w_k \text{ on } \Omega \text{ with } w_k = 0 \text{ on } \partial\Omega. \tag{3.21}$$

We distinguish two quite distinct cases: $\tau = 0$ and $0 < \tau < T$ with an assumed explicit bound M on the unknown initial data $\xi = u(0,\cdot)$ in $H_0^{2\vartheta}(\Omega)$ for the first case and in $\mathcal{H}$ for the second; in each case we take $\mathcal{P} = \mathcal{D} = \mathcal{H} := L^2(\Omega)$.

For the first case, $\mathbf{F}$ is just $\mathbf{H}(T)$ and $\mathbf{E}$ is the embedding: $H_0^{2\vartheta}(\Omega) \hookrightarrow L^2(\Omega) = \mathcal{H}$. If we use the norm

$$\|\xi\|_{2\vartheta} := \|(-\Delta)^\vartheta \xi\|_{\mathcal{H}} = \left[\sum_k \mu_k^{2\vartheta} c_k w_k\right]^{1/2} \qquad \left(\xi = \sum_k c_k w_k\right),$$

then $\mathbf{E}$ and $\mathbf{FE}$ commute and we have

$$\varphi_k = e^{-T\mu_k} \qquad \epsilon_k = \mu^{-\vartheta} \tag{3.22}$$

for use in the formulas above. Note that we here can obtain $\tilde{s}(\cdot)$ on $\mathbb{R}_+$ as the (concave) inverse of the convex monotone function $\lambda(\epsilon) := \epsilon \exp\left[-T/\epsilon^\vartheta\right]$. In principle the optimizations in (3.14) and (3.9) require only elementary calculus, but there is no simple formula so actual solution must be numeric. A straightforward, although somewhat messy, computation gives the asymptotic result (for very small δ) that one can approximate the optimal β by solving

$$\beta(\log\beta)^{\vartheta+1} = \vartheta T^\vartheta M/\delta$$

and still realize through use of (3.12) the optimal error estimate

$$\varepsilon(\delta;\mathcal{K}_M) = MT^\vartheta(\log 1/\delta)^{-\vartheta} + \ldots \tag{3.23}$$

where '...' denotes something 'negligible in comparison for δ sufficiently small'. We note that (3.23) is better than an order estimate for the error — it contains the correct constant, i.e., it is the first term in an asymptotic expansion for $\delta \to 0$ — but would still be useless for either our *a posteriori* or *a priori* interpretations until we would actually specify satisfactorily the meaning of 'sufficiently small'. For example, if we were to 'solve' (3.23) with some margin, prescribing a measurement accuracy $\delta = \exp\left[-T(2M/\hat{\varepsilon})^{1/\vartheta}\right]$, we would have a guarantee that we achieve the desired accuracy $\hat{\varepsilon}$ — once we would be in the domain of validity (as yet unspecified!) of the bound $\varepsilon(\delta;\mathcal{K}_M) \leq 2MT^\vartheta(\log 1/\delta)^{-\vartheta}$.

For the second case we follow [11]. Note that we then have $\mathbf{E} = \mathbf{H}(\tau)$ and $\mathbf{F} = \mathbf{H}(T-\tau)$ so

$$\varphi_k = e^{-(T-\tau)\mu_k}, \qquad \epsilon_k = e^{-\tau\mu_k}$$

so $\tilde{s}(\lambda) = \lambda^{\tau/T}$. Using this in (3.14), (3.15) gives, by elementary calculus, the explicit formula

$$\beta = \beta(\tilde{\nu}) := \frac{\tau}{T}(1/\tilde{\nu})^{1-\tau/T} \qquad (\tilde{\nu} = \delta/M) \tag{3.24}$$

for (3.12). We now recover the well-known 'log-convexity' estimate from (3.17), which here gives

$$\varepsilon(\delta;\mathcal{K}_M) = M^{1-\tau/T}\delta^{\tau/T}. \tag{3.25}$$

At this point we note that the construction (3.10) involves an infinite series expansion which cannot actually be implemented without some further approximation: we can only work computationally with a finite sum and also will typically have to compute the eigenvalues and eigenfunctions of (3.21). A 'brute force' approach to this error analysis would lead to completely unrealistic estimates but, as was noted in [11], the same analysis used here for (3.19) also applies to any plausible discretization. Such a discretization will certainly give a finite sum in (3.10) and (3.21) will be replaced by computation of the eigenvalues and eigenvectors of a finite (self-adjoint) matrix, so the total computation is implementable. Fortunately, the error can be estimated without bounding the errors in the spectral approximation, only considering the approximation error $\delta_*(\cdot)$ for the discretization of the forward problem (using the *a priori* bound on the initial data). When $\hat{x}$ is computed by this approximation procedure we thus get the composite error bound

$$\|u(\tau,\cdot)-\hat{x}\| \leq \delta_*(\tau) + M^{1-\tau/T}\left[\delta+\delta_*(T)\right]^{\tau/T}, \tag{3.26}$$

which, as expected, reduces to (3.25) as the computation is refined. [This assumes we have computed exactly the spectral data for the discretized problem and the coefficients for $\hat{y}$ with respect to those eigenvectors, but these additional sources of error can also be handled.] This treatment of discretization applies equally well to the first case ($\tau = 0$). ■

ACKNOWLEDGMENTS: I am pleased to thank H. Engl, H. Haario, W. Rundell, and UMBC for their different contributions, which made it possible for me to participate in the excellent and enjoyable conference at St. Wolfgangsee.

REFERENCES

[1] R.S. Anderssen, *The linear functional strategy for improperly posed problems*, in (J.R. Cannon and U. Horning, eds.) *Inverse Problems* [ISNM #77], Birkhäuser, Basel, 1987, pp. 11–30.

[2] W. Chewning and T.I. Seidman, *A convergent scheme for boundary controls for the heat equation,* SIAM J. Control Opt. **15** (1977), pp. 64–72.

[3] C.W. Groetsch, *The theory of Tikhonov regularization for Fredholm equations of the first kind*, Pitman, London, 1984.

[4] J. Hadamard, *Lectures on the Cauchy Problem in Linear Partial Differential Equations*, Yale Univ. Press, New Haven, 1923.

[5] K. Miller, *Least-squares methods for ill-posed problems with a prescribed bound,* SIAM J. Math. Anal. **1** (1970), pp. 52–74.

[6] T.I. Seidman, *The solution of singular equations, I: linear equations in Hilbert space*, Pac. J. Math. **61** (1975), pp. 513–520.

[7] T.I. Seidman, *Computational approaches to ill-posed problems: general considerations,* in *Proc. 1976 Conf. on Inf. Sci. and Systems*, Johns Hopkins Univ. (1976), pp. 258–262.

[8] T.I. Seidman, *Convergent approximation methods for ill-posed problems, Part I: general theory*, Control and Cybernetics **10** (1981), pp. 31–49.

[9] T.I. Seidman, *Some 'complexity' issues for ill-posed problems,* in *Inverse Problems in Partial Differential Equations* (D. Colton, R. Ewing, W. Rundell, eds.), SIAM (1990), pp. 17–35.

[10] T.I. Seidman, *The method of 'optimal filtering' for a class of ill-posed problems*, in *Wave Propagation and Inversion* (W. Fitzgibbon and M. Wheeler, eds.) SIAM (1992), pp. 108–123.

[11] T.I. Seidman, *Optimal filtering for the backward heat equation*, SIAM J. Num. Anal. (1994), to appear.

[12] T.I. Seidman and L. Eldén, *Optimal filtering for the sideways heat equation*, Inverse Problems **6** (1990), pp. 681–696.

[13] A.G. Werschulz, *What is the complexity of ill-posed problems?* Numer. Fnl. Anal. and Optim. **9** (1987), pp. 945–967.

Coefficient Recovery in a Parabolic Equation from Input Sources

Bruce D. Lowe[†] & William Rundell[†] *

1. Introduction.

In this talk we consider coefficient identification problems for the partial differential equation

$$\begin{aligned} \frac{\partial u}{\partial t} - \frac{\partial}{\partial x}\left(a(x)\frac{\partial u}{\partial x}\right) + q(x)u = f(x), \quad x \in (0,1)\,, t > 0 \\ u(0,t) = u(1,t) = 0, \quad t > 0 \\ u(x,0) = 0, \quad x \in (0,1), \end{aligned} \tag{1.1}$$

where it is assumed that $a > 0\,, a \in W^{1,\infty}(0,1)\,, q \in L^2(0,1)$ and $f \in L^2(0,1)$. One physical interpretation of the solution $u(x,t)$ of (1.1) is that it gives the temperature at position $x \in [0,1]$ along a rod at time $t \geq 0$. The coefficient a is the thermal conductivity, q is a control term and f is a time independent heat source which is distributed along the rod. It is assumed in (1.1) that the temperature at both ends of the rod is maintained at zero for all time and that the rod is initially at temperature zero. When the coefficients a and q are known and the source f is also known, the unique solution u of (1.1) is given by the expression

$$u(x,t) = \sum_1^\infty \frac{1 - e^{-\lambda_n t}}{\lambda_n}\left(\int_0^1 \phi_n f\right)\phi_n(x)\,, \tag{1.2}$$

where the eigenfunctions $\{\phi_n(x)\}_1^\infty$ and the eigenvalues $\{\lambda_n\}_1^\infty$ satisfy

$$\begin{aligned} \mathcal{L}\phi_n = -(a\phi_n')' + q\,\phi_n = \lambda_n\phi_n \\ \phi_n(0) = \phi_n(1) = 0 \\ \|\phi_n\|_{L^2} = 1\,, \phi_n'(0) > 0. \end{aligned} \tag{1.3}$$

This representation has been used as the starting point of several uniqueness results for the inverse problem of obtaining information about the coefficients from knowledge of additional data. For example, if overposed data consisting of the flux values at one of the boundaries or of the interior measurement of temperature at a point x_0, $0 < x_0 < 1$ is given as a function of time, then under certain circumstances one can utilize (1.2) to obtain spectral data for the operator $\mathcal{L}$ and reduce the original inverse problem to that for a classical inverse Sturm-Liouville problem [8], [9], [10]. This approach, while ultimately leading to a tractable, even numerically stable problem, suffers from the drawback that the transformation to the inverse Sturm Liouville problem is itself extremely ill conditioned. This is nothing else but a manifestation of the fact that recovering a coefficient depending on interior points only from boundary measurements is intrinsically an ill-posed problem. However, the technique of transforming to an inverse Sturm Liouville situation does not seem to be fruitful, and we shall instead attempt to solve the inverse problem in its original formulation.

† Department of Mathematics, Texas A&M University, College Station, TX 77843

⋆ Supported in part by the National Science Foundation under grant DMS-9202352

In this regard, the techniques used for reconstruction of the coefficients resemble that used for a particular case of the inverse Sturm Liouville problem in [2].

What we are proposing is that if one is able to change some of the interior conditions (and we use a variable source to achieve this) then recovery of coefficients from boundary measurements becomes a very reasonable problem. We thus envision that (at least in theory) an infinite number of inputs are given and the reconstruction is to be obtained from measurements of corresponding outputs, in our case these correspond to a single scalar quantity. In this regard our approach has some similarities to recent work in undetermined coefficient problems such as that using the so called Dirichlet to Neumann map (although the resulting mathematical issues are quite distinct) [1], [11].

For the first part of the paper we consider the simpler case of known thermal conductivity which we shall set to be $a(x) = 1$, but will later return to the more general situation. Thus we seek to determine the function $q(x)$ by giving overposed data on the forward solution $u(x,t)$ of (1.1). The overposed data are generated by performing the following experiment: we assume that we are given a basis $\{f_j\}_1^\infty$ of $L^2(0,1)$ and a single, fixed time $t^\star > 0$. For each j, we distribute at time $t = 0$ the input source $f(x) = f_j(x)$ along the rod and then at time $t^\star$, measure either the heat leaving the left end of the rod

$$\alpha_j := -\frac{\partial u_j(0,t^\star)}{\partial x} \tag{1.4}$$

or the net heat flux from the rod

$$\beta_j := \frac{\partial u_j(1,t^\star)}{\partial x} - \frac{\partial u_j(0,t^\star)}{\partial x}\,. \tag{1.5}$$

The procedure generates either the data $\{\alpha_j\}_1^\infty$ or $\{\beta_j\}_1^\infty$ from which to determine the coefficient $q(x)$. We begin by addressing the question of whether the data $\{\alpha_j\}_1^\infty$ or $\{\beta_j\}_1^\infty$ uniquely determines the coefficient.

2. Uniqueness Results.

It is useful to first consider the case when $t^\star = \infty$, that is, when measurements are made at steady state (assume that $q > -\lambda_1$). The question of uniqueness reduces to showing that $q(x)$ in the ordinary differential equation

$$\begin{aligned} -u_j'' + q(x)\,u_j &= f_j(x) \\ u_j(0) = u_j(1) &= 0 \end{aligned} \tag{2.1}$$

is determined by either the data (1.4) or (1.5), taken at $t^\star = \infty$. We denote by $u_j(x;q)$ the solution of the boundary value problem (2.1). We begin with the following uniqueness result [7].

Theorem 2.1. *Let $\{f_j\}_1^\infty$ be a complete set in $L^2(0,1)$ and assume that $q_1, q_2 \in L^2(0,1)$. If 0 is not a Dirichlet eigenvalue of the operator $\mathcal{L}_q := -d^2/dx^2 + q$ on (0,1) and if $u_j'(0;q_1) = u_j'(0;q_2)$ for all $j \geq 1$, then $q_1 = q_2$.*

Proof: Let $G(x,\xi) = G(x,\xi;q)$ denote the Green's function of $\mathcal{L}_q$, which for each $\xi \in (0,1)$ solves

$$\begin{aligned} -G_{xx}(x,\xi) + q(x)\,G(x,\xi) &= \delta_\xi(x)\,, \quad 0 < x < 1 \\ G(0,\xi) = G(1,\xi) &= 0. \end{aligned}$$

First we will show that the assumptions of the theorem imply that the corresponding pair of Green's functions satisfy

$$G_x(0,\xi;q_1) = G_x(0,\xi;q_2)\,. \tag{2.2}$$

This follows easily from the fact that the solutions $u_j(x;q_i)$ for $i = 1, 2$ of (2.1) satisfy

$$u_j(x;q_i) = \int_0^1 G(x,\xi;q_i) f_j(\xi)\, d\xi \,.$$

Consequently,

$$u_j'(0;q_i) = \int_0^1 G_x(0,\xi;q_i) f_j(\xi)\, d\xi \,,$$

where $G_x(0,\xi;q_i) \in H^2(0,1)$. Hence,

$$\int_0^1 [G_x(0,\xi;q_2) - G_x(0,\xi;q_1)] f_j(\xi)\, d\xi = u_j'(0;q_2) - u_j'(0;q_1) = 0$$

for all $j \geq 1$. Since $\{f_j\}_1^\infty$ is complete, equation (2.2) is immediate.

Defining $v_i(x) = -G_x(0,x;q_i)$, it is easily seen that these functions satisfy

$$\begin{aligned} -v_i'' + q(x)\, v_i &= 0, \quad 0 < x < 1 \\ v_i(0) &= -1 \\ v_i(1) &= 0. \end{aligned} \tag{2.3}$$

It follows from (2.2) that $v_1 = v_2$ on $(0,1)$. Moreover, v_1 can vanish at most at finitely many points in $[0,1]$, for otherwise, there would exist a point $x^* \in [0,1]$ for which $v_1(x^*) = v_1'(x^*) = 0$. Since v_1 satisfies a homogeneous differential equation on $(0,1)$, we conclude that v_1 must vanish identically on $(0,1)$, contradicting the initial condition $v_1(0) = -1$. Using the equations in (2.3), we have $v_1(q_1 - q_2) = 0$ almost everywhere on $(0,1)$. Since v_1 vanishes on at most a finite set, it follows that $q_1 = q_2$ almost everywhere on $(0,1)$.

Theorem 2.2. *Let $\{f_j\}_1^\infty$ be a complete set in $L^2(0,1)$ and assume that $q_1, q_2 \in L^2(0,1)$. If 0 is not a Dirichlet eigenvalue of $-d^2/dx^2 + q$ on $(0,1)$ and if $u_j'(1;q_1) - u_j'(0;q_1) = u_j'(1;q_2) - u_j'(0;q_2)$ for all $j \geq 1$, then $q_1 = q_2$.*

Proof: Proceeding as in the proof of the previous theorem, we obtain

$$G_x(1,\xi;q_1) - G_x(0,\xi;q_1) = G_x(1,\xi;q_2) - G_x(0,\xi;q_2)\,, \tag{2.4}$$

instead of (2.2). The function $w_i(x) = G_x(1,x;q_i) - G_x(0,x;q_i)$ satisfies

$$\begin{aligned} -w_i'' + q(x)\, w_i &= 0, \quad 0 < x < 1 \\ w_i(0) = w_i(1) &= -1 \end{aligned}$$

and the remainder of the argument follows directly as above.

Although either the data $\{\alpha_j\}_1^\infty$ or $\{\beta_j\}_1^\infty$ give uniqueness, we shall see below that the net flux data $\{\beta_j\}_1^\infty$ are preferable for numerical reconstruction. In the next section we will see that this is especially true in the case when the measurements are made at a finite time t^*. Prior to addressing this point, we make the following simple but important observation.

For each fixed q in (2.1), the mapping from the source f_j to the overposed data (either $u_j'(0)$ or $u_j'(1) - u_j'(0)$) is linear, this following from the linearity of the operator $\mathcal{L}_q$. Consequently, suppose that $\{f_k\}_1^\infty$ is a basis for $L^2(0,1)$ for which overposed data, say $\{u_k'(0)\}_1^\infty$ has been measured. If $\{\psi_j\}_1^\infty$ is another basis for $L^2(0,1)$ and $\psi_j(x) = \sum_1^\infty c_{jk} f_k(x)$, then if v_j denotes the solution of (2.1) with the input source $\psi_j(x)$, then $v_j'(0) = \sum_1^\infty c_{jk} u_k'(0)$. As a result of this we are able, without any loss of generality, to use any basis of sources.

Assume then that $\{\alpha_j\}_1^\infty$ and $\{\beta_j\}_1^\infty$ arise from the application in (2.1) of the input sources $f_j(x) = \sin j\pi x$. The function $y_j(x) = u_j(x;0)$ satisfies $-y_j'' = \sin j\pi x$ and $y_j(0) = y_j(1) = 0$, so that $y_j(x) = \sin j\pi x / j^2\pi^2$. By integrating (2.1) from $x = 0$ to $x = 1$ we obtain

$$u_j'(1;q) - u_j'(0;q) + \int_0^1 f_j(t)\,dt = \int_0^1 q(t)\,u_j(t;q)\,dt\,.$$

For $q(x)$ lying in a ball of sufficiently small radius about $q = 0$ we can make the approximation $u_j(x;q) \approx y_j(x)$ and so the above equation becomes

$$\int_0^1 q(t)\sin j\pi t\,dt \approx j^2\pi^2\left[u_j'(1;q) - u_j'(0;q) + \int_0^1 f_j(t)\,dt\right]\,.$$

Thus, the datum β_j yields

$$\begin{aligned}\int_0^1 q(t)\sin j\pi t\,dt &\approx j^2\pi^2\left[\beta_j + \int_0^1 f_j(t)\,dt\right]\\ &\approx j^2\pi^2\left(\beta_j + (1-(-1)^j)/j\pi\right)\,,\end{aligned} \tag{2.5}$$

an approximation for the jth Fourier sine coefficient of the function $q(x)$.

Consider now the case of boundary data given only at the left end point. We integrate (2.1) from 0 to s obtaining

$$u_j'(s;q) - u_j'(0;q) = -\int_0^s f_j(t)\,dt + \int_0^s q(t)u_j(t;q)\,dt.$$

Integrating from $s = 0$ to $s = 1$ and using (1.4) gives

$$u_j(1;q) - u_j(0;q) + \alpha_j = -\int_0^1\int_0^s f_j(t)\,dt\,ds + \int_0^1\int_0^s q(t)u_j(t;q)\,dt\,ds,$$

that is,

$$\int_0^1 (1-t)q(t)u_j(t;q)\,dt = \alpha_j + \int_0^1 (1-t)f_j(t)\,dt\,.$$

For $q(x)$ lying in a ball of sufficiently small radius about $q = 0$ we again make the approximation $u_j(x;q) \approx y_j(x)$ to obtain

$$\int_0^1 (1-t)q(t)\sin j\pi t\,dt \approx j^2\pi^2\left[\alpha_j + \int_0^1 (1-t)f_j(t)\,dt\right]\,. \tag{2.6}$$

We see that (2.6) yields an approximation to the jth Fourier sine coefficient of $(1-x)q(x)$.

Roughly speaking, for small q, the difference between the data $\{\beta_j\}_1^\infty$ and $\{\alpha_j\}_1^\infty$ is the difference between approximate Fourier sine reconstructions of $q(x)$ and $(1-x)q(x)$ respectively. In the latter situation, the information loss for $q(x)$ near $x = 1$ is consistent with all the measurements being taken at $x = 0$.

Before discussing the case when the measurements, either (1.4) or (1.5) are made at a finite time $t^\star$, we consider the simultaneous determination of $a(x)$ and $q(x)$ in (1.1) from steady state measurements. Consider the solution $v = v(x; a, q)$ of the ordinary differential equation

$$\begin{aligned} -(av')' + q\,v &= f \\ v(0) = v(1) &= 0\,. \end{aligned} \tag{2.7}$$

We have the following theorem, the details of which can be found in [6].

Theorem 2.3. *Let $\{f_j\}_1^\infty$ be a complete set in $L^2(0,1)$ and suppose that $a_i \in W^{1,\infty}(0,1)$ and $q_i \in L^2(0,1)$ for $i = 1, 2$. Assume that 0 is not a Dirichlet eigenvalue of $-\frac{d}{dx}\left(a\frac{d}{dx}\right) + q$ on $(0,1)$. Let $v_j(x; a_i, q_i)$ denote the solution of (2.7) with the input source $f = f_j(x)$. If for all $j \geq 1$, $a_1(0)v_j'(0; a_1, q_1) = a_2(0)v_j'(0; a_2, q_2)$ and $a_1(1)v_j'(1; a_1, q_1) = a_2(1)v_j'(1; a_2, q_2)$ and if in addition, $a_1(\tilde{x}) = a_2(\tilde{x})$ for some $\tilde{x} \in [0,1]$, then $a_1 = a_2$ and $q_1 = q_2$ on $[0,1]$.*

We now consider the situation when the measurements, either (1.4) or (1.5) are made at a finite time $t^\star$. Some results for this transient case were developed in [7]. For each input source $f_j(x)$ in a basis for $L^2(0,1)$, we define the function $u_j(x, t; q)$ to be the solution of

$$\begin{aligned} \frac{\partial u_j}{\partial t} - \frac{\partial^2 u_j}{\partial x^2} + q u_j &= f_j(x) \\ u_j(0, t; q) = u_j(1, t; q) &= 0 \\ u_j(x, 0; q) &= 0. \end{aligned} \tag{2.8}$$

In this situation there were differences in what we could prove depending on whether (1.4) or (1.5) were used. In either case our approach limited our results to "sufficiently small" coefficients, and perhaps more unusually, to "sufficiently large" times of measurement. Thus it should be viewed as a perturbation from the known steady state situation. The main results obtained are the following theorems:

Theorem 2.4. *Let $u_j(x, t; q_1)$ and $u_j(x, t; q_2)$ be solutions of (2.8) and suppose that $q_1 = q_2$ on the interval $[1-\delta, 1]$ for some $0 < \delta < 1$ and $\|q_i\|_{L^2} < \omega < \pi\delta$ for some ω. Then there exists a time $\bar{t}$, such that if for each $j \geq 1$,*

$$\frac{\partial u_j(0, t^\star; q_1)}{\partial x} = \frac{\partial u_j(0, t^\star; q_2)}{\partial x}$$

holds for some $t^\star > \bar{t}$, then $q_1 = q_2$ on $[0,1]$.

Theorem 2.5. *Let $u_j(x, t; q_1)$ and $u_j(x, t; q_2)$ be solutions of (2.8) and suppose that $\|q_i\|_{L^2} < \omega < \pi$ for some ω. Then there exists a time $\bar{t}$, such that if for each $j \geq 1$,*

$$\frac{\partial u_j(1, t^\star; q_1)}{\partial x} - \frac{\partial u_j(0, t^\star; q_1)}{\partial x} = \frac{\partial u_j(1, t^\star; q_2)}{\partial x} - \frac{\partial u_j(0, t^\star; q_2)}{\partial x}$$

holds for some $t^\star > \bar{t}$, then $q_1 = q_2$ on $[0,1]$.

3. Numerical Reconstruction.

We begin the discussion by considering a numerical method for the reconstruction of $q(x) \approx 0$ in the ordinary differential equation

$$\begin{aligned} -u_j'' + q(x)\, u_j &= f_j(x) \\ u_j(0) = u_j(1) &= 0\,, \end{aligned} \tag{3.1}$$

from finite data. We take $f_j(x) = \sin j\pi x$ and consider for now the data $\{\beta_j = u_j'(1) - u_j'(0)\}_1^N$. As indicated in (2.5), this data provides an approximation to the first N terms of the Fourier sine series of $q(x)$. Our method is to construct a function that lies in the span of $\{\sin j\pi x\}_1^N$ which is compatible with the overposed data. Setting $q(x) = \sum_1^N q_k \sin k\pi x$, we find a zero of the map $G : R^N \to R^N$ given by

$$G_j(\vec{q}) := 2j^2\pi^2(u_j'(1;\vec{q}) - u_j'(0;\vec{q}) - \beta_j), \quad j = 1,\ldots,N. \tag{3.2}$$

Here, $\vec{q} = (q_1,\ldots,q_N)$ and $u_j(x;\vec{q})$ is the solution of (3.1) with $q(x) = \sum_1^N q_k \sin k\pi x$ (the reason for the multiplicative factors $2j^2\pi^2$ will be apparent shortly). We have the following result:

Theorem 3.1. *For G given in (3.2),*

$$\frac{\partial G_j}{\partial q_k}(\vec{0}\,) = \delta_{jk}\,. \tag{3.3}$$

We write this as $G'_\infty(\vec{0})$ where the subscript refers to the fact that we are in the steady state case. The matrix is clearly uniformly invertible for all values of N. The finite dimensional Implicit Function Theorem guarantees that for each N there exists a unique zero of the map G in a neighborhood of the origin. We can attempt to reconstruct $q(x)$ by means of the quasi-Newton scheme:

$$\vec{q}_{n+1} = \vec{q}_n - (G'(\vec{0}\,))^{-1}G(\vec{q}_n\,) = \vec{q}_n - G(\vec{q}_n\,), \quad \vec{q}_0 = 0\,. \tag{3.4}$$

This will converge provided that $\|q(x)\|_{L^2}$ is sufficiently small.

If instead we have the data $\{\alpha_j = -u_j'(0)\}_1^N$, then we seek a zero of the map $F : R^N \to R^N$ given by

$$F_j(\vec{q}) := 2j^2\pi^2(u_j'(0;\vec{q}) + \alpha_j), \quad j = 1,\ldots,N, \tag{3.5}$$

by means of the quasi-Newton scheme

$$\vec{q}_{n+1} = \vec{q}_n - (F'(\vec{0}\,))^{-1}F(\vec{q}_n\,), \quad \vec{q}_0 = 0\,. \tag{3.6}$$

We have the following result:

Theorem 3.2. *For F given in (3.4),*

$$F'_\infty(\vec{0}) = \frac{\partial F_j}{\partial q_k}(\vec{0}) = \begin{cases} \frac{(-1)^{k+j}-1}{\pi^2}\left(\frac{1}{(k-j)^2} - \frac{1}{(k+j)^2}\right) & \text{if } j \neq k \\ -\frac{1}{2} & \text{if } j = k. \end{cases} \tag{3.7}$$

The analysis for the map F is not quite so straightforward, but it turns out that the Jacobian matrix $F'(\vec{0})$ is diagonally dominant for each N and hence invertible for each N. Indeed, if we denote by $[A_{jk}]$ the matrix representing the Jacobian $F'(\vec{0})$ and set $\sigma(n) = (1-(-1)^n)/\pi^2$, then

$$\begin{aligned}\sum_{k=1,k\neq j}^{N} |A_{jk}| &= \sum_{k=1,k\neq j}^{N} \sigma(k+j)\left(\frac{1}{(k-j)^2} - \frac{1}{(k+j)^2}\right) \\ &< \sum_{k=1,k\neq j}^{\infty} \frac{\sigma(k+j)}{(k-j)^2} = \sum_{k=1}^{j-1} \frac{\sigma(k+j)}{(k-j)^2} + \sum_{k=j+1}^{\infty} \frac{\sigma(k+j)}{(k-j)^2} \\ &< 2\sum_{s=1}^{\infty} \frac{1-(-1)^s}{\pi^2 s^2} = \frac{1}{2} = |A_{jj}|,\end{aligned}$$

and so A is diagonally dominant. However, it is not true that A is uniformly diagonally dominant for all values of N. Indeed, numerical computations show that the condition number $\kappa(A) \approx cN$ where the value of the constant c is very close to unity. For practical purposes we cannot use a large number of input sources, and in any case are unlikely to want to recover more than ten or twenty Fourier coefficients of the function $q(x)$ and so the relatively weak amount of ill-conditioning of the Jacobian matrix need not be of practical concern for this particular problem.

Note that in order to obtain a matrix that was uniformly invertible in the dimension N, it was necessary to define the map G with the jth component containing the multiplicative factor $2j^2\pi^2$. As intimated by the approximation given in (2.5), it can be established that [7]:

$$\beta_j = \frac{(-1)^j - 1}{j\pi} + \frac{c_j(q)}{j^2\pi^2},$$

where $\{c_j(q)\}_1^\infty$ is square summable. Let h_2^2 denote the space of sequences $\{x_n\}_1^\infty$ such that $\sum_1^\infty (n^2 x_n)^2 < \infty$. Consequently, the difference of $\{\beta_j\}_1^\infty$ and $\{[(-1)^j - 1]/j\pi\}_1^\infty$, the net flux data for $q = 0$ must converge in h_2^2.

This shows that the inverse problem with net flux data is indeed ill-posed, but the degree of ill-conditioning is rather mild.

For the situation of simultaneous measurements of both a and q from finite data consisting of the flux measurements at both ends of the rod, $\{a(0)u_j'(0; a, q) = \alpha_j\}_{j=1}^N$, and $\{a(1)u'(1; a, q) = \gamma_j\}_{j=1}^N$ we construct approximations $a_N(x)$ and $q_N(x) = a_N(x)r_N(x)$ for $a(x) \approx 1/a_0$ and $q(x) \approx 0$ of the form

$$\frac{1}{a_N(x)} = a_0 + \sum_{k=1}^{N} a_k \cos \pi k x$$

$$r_N(x) = \sum_{k=1}^{N} c_k \sin \pi k x$$

where the unknown coefficients $a_1, \ldots, a_N$ and $c_1, \ldots c_N$ are determined from the data $\{\alpha_j, \gamma_j\}_{j=1}^N$ and the assumed known value of a_0 (this requirement cannot be dropped, [*]) by finding a zero of the map $F : R^{2N} \to R^{2N}$ given by

$$\begin{aligned}F_j(\vec{a}, \vec{c}) &:= 2\pi^2 j^2 \left(a(0)u_j'(0; a, q) - \alpha_j\right) \\ F_{j+N}(\vec{a}, \vec{c}) &:= 2\pi^2 j^2 \left(a(1)u'(1; a, q) - \gamma_j\right) \qquad j = 1, \ldots, N.\end{aligned}$$

The construction again employs a quasi-Newton method from the initial guess $(\vec{a}, \vec{c}) = (\vec{0}, \vec{0})$. This is feasible since the Jacobian Matrix $F'(\vec{0}, \vec{0})$ can be shown, for sufficiently small values of both $\vec{a}$ and $\vec{c}$ to be invertible [*].

In the transient case

$$\begin{gathered}\frac{\partial u_j}{\partial t} - \frac{\partial^2 u_j}{\partial x^2} + q u_j = \sin j\pi x \\ u_j(0,t) = u_j(1,t) = 0 \\ u_j(x,0) = 0\,,\end{gathered} \tag{3.8}$$

when measurements are taken at a finite time $t^\star$, the previous quasi-Newton method is easily modified.

We just target on either the overposed data

$$\left\{-\frac{\partial u_j(0,t^\star)}{\partial x}\right\}_1^N \tag{3.9}$$

or

$$\left\{\frac{\partial u_j(1,t^\star)}{\partial x} - \frac{\partial u_j(0,t^\star)}{\partial x}\right\}_1^N . \tag{3.10}$$

The corresponding mappings G and F given by (3.2) and (3.5) differ only in the fact that $u(x,t)$ is evaluated at a finite value $t^\star$ for t. The resulting Jacobian matrices now depend on the time $t^\star$ of the flux measurement and these have the following properties.

Theorem 3.4. *The Jacobian matrices can be written in the form*

$$\begin{gathered}G'_{t^\star}(\vec{0}) = G'_\infty(\vec{0}) + T_1(t^\star) \\ F'_{t^\star}(\vec{0}) = F'_\infty(\vec{0}) + T_2(t^\star)\end{gathered} \tag{3.11}$$

where the $N \times N$ matrices T_1 and T_2 have entries that tend to zero as $t^\star \to \infty$.

From this it readily follows that both $G'_{t^\star}(\vec{0})$ and $F'_{t^\star}(\vec{0})$ are diagonally dominant, and hence invertible, provided that $t^\star$ is sufficiently large, a result very much in keeping with the uniqueness theorems of the previous section. It is not true that these matrices, even $G'_{t^\star}(\vec{0})$, are diagonally dominant for all values of $t^\star$; however numerical computations indicated that while the condition number increased as $t^\star$ approached zero, the matrices appeared to be nonsingular for al values of $t^\star$. In practice we found that the convergence rate of the schemes decreased as the time $t^\star$ decreased. For the single point flux measurement (3.9) we found that the numerical scheme diverged for $t^\star$ less than a certain value (which was approximately 0.35); while the scheme corresponding to the net flux (3.10) appeared to converge for all $t^\star > 0$ (albeit quit slowly for very small $t^\star$).

4. Generalizations.

We consider here two possible directions in which these results might be extended; to the recovery of both coefficients (1.1) and to the recovery of one or more coefficient in higher space dimensions.

If $q(x)$ is known, then the problem of determining $a(x)$ from either the value of left flux or the net flux measurements at a fixed time proceeds in a similar way and leads to identical results to that of the recovery of q case.

We have been unable to extend theorem 2.3 to the finite time case. However, one can show that the Jacobian matrix for the transient case can be written as a sum of the Jacobian matrix for steady state case plus a time-dependent perturbation that tends to zero as $t \to \infty$. Numerical results were not so encouraging here, for our scheme would only converge when $t^\star$ was so large that the perturbation matrix was very small.

Let us now consider the case of the steady state problem in two spatial dimensions. Let Ω be a bounded domain in R^2 with smooth boundary $\partial\Omega$. We consider the recovery of $q(x)$ in the elliptic equation

$$\begin{aligned} -\Delta u_j + q(x)u_j &= f_j \quad \text{in } \Omega \\ u_j|_{\partial\Omega} &= 0 \end{aligned} \tag{4.1}$$

from the application of a basis $\{f_j\}_1^\infty$ of $L^2(\Omega)$ of input sources. In particular, for $q(x) \geq 0$, equation (4.1) can be viewed as the steady state of

$$\begin{aligned} v_t - \Delta v + q(x)v &= f_j(x), \quad \Omega \times (0,\infty) \\ v(x,t)|_{\partial\Omega} &= 0, \quad t > 0 \\ v(x,0) &= 0, \quad x \in \Omega. \end{aligned}$$

The natural generalization of the steady state overposed data $\{u_j'(1) - u_j'(0)\}_1^\infty$ in equation (2.1) to higher dimensions is

$$\left\{\beta_j := \int_{\partial\Omega} \frac{\partial u_j}{\partial n}\, ds\right\}_1^\infty . \tag{4.2}$$

We have the following uniqueness theorems for $q(x) \in C^1(\overline{\Omega})$ [6]:

Theorem 4.1. *If $q(x)$ in equation (4.1) satisfies the condition $q(x) \geq 0$, then $q(x)$ is uniquely determined by the data (4.2).*

Theorem 4.2. *If $q(x)$ in equation (4.1) satisfies $|q|_{L^\infty} < \min\left\{\lambda_1, \frac{1}{2(e^d - 1)}\right\}$, where λ_1 is the smallest Dirichlet eigenvalue of $-\Delta$ on Ω and the domain Ω lies between parallel lines distance d apart, then $q(x)$ is uniquely determined by the data (4.2).*

For a fixed q, the mapping from the source f_j to the data $\int_{\partial\Omega} \frac{\partial u_j}{\partial n}\, ds$ is linear. We may therefore choose without loss of generality any basis of input sources. We select an orthonormal basis $\{f_j\}_1^\infty$ of Dirichlet eigenfunctions of $-\Delta$ on Ω, where f_j corresponds to the eigenvalue λ_j and $\lambda_{j+1} \geq \lambda_j$ for all $j \geq 1$. Note that when $q(x) = 0$, $u_j(x) = \lambda_j^{-1} f_j(x)$. An integration of (4.1) over Ω and an application of the divergence theorem gives

$$-\int_{\partial\Omega} \frac{\partial u_j}{\partial n}\, ds + \int_\Omega q(x)u_j(x)\, dx = \int_\Omega f_j(x)\, dx .$$

For $q(x)$ lying in a ball of sufficiently small radius about $q = 0$ we can make the approximation $u_j(x) \approx \lambda_j^{-1} f_j(x)$ and so the above equation becomes

$$\int_\Omega q(x) f_j(x)\, dx \approx \lambda_j \left(\beta_j + \int_\Omega f_j(x)\, dx\right) . \tag{4.3}$$

Consequently, the datum β_j yields an approximation to the jth Fourier coefficient of the function $q(x)$ in the basis $\{f_j\}_1^\infty$. This generalizes the approximation given in (2.5) to higher dimensions.

It is also the case that $q(x)$ can be uniquely determined by the data

$$\left\{\alpha_j := \int_\Gamma \frac{\partial u_j}{\partial n}\, ds\right\}_1^\infty , \tag{4.4}$$

for any arc $\Gamma \subset \partial\Omega$ [7].

We observed in the one-dimensional case that associated with left flux measurements $\{u_j'(0)\}$ of (2.1) is some difficulty in resolving $q(x)$ near $x = 1$. A similar phenomenon is encountered in two dimensions. It becomes more difficult to resolve $q(x)$ as one moves away from the arc Γ where the data are collected.

Let w be the solution of

$$\begin{aligned} \Delta w &= 0 \\ w|_\Gamma &= 1 \\ w|_{\partial\Omega\sim\Gamma} &= 0\,. \end{aligned}$$

Then it follows that $w > 0$ on Ω. Moreover, using (4.1) we obtain that

$$\int_\Omega q(x)w(x)u_j(x)\,dx = \int_\Omega w(x)f_j(x)\,dx + \alpha_j\,.$$

Again using that $u_j(x) \approx \lambda_j^{-1} f_j(x)$ for $q(x)$ lying in a ball of sufficiently small radius about $q = 0$, we have

$$\int_\Omega q(x)w(x)f_j(x)\,dx \approx \lambda_j \left(\int_\Omega w(x)f_j(x)\,dx + \alpha_j \right)\,. \tag{4.5}$$

Hence, the datum α_j yields an approximation for the jth Fourier coefficient of $q(x)w(x)$ in the basis $\{f_j\}_1^\infty$. As x approaches $\partial\Omega \sim \Gamma\,, w(x)$ approaches 0 and an information loss in $q(x)$ results. The approximation (4.5) generalizes (2.6) to the two dimensional setting.

References.

[1] S. Avdonin and T.I. Seidman, Identification of $q(x)$ in $u_t = \Delta u - qu$ from boundary measurements, preprint.

[2] B.D. Lowe, M. Pilant and W. Rundell, The recovery of potentials from finite spectral data, *SIAM J. Math. Anal.*, **23**, (1992), 482–504.

[3] B.D. Lowe and W. Rundell, An inverse problem for a Sturm-Liouville operator, *J. Math. Anal. Appl.*, **181**, (1994), 188–199.

[4] B.D. Lowe and W. Rundell, The determination of multiple coefficients in a second order differential equation from input sources, *Inverse Problems*, **9**, (1993), 469–482.

[5] B.D. Lowe and W. Rundell, The determination of a coefficient in a parabolic equation from input sources, *IMA J. Appl. Math.*, **52**, (1994), 31–50.

[6] B.D. Lowe and W. Rundell, Unique recovery of a coefficient in an elliptic equation from input sources, submitted for publication.

[7] B.D. Lowe and W. Rundell, Numerical recovery of a coefficient in an elliptic equation from partial flux data, in preparation.

[8] A. Pierce, Unique identification of eigenvalues and coefficients in a parabolic problem, *SIAM J. Contr. Opt.*, **17**, 1979, 494–499.

[9] T. Suzuki, Uniqueness and nonuniqueness in an inverse problem for the parabolic equation, *J. Diff. Eqns.*, **47**, 1983, 296–316.

[10] T. Suzuki and R. Murayama, A uniqueness theorem in an identification problem for coefficients of parabolic equations, *Proc. Japan Acad. Ser. A. Math. Sci.*, **56**, 1980, 259–263.

[11] J. Sylvester and G. Uhlmann, The Dirichlet to Neumann map and applications, in *Inverse Problems in Partial Differential, SIAM Publications,* 1990, 99–138.

Numerical Solution of the Sideways Heat Equation

Lars Eldén[*]

Abstract

Numerical procedures are discussed for solving a Cauchy problem for the heat equation in the quarter plane, where data are given at $x = 1$ and a solution is sought in the interval $0 < x < 1$. This *sideways heat equation* is a model of a problem, where one wants to determine the temperature on both sides of a thick wall, but where one side is inaccessible to measurements. A singular value analysis is made, which gives information about the character of the ill-posedness of the problem. An efficient algorithm for implementing Tikhonov regularization is described. The problem is analysed using the Fourier transform, and an approximation method based on a hyperbolic equation is discussed. Difference methods are often used for solving the sideways heat equation; it is shown that time differencing prevents high frequency noise from blowing up. A few numerical examples are given, which demonstrate the regularizing effect of time differencing.

1 Introduction

Consider, first, the *direct* problem for the heat equation in the quarter plane ($t \geq 0$, $x \geq 0$):

$$(1) \qquad \begin{cases} \kappa u_{xx} = u_t, & x \geq 0, \quad t \geq 0, \\ u(x,0) = 0, & x \geq 0, \\ u(0,t) = f(t) \end{cases}$$

The functions $f(\cdot)$ and $u(x,\cdot)$ are to be in $\mathcal{X} := L^2(-\infty,\infty)$, with the understanding that f and u vanish for $t < 0$. If we include in the problem specification that u is to be bounded as $x \to \infty$, then it is well-known (cf., e.g., [2]) that for each $f \in \mathcal{X}$, (1) has a unique solution u with $u(x,\cdot) \in \mathcal{X}$ for each $x \geq 0$. Thus we have a well-defined continuous 'input–output map'

$$(2) \qquad f \longmapsto g \quad (g(t) := u(1,t)) \; : \mathcal{X} \to \mathcal{X}$$

induced by (1).

Note that (2) expresses the fact that one has well-posed propagation of data 'to the right'. We consider here the *inverse problem* of propagating data 'to the left'. We assume that the 'initial' data function f for (1) is unknown and that we wish to determine $u(x,\cdot)$ for $0 \leq x < 1$ from the specification of $g(\cdot) := u(1,\cdot)$. Physically, this corresponds to a situation in which the end-point $x = 0$ is 'inaccessible' but for which one can make (internal) measurements at $x = 1$. Of course, since such physical measurements can never be infinitely precise, we would actually have as data some function $g_m \in L^2$, for which

$$(3) \qquad \|g_m - g\| = \|g_m - u(1,\cdot)\| \leq \epsilon,$$

[*]Department of Mathematics, Linköping University, S-581 83 Linköping, Sweden

where the constant $\epsilon > 0$ represents a bound on the measurement error. For $x \geq 1$ one can solve the direct problem (1) using g_m as data. For $0 < x < 1$ we have the *sideways heat equation*

$$(4) \qquad \begin{cases} \kappa u_{xx} = u_t, & x \geq 0, \quad t \geq 0, \\ u(x,0) = 0, & x \geq 0, \\ u(1,t) = g_m(t), & t \geq 0. \end{cases}$$

In the literature this is often referred to as the *Inverse Heat Conduction Problem (IHCP)*. Note that, although we seek to recover u only for $0 \leq x < 1$, the problem specification includes the heat equation for $x > 1$ together with the boundedness at infinity. Since we can obtain u for $x > 1$, also $u_x(1,\cdot)$ is determined. Thus we can consider (4) as a *Cauchy problem* with appropriate Cauchy data $[u, u_x]$ given on the line $x = 1$.

The problem (4) is ill-posed: if the solution exists, then it does not depend continuously on the data g_m. However, if we impose an a priori bound on the solution at $x = 0$, and, in addition, we do not try to match the data exactly[1], and instead consider the problem,

$$(5) \qquad \begin{cases} \kappa u_{xx} = u_t & x \geq 0, \quad t \geq 0, \\ u(x,0) = 0, & x \geq 0, \\ \|u(0,\cdot)\| \leq M, & \\ \|u(1,\cdot) - g_m(\cdot)\| \leq \epsilon, & \end{cases}$$

then we have stability in the following sense: any two solutions of (5), u_1 and u_2, satisfy

$$(6) \qquad \|u_1(x,\cdot) - u_2(x,\cdot)\| \leq 2M^{1-x}\epsilon^x, \qquad 0 \leq x \leq 1.$$

This was proved by Levine in [24] (it is also implicit in the earlier paper by Carasso[3]). For this reason we refer to (5) as the *stabilized problem.* It is noted in [33] that the error estimate (6) is sharp. Further stability results are given in [22], [15]. An interesting new technique for improving the stability of certain ill-posed problems is presented in [6].

In this paper we will consider numerical procedures for approximating solutions of the sideways heat equation (4) and the stabilized problem (5). We restrict the presentation to the quarter plane problem, but many of the results are also valid for the case of a finite interval in space. In part the analysis will be for an infinite time interval, but the numerical procedures discussed are for a finite interval. We will mainly deal with the heat equation (4). In actual applications, equations with non-constant coefficients and non-linearities are of interest; we will indicate the applicability of the methods discussed to more general equations.

Numerical procedures for (4) can roughly be divided into two classes. The first class consists of methods that are based on an operator equation formulation of the problem, where (4) is written

$$Kf = g,$$

where K is an integral operator. Then this equation is discretized and methods for solving discrete ill-posed linear problems are applied, such as truncated singular value decomposition (SVD) and Tikhonov regularization. In Section 2 we consider the SVD and analyse the sideways heat equation in terms of approximate singular functions. This analysis is of interest also for other solution procedures. Tikhonov regularization is one of the most important methods for solving ill-posed problems; we discuss the application of this method to the sideways heat equation in Section 3.

[1]Since we are only given g_m, satisfying (3), it seems inappropriate to require that the given data be matched more accurately than is warranted by the precision of the data.

The second class of numerical procedures is based the formulation of the problem as a partial differential equation. Here it is useful to investigate the problem using Fourier transform techniques. This is done in Section 4. Also, it is natural to try to approximate a solution of the stabilized problem (5), since in some cases it is possible to obtain error estimates for the numerical procedure, which, in the optimal cases, are of the type (6). In one important subclass of such methods, the parabolic equation is perturbed by adding an extra term, which makes the corresponding Cauchy problem well-posed. In Section 5 we discuss such a method, where the equation is made hyperbolic.

Space-marching difference schemes applied to sideways parabolic equations have been studied by several authors. In most cases such methods are combined with data filtering, to remove measurement errors (noise). It has been shown that such schemes themselves have a regularizing effect [4, 5]. In order to differentiate between effects of space and time discretizations, in Section 6 we consider time discretization alone, and show that time discretization prevents high frequencies in the solution from blowing up. A few numerical examples are given in Section 7, which illustrate the behaviour of difference schemes.

Beck's method (see e.g. [1], [23] and articles in this volume) can be considered as a hybrid method, in that it is based on an integral equation formulation, but the approximation is computed in a sequential procedure.

As our presentation will focus on the numerical solution of the sideways heat equation, we will not discuss the more theoretical aspects of this and related problems. In the recent papers by Dinh Nho Hao and Gorenflo [7] and Reinhardt and Seiffarth [31], comprehensive bibliographies are given. For a discussion of the theory of regularization methods and applications, see Engl [14], and for numerical algorithms for inverse problems in general, see the survey by Hanke and Hansen [20].

We restrict the presentation to the case when we want to determine only the temperature u for $0 \leq x < 1$. Most of what is written can be modified to apply also to the problem of computing the heat flux u_x.

2 Singular Value Analysis

It is well-known (see e.g. [2, Chapter 4], [3]), that the unique bounded solution of (1) is given by

$$u(x,t) = \int_0^t \hat{k}(x, t-\tau) f(\tau) d\tau,$$

where the kernel function is defined by

$$\hat{k}(x,t) = \frac{x}{2t^{3/2}\sqrt{\kappa\pi}} \exp\left[\frac{-x^2}{4\kappa t}\right]. \tag{7}$$

Therefore, given the data $g_m(\cdot) \approx u(1,\cdot)$ on the interval $[0,1]^2$, the inverse problem (4) of determining f can be written [17, p. 16]

$$g_m(t) = Kf(t) = \int_0^t k_\kappa(t-\tau) f(\tau) d\tau, \qquad 0 \leq t \leq 1, \tag{8}$$

where $k_\kappa(t) = \hat{k}(1,t)$. (A slightly different kernel function with the same basic properties as the one in (7), is obtained in the case of a finite space interval [8], [26].)

[2] For notational convenience we assume that any interval $[0,T]$, where g_m is measured, has been normalized to $[0,1]$.

We will now consider the Volterra integral equation of the first kind (8). One standard method to deal with such equations is to transform it to a second kind Volterra equation by differentiation. Using the same argument as in [8], we see that this is impossible here, since k_κ and all derivatives of k_κ are equal to zero at $t = 0$.

The kernel function k_κ is an impulse response for the map (2): if we take f equal to a Dirac delta function with its mass concentrated at $t = 0$, then by (2) it is mapped to k_κ. The parameter κ is a measure of the heat conduction properties of the medium, and, therefore, the larger the value of κ, the closer to a Dirac delta function will k_κ be. In Figure 1 we illustrate the kernel function for a few values of κ.

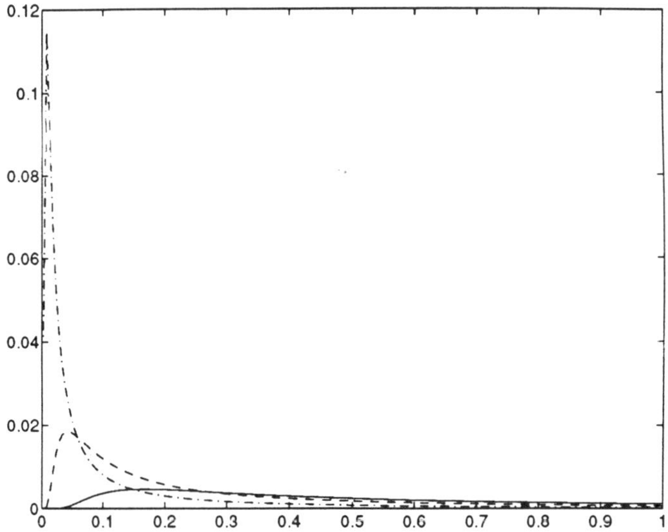

FIG. 1. *Kernel function $k_\kappa(t)$ for $0 \leq t \leq 1$ and with $\kappa = 1$ (solid), $\kappa = 2$ (dashed), and $\kappa = 5$ (dashed-dotted).*

Consider for a moment the map (2) as the propagation of a signal (at $x = 0$) through a medium to a recording device (at $x = 1$). From Figure 1 we see that the signal is diffused by the medium. Further, even if the speed of propagation of information in a parabolic equation is infinite (the characteristic lines are parallel to the x axis), there is a time delay before most of the information in the signal reaches the recording device. This observation is important in the construction of numerical algorithms for solving the sideways heat equation (4), and we will refer to this as a *causality principle.*

We will now discuss briefly and rather informally some properties of K in terms of the singular value expansion (SVE) of the operator. For a more extensive treatment of the SVE, see [14],[34]. We consider the operator K in the Hilbert space with inner product

$$(f, g) = \int_0^1 f(t)g(t)dt.$$

The Volterra equation (8) can be written as a Fredholm equation

$$g_m(t) = Kf(t) = \int_0^1 k_\kappa(t - \tau)f(\tau)d\tau, \qquad 0 \leq t \leq 1,$$

if we define $k_\kappa(t) \equiv 0$ for $t < 0$. A singular system $(\sigma_i; u_i, v_i)$ for the operator K is defined by the relations

$$\begin{aligned} K v_i &= \sigma u_i, \\ K^* u_i &= \sigma_i v_i, \qquad\qquad i = 1, 2, \ldots \\ \sigma_i &\geq \sigma_{i+1} \geq 0, \end{aligned}$$

where K^* is the adjoint of K and $(v_i)_1^\infty$ and $(u_i)_1^\infty$ are the orthonormal systems of eigenfunctions of K^*K and KK^*, respectively. The σ_i are called *singular values*, and u_i and v_i *singular functions*. The SVE of K can then be written

$$Kf = \sum_1^\infty \sigma_i (f, v_i) u_i. \tag{9}$$

The fact that the kernel function is smooth implies that

$$\lim_{n \to \infty} \sigma_n = 0,$$

[34, p. 148]. Furthermore, in [3] it shown that that for the operator under consideration, $\sigma_i > 0$ for all i. We now see that the operator equation (8) is ill-posed: we can write the solution formally

$$f = \sum_1^\infty \frac{(g_m, u_i)}{\sigma_i} v_i,$$

and for this series to converge, the *Picard criterion* [34, p. 164],

$$\sum_1^\infty \frac{(g_m, u_i)^2}{\sigma_i^2} < \infty,$$

must be satisfied. This is an extremely restrictive condition, because of measurement errors in g_m.

The rate of decay of the singular values (and thus the degree of ill-posedness) depends on the parameter κ: the smaller κ, the faster the singular values tend to zero. This is illustrated in Figure 2. Note that our plots are based on approximations from discretizations of the integral operator, since explicit expressions for the singular values and singular functions are not known to us.

In the lower plot of Figure 2 it is seen that for $\kappa = 2$ the last two singular values are exceptionally small. In Figure 3 we show that there is a qualitative difference between the corresponding singular vectors (that approximate the singular functions) and those corresponding to larger singular values.

We see that the first singular vectors v_1 and v_2 are smooth, while v_{126} oscillates rapidly. On the other hand, v_{128} have very small components up to the last one, which is approximately equal to 1 (v_{127} is similar). Thus v_{128} approximates a Dirac delta function with its mass at $t = 1$. These observations indicate that the ill-conditioning is due to two different factors:

- K is a smoothing operator,

- the solution $f(t)$ can not be reconstructed for t close to 1.

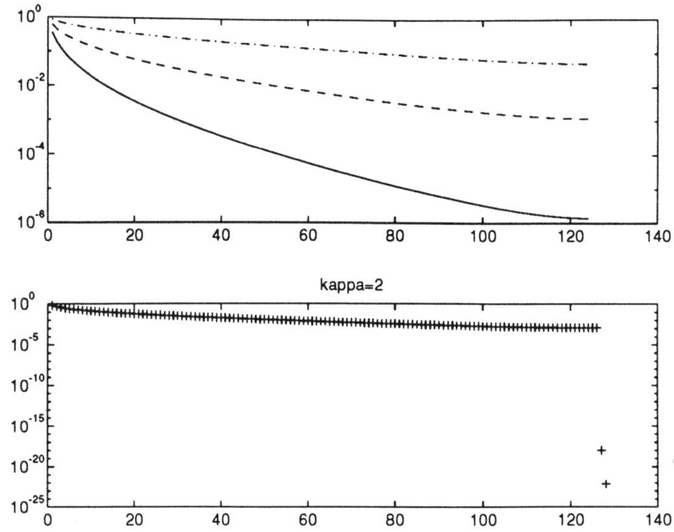

FIG. 2. *The upper plot shows first 124 singular values of a* 128×128 *discretization of the integral operator* K *for* $\kappa = 1$ *(solid),* $\kappa = 2$ *(dashed), and* $\kappa = 5$ *(dashed-dotted). In the lower plot we have plotted all the 128 singular values for the case* $\kappa = 2$.

The first cause of ill-conditioning is typical in ill-posed problems, and it can actually be seen in Figure 1. It is important to note that the second type of ill-conditioning occurs only if the dimension of the discretized problem is large enough. For example, with $\kappa = 5$ and a 128×128 discretization, the smallest singular value is approximately equal to 0.049. We now show that the second type of ill-conditioning is not just an anomaly created in the discretization of the integral operator. It is shown in [8] that the singular functions have the symmetry property

$$u_i(t) = \pm\, v_i(1-t), \quad 0 \le t \le 1,$$

which, using the definition of the SVE, implies that

$$\int_0^t k_\kappa(t-\tau) v_i(\tau) d\tau = \pm\, \sigma_i v_i(1-t).$$

Since $\sigma_i > 0$, we can conclude that

$$v_i(1) = 0, \quad \text{for all } i. \tag{10}$$

Now, if we take f to a Dirac delta function with its mass at $t = 1$, then from (9) and (10), we see that $Kf = 0$.

It is obvious that any numerical procedure for solving the sideways heat equation must take both sources of ill-conditioning into account.

In view of the above, it is natural to consider solving the sideways heat equation, using the the singular value decomposition (SVD) of a discretization of the operator K, and truncating it to alleviate the ill-conditioning associated with small singular values. If the solution is smooth, then a good approximation of it will be spanned by the singular vectors

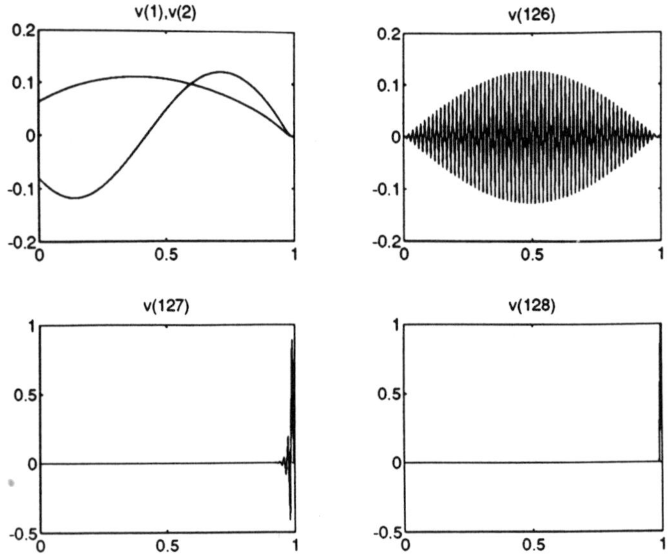

FIG. 3. *The singular vectors* v_1, v_2, v_{126}, v_{127}, *and* v_{128} *of the* 128×128 *discretization of the operator* K *for* $\kappa = 2$.

v_i corresponding to large singular values. This is a standard method for solving discrete, ill-posed problems, see e.g. [21],[37], and it is a powerful method in the sense that the SVD gives much information about the problem. On the other hand, it assumes that the kernel function in the integral equation is known, or can be computed. Another disadvantage of this method, as compared to other methods described below, is that it is considerably more expensive in terms of the number of floating point operations (flops): the computation of the SVD of an $N \times N$ matrix requires $O(N^3)$ flops, whereas the method described in Section 3 requires $O(N^2)$ flops.

3 Tikhonov Regularization

Tikhonov regularization is a method of wide applicability in ill-posed problems, and the literature is vast. We here mention only the book by Tikhonov and Arsenin [36] and the recent survey by Engl [14]. In this section we will consider a discretization of the integral equation (8), which we write

$$Kf = g_m, \tag{11}$$

where K is a square matrix, $K \in \mathbf{R}^{N \times N}$, and f and g_m are vectors. For discretizations of Volterra integral equations, see [25]. Due to the fact that the integral operator is a convolution operator of Volterra type, the matrix K becomes a lower triangular Toeplitz matrix (constant along the diagonals)

$$K = \begin{pmatrix} k_1 & & & & & \\ k_2 & k_1 & & & & \\ k_3 & k_2 & k_1 & & & \\ \vdots & & \ddots & \ddots & & \\ k_{N-1} & \cdots & \cdots & k_2 & k_1 & \\ k_N & k_{N-1} & \cdots & \cdots & k_2 & k_1 \end{pmatrix} \tag{12}$$

As seen in Section 2, the matrix K is extremely ill-conditioned if the dimension n is taken large enough, and we refer to (11) as a *discrete ill-posed problem.* If the linear system (11) is solved by forward substitution, the solution obtained will in general be meaningless, often wildly oscillating.

In Tikhonov regularization applied to a discrete ill-posed problem, the linear system (11) is replaced by a linear least squares problem,

$$\min_f \{\|Kf - g_m\|^2 + \mu^2\|Lf\|^2\}, \tag{13}$$

where the norm is the Euclidean vector norm $\|x\| = (x^T x)^{1/2}$, and L is a discretization of a differential operator, see below. The value of μ, the *regularization parameter*, determines the degree of smoothness of the solution. In general, it is a non-trivial problem to choose a suitable value of μ. If the measurement errors in g_m are small, then a small value of μ can be taken, but if the errors are large, then μ must be chosen large to suppress the influence of the errors. Here we do not go into details, but note that, in most cases, it is necessary to solve (13) for several values of the regularization parameter, and refer to [20] for a discussion of different methods for choosing μ based on the data.

The matrix L is often taken equal to the identity matrix or as a discretization of a differentiation operator. For example, the band matrix

$$L = \begin{pmatrix} 1 & -1 & & & \\ & 1 & -1 & & \\ & & \ddots & \ddots & \\ & & & 1 & -1 \end{pmatrix}, \qquad L \in \mathbf{R}^{(N-1)\times N}, \tag{14}$$

corresponds to the discretization of the first derivative of the solution. Note that L is (essentially) a lower triangular matrix with Toeplitz structure. This is the case also when L corresponds to a higher derivative.

For reasons of numerical accuracy it is advisable not to use the method of normal equations in the numerical solution of (13). Instead one should use a method based on an orthogonal decomposition of the matrix, see the discussion in [16, p. 230]. For ill-posed problems this is particularly important in the case when the perturbations of the data are small, and one hopes for an accurate solution, see an example in [9]. It is obvious that the following is an equivalent way of writing (13):

$$\min_f \left\| \begin{pmatrix} K \\ \mu L \end{pmatrix} f - \begin{pmatrix} g_m \\ 0 \end{pmatrix} \right\|. \tag{15}$$

If we can determine an orthogonal matrix Q, such that

$$Q^T \begin{pmatrix} K \\ \mu L \end{pmatrix} = \begin{pmatrix} K_\mu \\ 0 \end{pmatrix}, \quad Q^T \begin{pmatrix} g_m \\ 0 \end{pmatrix} = \begin{pmatrix} \bar{g}_1 \\ \bar{g}_2 \end{pmatrix}, \tag{16}$$

where K_μ is lower triangular, then, since

$$\left\| \begin{pmatrix} K \\ \mu L \end{pmatrix} f - \begin{pmatrix} g_m \\ 0 \end{pmatrix} \right\|^2 = \left\| \begin{pmatrix} K_\mu \\ 0 \end{pmatrix} f - \begin{pmatrix} \bar{g}_1 \\ \bar{g}_2 \end{pmatrix} \right\|^2 = \|K_\mu f - \bar{g}_1\|^2 + \|\bar{g}_2\|^2,$$

we get the solution of (15) by solving the lower triangular system

$$K_\mu f = \bar{g}_1.$$

It is not necessary to compute the orthogonal matrix Q explicitly, but only apply the transformation to the matrix and right hand side.

We now show that it is possible to compute the decomposition (16) efficiently when K and L are triangular Toeplitz matrices, using an algorithm from [10]. To describe the algorithm we use a small example with $N = 4$, and we assume that both K and L are square (this is no essential restriction). We assume that the reader is familiar with Givens rotations (see, e.g., [16, Chap. 5]). Consider the augmented matrix

$$\begin{pmatrix} K & g_m \\ \mu L & 0 \end{pmatrix} = \begin{pmatrix} k_1 & & & & g_1 \\ k_2 & k_1 & & & g_2 \\ k_3 & k_2 & k_1 & & g_3 \\ k_4 & k_3 & k_2 & k_1 & g_4 \\ l_1 & & & & 0 \\ l_2 & l_1 & & & 0 \\ l_3 & l_2 & l_1 & & 0 \\ l_4 & l_3 & l_2 & l_1 & 0 \end{pmatrix},$$

where l_i is equal to μ times a component of L. In the first step of the algorithm we zero all the diagonal elements of L by N rotations in the planes $(1, N+1), (2, N+2), \cdots, (N, 2N)$. Because of the Toeplitz structure all transformations involve the same angle of rotation. The result is

$$\begin{pmatrix} k'_1 & & & & g'_1 \\ k'_2 & k'_1 & & & g'_2 \\ k'_3 & k'_2 & k'_1 & & g'_3 \\ k'_4 & k'_3 & k'_2 & k'_1 & g'_4 \\ 0 & & & & g'_5 \\ l'_2 & 0 & & & g'_6 \\ l'_3 & l'_2 & 0 & & g'_7 \\ l'_4 & l'_3 & l'_2 & 0 & g'_8 \end{pmatrix}.$$

Note that the Toeplitz structure is not destroyed. Obviously only about $4N$ elements need be computed.

In the next step, the subdiagonal of the lower matrix is zeroed by $N-1$ rotations in the planes $(1, N+2), (2, N+3), \cdots, (N-1, 2N)$. Again the same angle of rotation can be used in all transformations, and in this step $4(N-1)$ elements are computed.

It is now obvious how the algorithm proceeds. The total operation count for the computation of K_μ and $\bar{g}_1$ is $4N^2$ multiplications, and thus Tikhonov regularization using this algorithm is one order of magnitude faster than the truncated SVD algorithm. However, it has the same disadvantage as the latter method in that it assumes that we know the kernel function explicitly or can compute it.

Tikhonov regularization can also be implemented using the Fast Fourier Transform [3]. We consider this in the next section.

4 Fourier Transform Techniques

As in [3] we will now analyse the Cauchy problem by taking the Fourier transform with respect to the time variable. For this presentation we will consider functions defined for all real real t, but equal to zero for negative t. Also, to avoid too much notation we will take $\kappa = 1$. The Fourier transform is defined

$$Z(s) = [\mathbf{FT}(z)](s) := \frac{1}{\sqrt{2\pi}} \int_{-\infty}^{\infty} z(t) e^{-ist} dt, \qquad -\infty < s < \infty,$$

for $z(\cdot) \in \mathcal{X} = L^2(\mathbf{R})$. In particular we will use the identity

$$[\mathbf{FT}(z_t)](s) = isZ(s). \tag{17}$$

In working with the Fourier transform we will consistently use s for the transform variable, and for transforms of functions we will use the corresponding capital letters so $F = \mathbf{FT}(f)$, $U(x,\cdot) = \mathbf{FT}(u(x,\cdot))$, etc.

Using (17), the quarter plane problem (1) corresponds to a boundary value problem for an ordinary differential equation in the variable $x \in \mathbf{R}_+$, parameterized by $s \in \mathbf{R}$:

$$U_{xx} = isU, \qquad U|_{x=0} = F, \qquad U|_{x\to\infty} \text{ bounded}. \tag{18}$$

The unique solution of (18) is easily seen to be

$$U(x,s) = e^{-\theta(s)x}F(s) = e^{\theta(s)(1-x)}G(s), \tag{19}$$

where $\theta(s)$ is the principal value of $\sqrt{is}$:

$$\theta(s) = \begin{cases} (1+i)\sigma, & s \geq 0, \\ (1-i)\sigma, & s < 0, \end{cases} \qquad \text{with } \sigma = \sigma(s) := \sqrt{|s|/2}. \tag{20}$$

Using the notation $\bar{u}$ (with Fourier transform $\bar{U}$) to distinguish between the exact solution u of (1) with unperturbed data and the solution $\bar{u}$ of (4) with data g_m, the transformed Cauchy problem corresponding to the sideways heat equation (4) becomes

$$\bar{U}_{xx} = is\bar{U}, \qquad \bar{U}|_{x=1} = G_m, \qquad \bar{U}|_{x\to\infty} \text{ bounded}. \tag{21}$$

We can write the formal solution of (21) explicitly as

$$\bar{U} = \bar{U}(x,s) = e^{\theta(s)(1-x)}G_m(s), \tag{22}$$

for all x, s, so in the time domain this solution is given by

$$\bar{u}(x,t) = \frac{1}{2\pi}\int_{-\infty}^{\infty} e^{ist}\left[e^{\theta(s)(1-x)}\int_{-\infty}^{\infty} g_m(r)e^{-isr}dr\right]ds. \tag{23}$$

The real part of $\theta(s)$ is positive so the solution operator $g_m \mapsto \bar{u}(x,\cdot)$ 'defined' by (23) is unbounded for $0 \leq x < 1$. High frequency perturbations of the data are blown up and the solution will exhibit wild oscillations. Thus, again we see that the problem (4) is *ill-posed* when we want to solve it for $0 \leq x < 1$.

Tikhonov regularization in a Fourier setting has been considered by several authors, e.g. Manselli and Miller [26], Murio [27], Carasso [3]. Here we solve the minimization problem

$$\min_w[\|Kw - g_m\|^2 + \mu^2\|w\|^2],$$

and get the solution [3],

$$w(t) = \frac{1}{2\pi}\int_{-\infty}^{\infty} \frac{e^{ist}e^{\theta(s)}G_m(s)}{1+\mu^2 e^{\sqrt{2|s|}}}ds.$$

The exponential term in the denominator damps high frequencies, and if μ is taken small, then low frequencies are only slightly perturbed compared to the "true" solution operator (23). In fact, we can introduce x-dependence here and define

$$w(x,t) = \frac{1}{2\pi}\int_{-\infty}^{\infty} \frac{e^{ist}e^{(1-x)\theta(s)}G_m(s)}{1+\mu^2 e^{\sqrt{2|s|}}}ds, \tag{24}$$

as an approximation of a solution of the stabilized problem (5). Carasso [3] shows that with $\mu = \epsilon/M$, the following error estimate holds,

$$\|w(x,\cdot) - u(x,\cdot)\| \leq 2M^{1-x}\epsilon^x, \tag{25}$$

where $u(x,t)$ denotes the solution of (4) with *exact data*. In view of (6) we see that (24) defines an optimal approximation.

Another optimal method is given in [33], where the approximate solution of (5) is given by

$$v(x,t) = \frac{1}{2\pi}\int_{-\infty}^{\infty} e^{ist}\exp[\int_x^1 \rho(\xi,s)d\xi]G_m(s)ds,$$

where $\rho(x,s)$ is defined by

$$\begin{aligned} \rho(x,s) &= \min\{\beta(x),\sigma\} + i(\operatorname{sgn} s)\sigma, \quad \sigma = \sqrt{|s|/2}, \\ \beta(x) &= \log(M/\epsilon) - 1/x. \end{aligned}$$

The two optimal methods discussed in this section both rely on the fact that an explicit solution of the ordinary differential equation (18) can be obtained. For more general equations it is not obvious how to use these methods. However, Fourier techniques, and in particular the Fast Fourier Transform (FFT), in combination with Tikhonov regularization, can be used very efficiently to filter out high frequency components in the *data* g_m[3]. Then the filtered data are used in a space-marching procedure based on finite differences for solving the sideways problem.

5 Hyperbolic Approximation

It was suggested by Weber [38] to approximate the parabolic equation (4), for which the Cauchy problem is ill-posed, by a *hyperbolic* equation, for which the Cauchy problem is *well-posed.* This method was further analysed in [12], and, independently, in [30].

A general hyperbolic approximation of the heat equation $u_{xx} = u_t$ (also in this section we assume that $\kappa = 1$) is given by

$$u_{xx} = u_t - 2\delta u_{xt} + \beta u_{tt}. \tag{26}$$

It is shown in [12] that considering the mapping (2) as a signal propagation problem (cf. [32] and Section 2), and invoking a causality principle (see Section 2), one is lead to choosing the parameters δ and β so that the two characteristic lines of the hyperbolic equation (26) have positive and negative slope respectively. Furthermore, the error analysis in [12] shows that nothing is gained by taking δ different from zero. Thus we are led to the following hyperbolic equation, which is to be considered as a Cauchy problem in the space variable,

$$\begin{cases} v_{xx} = v_t + \beta v_{tt}, & \beta \geq 0, \\ v(1,t) = g(t), & t \geq 0 \\ v_x(1,t) = h(t), & t \geq 0 \\ v(x,0) = 0, & 0 \leq x \leq 1. \end{cases} \tag{27}$$

The extra condition for v_x at $x = 1$ can be obtained by solving the parabolic equation (1) for $x > 1$.

The characteristic lines of the hyperbolic equation are illustrated in Figure 4. The slope of the characteristic lines is determined by the value of β: the smaller β, the more

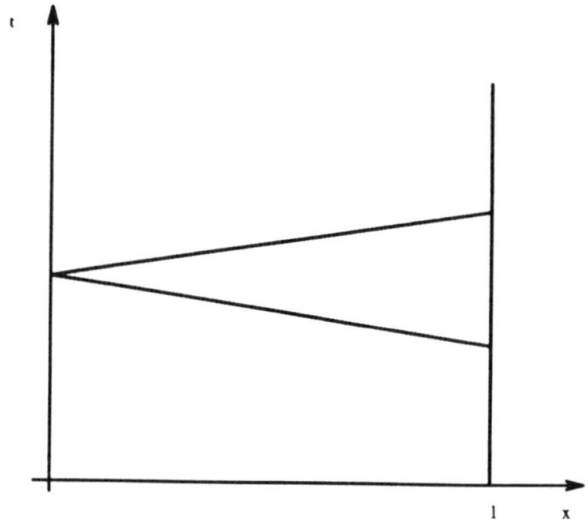

FIG. 4. *Characteristic lines of the hyperbolic equation.*

the characteristics approach a line parallel to the x axis. The interval on the line $x = 1$ intercepted by the characteristics is called the *domain of dependence* of the starting point on the t axis.

Taking the Fourier transform of the hyperbolic equation (27) we get

$$V_{xx} = (is - \beta s^2)V. \tag{28}$$

The characteristic equation for (28) has the roots

$$\rho_{1,2} = \pm\sqrt{|s|}(1 + \beta^2 s^2)^{1/4} e^{i\,\mathrm{sgn}(s)(\alpha/2+\pi/4)}, \quad \alpha = \tan^{-1}|\beta s|.$$

The real part of $\exp[i\,\mathrm{sgn}(s)(\alpha/2 + \pi/4)]$ is positive, and therefore, due to the boundedness requirement in (21), it is the positive root that approximates the function $\theta(s)$ defined by (20). Thus, we consider the solution of (28),

$$V = e^{\rho(s)(1-x)}G, \tag{29}$$

where

$$\rho(s) = \sqrt{|s|}(1 + \beta^2 s^2)^{1/4} e^{i\,\mathrm{sgn}(s)(\alpha/2+\pi/4)}. \tag{30}$$

We see that for small values of $|\beta s|$, the angle α is small, and ρ is close to θ, which implies that for small $|s|$ the solution (29) approximates (19). It is also easy to show that

$$\mathrm{Re}(\rho) \to \frac{1}{2\sqrt{\beta}}, \quad \text{as } |s| \to \infty,$$

which means that the solution operator defined by (29) is bounded.

The approximation properties of the hyperbolic approximation can be stated as follows. First, choose[3]

$$\beta = \frac{1}{4(\log(M/\epsilon))^2}.$$

[3] It is noted in [12] that this choice is probably not optimal.

Let x be a fixed point in the interval $(0,1)$, and let ϵ be a measure of the accuracy of the data (3). Then for small enough ϵ,

$$\|u(x,\cdot) - v(x,\cdot)\| \sim \frac{1-x}{x^3} \frac{M}{(\log(M/\epsilon))^2}, \tag{31}$$

where v is obtained from the hyperbolic approximation and u is the solution of (4) with exact data. Obviously, this is a much worse error estimate than the optimal one, in the sense that as ϵ tends to zero, the bound (31) tends to zero much more slowly than (6).

An error estimate for the hyperbolic approximation is also derived in [30]; however, the estimate is not explicit, and it is not clear how to choose the parameter β. One important observation made in [30] is that the stability of the hyperbolic approximation method can be improved by filtering the noisy data. The numerical solution procedure suggested in [30] is based on writing the hyperbolic equation as a Volterra integral equation of the second kind, and solving a discretization of this equation (note that this is similar to using Tikhonov regularization for the integral equation (8)).

It is interesting to note that under some circumstances hyperbolic equations are used to model heat conduction also for well-posed problems, see [35]. The introduction into the parabolic equation of the term v_{tt} can be thought of as a singular perturbation. Different singular perturbations are analysed in [11] and [29].

6 Time Discretization

We have indicated earlier that filtering the data g_m enhances the stability of the problem[4]. Let w denote the solution of the sideways heat equation with filtered data,

$$\begin{cases} w_{xx} = w_t, & x \geq 0, \quad t \geq 0, \\ w(x,0) = 0, & x \geq 0, \\ w(1,t) = g_{filt}(t), & t \geq 0, \end{cases} \tag{32}$$

where the filtering of g_m to get g_{filt} is assumed to have been done by either the Tikhonov based approach of Carasso, Section 4, or by mollification, see [26], [27], [30], [18]. In both cases we have continuous dependence on the data, for the Carasso method (25), and for mollification, see [30]. Now it may appear as if we can safely discretize and solve (32) using standard methods for well-posed partial differential equations in a space-marching scheme (solving the equation sideways).

It turns out that the situation is not quite that simple because *(32) is still an ill-posed problem*: filtering has reduced the high frequency components in the data function, and thus reduced the noise level, but rounding errors introduced in the computations will be magnified as in any ill-posed problem.

Different finite difference discretizations of (32) are analysed and compared by Carasso in [4, 5]. The stability aspects of finite difference schemes in connection with mollification are considered in a series of papers by Murio, Guo and Roth[28, 19, 29]. In these papers successful computations with space-marching finite difference schemes for are reported, and at first sight, these reports seem to contradict our statement above that (32) is an ill-posed problem. However, the finite difference discretization itself has a regularizing effect.

In order to study the properties of discretizations, it is useful to study discretization in time and space separately. Here we only consider discretization in time, and we show

[4] As is pointed out in [5], filtering induces a loss of high frequency information in the data function, and thus makes it impossible to reconstruct sharp features in the solution.

that *time discretization prevents high frequency components in the solution from blowing up*. We will also state an approximation theorem that will give insight into the problem of choosing the step length in the discretization.

As an approximation of the heat equation we will consider the following *differential-difference equation*,

$$w_{xx}(x,t) = \frac{1}{k}(w(x,t+k) - w(x,t)), \tag{33}$$

where we have replaced the time derivative by a *forward difference* with step length k. Similarly, we can approximate the time derivative by a *central difference*,

$$w_{xx}(x,t) = \frac{1}{2k}(w(x,t+k) - w(x,t-k)), \tag{34}$$

or a *backward difference*,

$$w_{xx}(x,t) = \frac{1}{k}(w(x,t) - w(x,t-k)). \tag{35}$$

We immediately see that the backward difference approximation (35) does not take into account the causality principle (see Section 2) in a proper way, since the value of the solution at a point (x_0, t_0) is based on what happens *before* $t = t_0$.

Since at present we are primarily interested in the stability properties of the approximations (33)–(35), and since we will use Fourier transforms, we here assume that data are given in the infinite interval $t \geq 0$,

$$\begin{cases} w(x,0) = 0, & x \geq 0, \\ w(1,t) = g_m(t), & t \geq 0. \end{cases} \tag{36}$$

Furthermore, we assume that the solution w is bounded as $x \to \infty$.

Proceeding formally, taking the Fourier transform of (33), (36), we get

$$\begin{cases} W_{xx}(x,s) = \frac{1}{k}(e^{isk} - 1)W(x,s), & 0 \leq x \leq 1, \\ W(1,s) = G_m(s). \end{cases} \tag{37}$$

It is seen that for small $|sk|$ we have $\frac{1}{k}(e^{isk} - 1) \approx is$, so that (37) is an approximation of (21). The solution of (37) is given by

$$W(x,s) = e^{\rho_f(s)(1-x)} G_m(s), \qquad \rho_f(s) = \left(\frac{1}{k}(e^{isk} - 1)\right)^{1/2}, \tag{38}$$

where $\rho_f(s)$ is the principal square root. Since $\rho_f(s)$ is a bounded periodic function of s, we see that (38) defines a bounded solution operator.

Similarly, we can take the Fourier transform of (34) and (35), and solve the corresponding equations. For instance, the equation obtained for the central difference is

$$\begin{cases} W_{xx}(x,s) = \frac{i}{k}\sin(ks)W(x,s), & 0 \leq x \leq 1, \\ W(1,s) = G_m(s), \end{cases} \tag{39}$$

with solution

$$W(x,s) = e^{\rho_c(s)(1-x)} G_m(s), \qquad \rho_c(s) = \left(\frac{i}{k}\sin(ks)\right)^{1/2}. \tag{40}$$

To illustrate the approximation properties of time discretizations, we cite the following error estimate for the central difference approximation[13].

THEOREM 6.1. *Let w be the solution of (34), with data (36), i.e., the function whose Fourier transform is given by (40), with*

$$k = \frac{1}{2(\log(M/\epsilon))^2},$$

and let u be any solution of (5). If x is fixed in the interval $(0,1)$, *then, asymptotically as* $\epsilon \to 0$,

$$\|u(x,\cdot) - w(x,\cdot)\| \sim \frac{M}{(\log(M/\epsilon))^4} \tag{41}$$

This shows that for fixed $x \in (0,1]$, the solution of (34) tends to a solution of (5) as $\epsilon \to 0$. However, the rate of convergence is not optimal (in the sense that the error estimate (41) is worse than the estimate (6) for (5)).

In order to further illustrate the approximation properties of the finite difference approximations, we study the *amplification factors* of the solution operators (such factors are defined in [4, 5] for the case of a discrete equation in both time and space). For the analytic solution of the sideways heat equation, the solution operator is defined by (22), and its amplification factor is

$$A(s) = |e^{\theta(s)}| = e^{\mathrm{Re}\theta(s)}. \tag{42}$$

The amplification factors for the solutions of (33),(34),(35) are defined analogously. The amplification factor shows how the different frequencies are amplified by the solution operator. Obviously, the amplification factor (42) for the analytic solution is an unbounded function of the frequency s. The amplification factors for the approximate solutions should approximate (42) for low frequencies, and stay bounded for high frequencies (in order not to amplify high frequency noise).

Figure 5 shows that the backward difference is a bad approximation of the analytic solution operator, as many frequencies are amplified too much. Further, the figure shows how the approximation properties of the approximation are determined by the choice of step length: the larger we choose k, the sooner the amplification curve for the approximation deviates from that of the analytic solution operator (cf. Figure 2 in [5]). These results confirm those in [4], [5], where the importance of the time discretization is stressed.

At this point it is useful to give the corresponding graph of the amplification factor of the hyperbolic approximation. Thus, we consider the hyperbolic equation

$$v_{xx} = v_t + \beta v_{tt},$$

discretized in time by central differences

$$\begin{aligned} w_{xx}(x,t) &= \frac{1}{2k}(w(x,t+k) - w(x,t-k)) \\ &+ \frac{\beta}{k^2}(w(x,t+k) - 2w(x,t) + w(x,t-k)), \end{aligned}$$

which, taking Fourier transforms, becomes

$$W_{xx}(x,s) = \left(\frac{i}{k}\sin(ks) + \frac{\beta}{k^2}(e^{isk} - 2 + e^{-isk})\right) W(x,s). \tag{43}$$

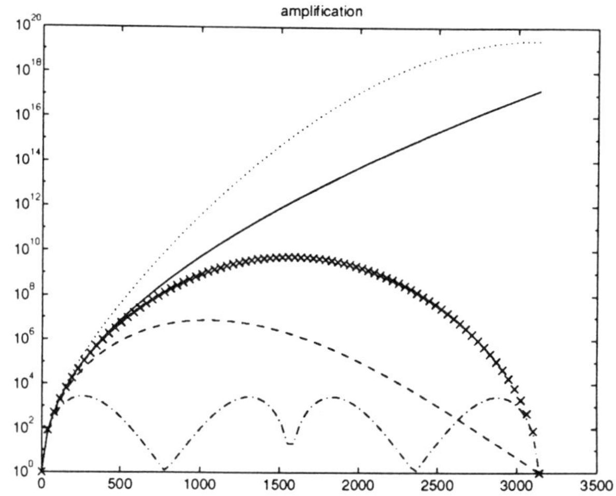

FIG. 5. *The amplification factor is plotted as a function of frequency for the analytic solution (solid line), forward time difference (dashed), central difference ('×'), and backward difference (dotted). The step length k is chosen equal to 10^{-3}. To show the dependence on the step length we have also plotted the amplification factor for the forward approximation with step length $4k$ (dashed-dotted). Note that the amplification factors of the approximations are periodic.*

In Figure 6 we have plotted the amplification factor for (43) for a few values of β. The dependence of the amplification of the value of β is clearly seen. Note that for $\beta = 0$ the discretized hyperbolic approximation coincides with (34).

It is obvious that the time derivative can be discretized with a certain step length k (as required from Theorem 6.1), based on a priori knowledge about the noise level of the data, even if the approximate solution is considered on a finer grid with step size $k_0 < k$. This is done simply by choosing k equal to a integer multiple of k_0, see Section 7.

We conclude this section by summarizing the investigations:

- Time discretization has a 'regularizing effect': high frequency noise is prevented from blowing up.
- The time difference should have a substantial forward component (cf. [4], [5]).
- Theorem 6.1 gives an indication how to choose the step length k.

We believe that the insight into the effects of time discretization will be useful in the study of the required properties of *space discretizations*. We intend to pursue this elsewhere.

7 Numerical Experiments

In the numerical experiments we illustrate some properties of discretizations of the sideways heat equation on the unit square, using a rectangular grid with step sizes $h = 1/M$ and $k = 1/N$ in space and time, respectively. Thus we have grid points

$$(x_m, t_n) = (mh, nk), \qquad m = 0, 1, \ldots, M, \quad n = 0, 1, \ldots, N.$$

In the tests presented below, we use $M = 100$, and $N = 256$.

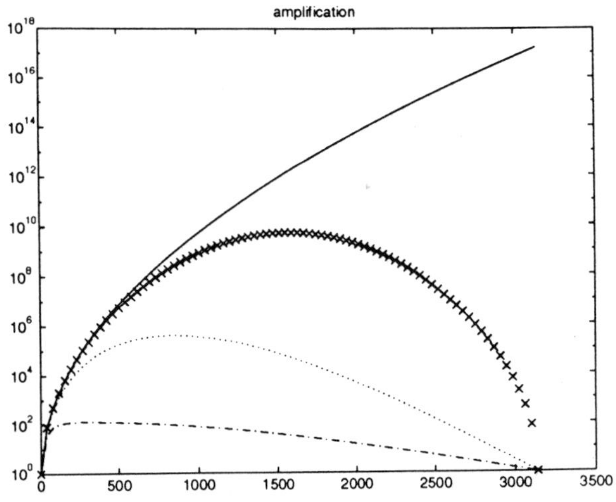

FIG. 6. *The amplification factor is plotted as a function of frequency for the analytic solution (solid line), and the hyperbolic approximation discretized by central differences. The 'regularization parameter' β is chosen equal to 0 ('×'), 10^{-3} (dotted), and 10^{-2} (dashed-dotted). The step length k is equal to 10^{-3}.*

The heat equation is written as a system of two first order equations

$$\begin{cases} w_x = q \\ q_x = w_t, \end{cases} \tag{44}$$

which we discretize in the point (x_m, t_n) by forward differences in both time and space,

$$\begin{cases} w_{m-1}^n = w_m^n - hq_m^n, \\ q_{m-1}^n = q_m^n - \frac{h}{\nu k}(w_m^{n+\nu} - w_m^n), \end{cases} \tag{45}$$

where ν is an integer (note that we are solving the problem from right to left). For ν points close to $t = 1$ it is not possible to use the forward difference for the time derivative. For those points we use the same value as in the last point that can be computed from the difference formula.

In a real application, we can obtain data for (44) by solving the heat equation in the quarter plane $x \geq 1$, $t \geq 0$. For our test examples we generate data using a discretization of the integral equation (11). A solution vector f is selected and a data vector g is created by multiplying f by the matrix K. Then a 'noisy' data vector g_m is created by adding a uniformly distributed perturbation from the interval $(0, \alpha)$, for different values of α. To obtain data for the component q in (44), we differentiate the kernel function, discretize it, and multiply the solution vector f by the corresponding matrix. Random errors from the same distribution are added.

Given data in the grid points at $x = 1$, we can compute approximations of w and q at $x = 1 - h, 1 - 2h$, etc from (45). Thus, (45) can be used as a space-marching procedure.

The first two examples show the regularizing effect of the time discretization for two different values of ν. We added perturbations from $(0, 10^{-3})$ to the data, and solved the problem using $\nu = 4$ and 2, without prefiltering the data. The results in Figure 7 and

8 show that the longer step length in the time difference damps the high frequency noise more efficiently.

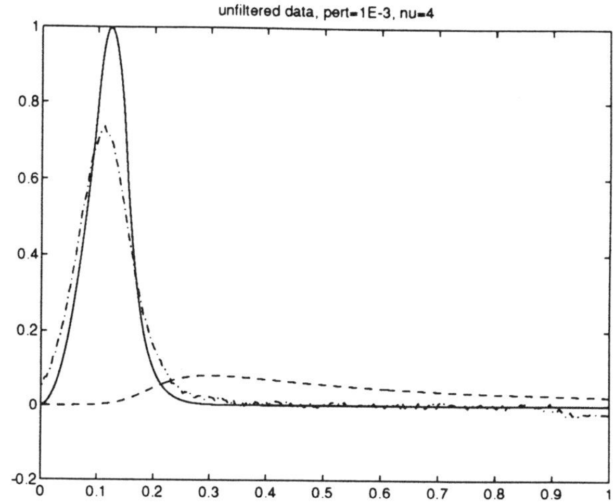

FIG. 7. *The graph shows the results of solving the sideways heat equation using difference approximations. The dashed line is the data vector g_m, and the solid line the correct solution. A perturbation in $(0, 10^{-3})$ was added to the data. The dashed-dotted line shows the result obtained from the difference method using $\nu = 4$.*

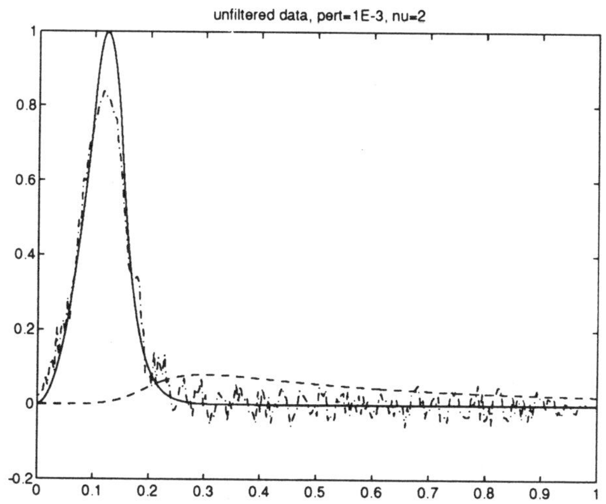

FIG. 8. *Results from the difference method using $\nu = 2$.*

We were not able to obtain acceptable results for higher noise levels without prefiltering the data. In Figure 9 we show the results from an experiment where a random perturbation from $(0, 10^{-2})$ was added to the data, and prefiltering was done using the Tikhonov regularization scheme of Carasso[3] (implemented using FFT). Then the problem was solved using (45) with $\nu = 1$. It is seen that the difference method gave a good approximation of

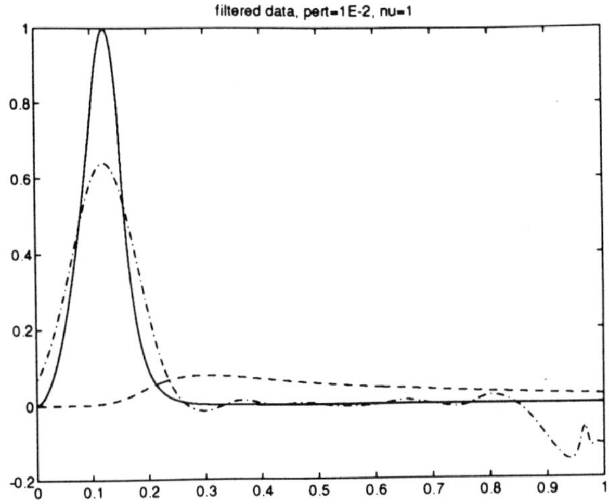

FIG. 9. *Results from the difference method applied to a problem, where the data were contaminated by noise from* $(0, 10^{-2})$ *and prefiltering was done. The step length in the time difference was* $\nu = 1$.

the peak position, but the amplitude of the peak is incorrect. This is due to prefiltering and time differencing, which both induce a loss of high frequency information. Observe that the accuracy of the solution is bad close to $t = 1$. This is because the problem of recovering the solution here is more ill-conditioned than in the rest of the interval (cf. Section 2).

8 Conclusion

In this paper we have discussed some methods for solving numerically the sideways heat equation. To a large extent, we have been able to use analytic techniques to investigate the properties of this model problem.

Procedures based on an integral equation formulation of the problem, such as the SVD and Tikhonov regularization methods, are well understood and reliable. However, their applicability is limited in the sense that they cannot be used for the more general problems occurring in applications, e.g. equations with nonlinearities, since in those cases the integral operator is not known explicitly.

Methods based on finite difference approximations of the partial differential equation can easily be used for nonlinear problems. However, the properties of these methods are not well understood.

In this paper we have studied some properties of time differencing in the sideways heat equation, and we have shown that time differencing has a regularizing effect. We hope that these investigations will be useful also for the study of more general problems.

References

[1] J. V. Beck and B. Blackwell, *Inverse problems*, in Handbook of Numerical Heat Transfer, W. J. Minkowycz, E. M. Sparrow, G. E. Scheider, and R. H. Pletcher, eds., J. Wiley and Sons, New York, 1988, Ch. 19, pp. 787–834.

[2] J. R. Cannon, *The One-Dimensional Heat Equation*, Addison-Wesley, Reading, MA, 1984.

[3] A. S. Carasso, *Determining surface temperatures from interior observations*, SIAM J. Appl. Math., 42 (1982), pp. 558–574.

[4] ——, *Space marching difference schemes in the nonlinear inverse heat conduction problem*, Inverse Problems, 8 (1992), pp. 25–43.

[5] ——, *Slowly divergent space marching schemes in the inverse heat conduction problem*, Numerical Heat Transfer, part B, 23 (1993), pp. 111–126.

[6] ——, *Overcoming Hölder continuity in ill-posed continuation problems*, SIAM J. Numer. Anal., 31 (1994).

[7] Dinh Nho Hao and R. Gorenflo, *A noncharacteristic Cauchy problem for the heat equation*, Acta Appl. Math., 24 (1991), pp. 1–27.

[8] L. Eldén, *The numerical solution of a non-chracteristic Cauchy problem for a parabolic equation*, in Numerical Treatment of Inverse Problems in Differential and Integral Equations, Proceedings of an International Workshop, Heidelberg, 1982, P. Deuflhard and E. Hairer, eds., Birkhäuser, Boston, 1983, pp. 246–268.

[9] ——, *An algorithm for the regularization of ill-conditioned, banded least squares problems*, SIAM J. Sci. Stat. Comput., 5 (1984), pp. 237–254.

[10] ——, *An efficient algorithm for the regularization of ill-conditioned least squares problems with triangular Toeplitz matrix*, SIAM J. Sci. Stat. Comput., 5 (1984), pp. 229–236.

[11] ——, *Approximations for a Cauchy problem for the heat equation*, Inverse Problems, 3 (1987), pp. 263–273.

[12] ——, *Hyperbolic approximations for a Cauchy problem for the heat equation*, Inverse Problems, 4 (1988), pp. 59–70.

[13] ——, *Numerical solution of the sideways heat equation by difference approximation in time*, tech. rep., Department of Mathematics, Linköping University, Sweden, 1994.

[14] H. W. Engl, *Regularization methods for the stable solution of inverse problems*, Surv. Math. Ind., 3 (1993), pp. 71–143.

[15] H. W. Engl and P. Manselli, *Stability estimates and regularization for an inverse heat conduction problem in semi–infinite and finite time intervals*, Numer. Funct. Anal. Optimiz., 10 (1989), pp. 517–540.

[16] G. H. Golub and C. F. Van Loan, *Matrix Computations. 2nd ed.*, Johns Hopkins Press, Baltimore, MD., 1989.

[17] C. W. Groetsch, *Inverse Problems in the Mathematical Sciences*, Vieweg, Braunschweig, 1993.

[18] L. Guo and D. Murio, *A mollified space-marching finite-difference algorithm for the two-dimensional inverse heat conduction problem with slab symmetry*, Inverse Problems, 7 (1991), pp. 247–259.

[19] L. Guo, D. A. Murio, and C. Roth, *A mollified space marching finite differences algorithm for the inverse heat conduction problem with slab symmetry*, Computers Math. Applic., 19 (1990), pp. 75–89.

[20] M. Hanke and P. C. Hansen, *Regularization methods for large-scale problems*, Surv. Math. Ind., 3 (1993), pp. 253–315.

[21] R. J. Hanson, *A numerical method for solving Fredholm integral equations of the first kind using singular values*, SIAM J. Numer. Anal., 8 (1971), pp. 616–622.

[22] P. Knabner and S. Vessella, *Stability estimates for ill-posed Cauchy problems for parabolic equations*, in Inverse and Ill-Posed Problems, H. W. Engl and C. W. Groetsch, eds., St. Wolfgang, Austria, 1987, Academic Press, pp. 351–368.

[23] P. Lamm, *Future-sequential regularization methods for ill-posed Volterra equations. Applications to the inverse heat conduction problem*, J. Math. Anal. Appl., to appear, (1994).

[24] H. Levine, *Continuous data dependence, regularization, and a three lines theorem for the heat equation with data in a space like direction*, Ann. Mat. Pura Appl. (IV), CXXXIV (1983), pp. 267–286.

[25] P. Linz, *A survey of methods for the solution of Volterra integral equations of the first kind*, in The Application and Numerical Solution of Integral Equations, R. S. Anderssen, F. R. de Hoog, and M. A. Lukas, eds., Sijthoff & Noordhoff, Alphen aan den Rijn, 1980, pp. 183–194.

[26] P. Manselli and K. Miller, *Calculation of the surface temperature and heat flux on one side of a

wall from measurements on the opposite side, Ann. Mat. Pura Appl., 123 (1980), pp. 161–183.

[27] D. A. Murio, *The mollification method and the numerical solution of an inverse heat conduction problem*, SIAM J. Sci. Stat. Comput., 2 (1981), pp. 17–34.

[28] ——, *The mollification method and the numerical solution of the inverse heat conduction problem by finite differences*, Computers Math. Applic., 10 (1989), pp. 1385–1396.

[29] D. A. Murio and L. Guo, *A stable space marching finite differences algorithm for the inverse heat conduction problem with no initial filtering procedure*, Computers Math. Applic., 19 (1990), pp. 35–50.

[30] D. A. Murio and C. C. Roth, *An integral solution for the inverse heat conduction problem after the method of Weber*, Comput. Math. Appl., 15 (1988), pp. 39–51.

[31] H.-J. Reinhardt and F. Seiffarth, *On the approximate solution of illposed Cauchy problems for parabolic differential equations*, in Inverse Problems: Principles and Applications in Geophysics, Technology, and Medicine, G. Anger, R. Gorenflo, H. Jochmann, H. Moritz, and W. Webers, eds., Akademie Verlag, August 1993, pp. 284–298.

[32] T. Seidman, *Recovery of a diffused signal*, in Seminaires IRIA, Analyse et Controle de Systemes, INRIA, Rocquencourt, 1980, pp. 71–82.

[33] T. Seidman and L.Eldén, *An optimal filtering method for the sideways heat equation*, Inverse Probl., 6 (1990), pp. 681–696.

[34] F. Smithies, *Integral Equations*, Cambridge University Press, Cambridge, 1965.

[35] J. C. Tannehill, *Hyperbolic and hyperbolic-parabolic systems*, in Handbook of Numerical Heat Transfer, W. J. Minkowycz, E. M. Sparrow, G. E. Scheider, and R. H. Pletcher, eds., J. Wiley and Sons, New York, 1988, ch. 12, pp. 463–517.

[36] A. N. Tikhonov and V. Y. Arsenin, *Solutions of Ill-posed Problems*, Winston, Wiley, New York, 1977.

[37] J. M. Varah, *On the numerical solution of ill-conditioned linear systems with applications to ill-posed problems*, SIAM J. Numer. Anal., 10 (1973), pp. 257–267.

[38] C. F. Weber, *Analysis and solution of the ill-posed inverse heat conduction problem*, Int. J. Heat Mass Transfer, 24 (1981), pp. 1783–1792.

NUMERICAL EXPERIMENTS IN MULTIDIMENSIONAL IHCP ON BOUNDED DOMAINS

DIEGO A. MURIO, YINGHUI LIU AND HONGCAI ZHENG
DEPARTMENT OF MATHEMATICAL SCIENCES
UNIVERSITY OF CINCINNATI
CINCINNATI, OH 45221-0025

Abstract. Space marching implementations of the Mollification Method are described to numerically recover the temperature and heat flux histories at interior and/or boundary points of finite multidimensional bodies when the temperature and heat flux functions are approximately measured at one portion of their boundary.

Keywords: *Ill-posed problems, multidimensional inverse heat conduction, discrete mollification, singular perturbation, finite differences.*

1. Introduction. Theoretical concepts and computational implementations related to the inverse heat conduction problem (IHCP) have been mostly restricted to one-dimensional models. The difficulties of the multidimensional IHCPs are more pronounced and very few results are available in these cases. See D. A. Murio [8] and the references therein for a description of one-dimensional algorithms and a historical perspective.

The first analytical solution – requiring exact data – that is applicable to two-dimensional conduction systems for geometries of arbitrary shape was introduced by M. Imber [5]. Most of the literature related to the numerical treatment of the two-dimensional IHCP is based on different ways of combining finite elements realizations with the Future Temperatures Method of J. V. Beck [3]. For some of the early applications of these ideas, see B. R. Bass and L. J. Ott [1]. More numerical experimentation can be found in T. Yoshimura and K. Ituka [10]. An elaborated and comprehensive exposition of the method was presented later by J. Baumeister and H. J. Reinhardt [2]. See also N. Zabaras and J. C. Liu [11]. More recently, space marching implementations of the mollification method have been introduced by Y. Liu and D. A. Murio [6], [7]. Theoretical and computational results involving the numerical solution of the three-dimensional IHCP can be found in D. A. Murio and H. Zheng [9], and H. Zheng and D. A. Murio [12].

In section 2 of this paper we briefly review the space marching implementation of the mollification method for the three-dimensional inverse heat conduction problem in an infinite slab – including the numerical procedure and the corresponding numerical analysis – as introduced by D. A. Murio and H. Zheng [9], generalizing the earlier two-dimensional presentation of L. Guo and D. A. Murio [4]. In section 3 we initially concentrate our attention on the actual application of the fully explicit space-marching algorithm to the more realistic situation related with a bounded rectangular space domain in thc (x, y) planc. Wc study the task of recovering transient temperatures an heat fluxes at "interior" space domains for times greater than a certain $t_{min} > 0$; a singular perturbation technique is introduce and analyze to enhance the stability of the numerical method. The question of attempting to recover the initial temperature distribution on the interior domains and the testing of numerical strategies for the approximate reconstruction of the temperature functions at the boundaries of the whole original bounded domain – a more difficult task – are briefly described in section 4 as reported in Y. Liu and D. A. Murio [7]. In all cases, we provide several model examples to illustrate the results of the numerical experiments. Finally, in section 5, following H. Zheng and D. A. Murio [12], we reproduce the numerical scheme for the three-dimensional "finite cube" case and present some numerical examples of interest.

2. Three dimensional IHCP in a slab. We consider a three-dimensional IHCP on a semi-infinite slab in which the temperature and heat flux histories $F(y,z,t)$ and $Q(y,z,t)$ respectively, on the left-hand surface ($x = x_0 = 0$) are approximately measured, and the temperature and heat flux histories $f(y,z,t,)$ and $q(y,z,t,)$ respectively, on the right-hand surface ($x = x_1 > x_0$) are desired and unknown. We assume linear heat conduction with constant coefficients. Without loss of generality, we also assume zero initial temperature and normalize the problem by dimensionless quantities. The mathematical description of the problem is as follows.

The unknown temperature $u(x,y,z,t)$ satisfies:

(1)
$$\begin{aligned}
&u_t(x,y,z,t) = u_{xx}(x,y,z,t) + u_{yy}(x,y,z,t) + u_{zz}(x,y,z,t), && (1a)\\
&0 < x < x_1, \quad y > 0,\ z > 0,\ t > 0, &&\\
&u(0,y,z,t) = F(y,z,t), \quad y > 0,\ z > 0,\ t > 0, && (1b)\\
&\text{with corresponding approximate data function } F_m(y,z,t) &&\\
&-u_x(0,y,z,t) = Q(y,z,t), \quad y > 0,\ z > 0,\ t > 0, && (1c)\\
&\text{with corresponding approximate data function } Q_m(y,z,t), &&\\
&u(x,y,z,0) = 0, \quad 0 \le x \le x_1,\ y \ge 0,\ z \ge 0, && (1d)\\
&u(x,0,z,t) = 0, \quad 0 \le x \le x_1,\ z \ge 0,\ t \ge 0, && (1e)\\
&u(x,y,0,t) = 0, \quad 0 \le x \le x_1,\ y \ge 0,\ t \ge 0, && (1f)\\
&u(x_1,y,z,t) = f(y,z,t), \quad y > 0,\ z > 0, t > 0, && (1g)\\
&\text{the desired but unknown temperature function,} &&\\
&-u_x(x_1,y,z,t) = q(y,z,t), \quad y > 0\ ,\ z > 0,\ t > 0, && (1h)\\
&\text{the desired but unknown heat flux function,} &&
\end{aligned}$$

We consider all the functions involved – and their extensions – to be L_2 functions in R^3 and use the corresponding L_2 norm, as defined below, to measure errors:

$$\|g\| = \left(\int_{R^3} |g(y,z,t)|^2\, dydzdt\right)^{\frac{1}{2}}.$$

We hypothesize that the exact data functions $F(y,z,t)$ and $Q(y,z,t)$ and the measured data functions $F_m(y,z,t)$ and $Q_m(y,z,t)$ satisfy the L_2 data error bounds

$$\|F - F_m\| \le \epsilon \text{ and } \|Q - Q_m\| \le \epsilon.$$

The Fourier analysis of system (1), presented in D. A. Murio and H. Zheng [9], demonstrates that if w_1, w_2 and w_3 represent the Fourier transform variables associated with y, z and t respectively, attempting to solve problem (1) – obtaining $f(y,z,t)$ and $q(y,z,t)$ from $F(y,z,t)$ and $Q(y,z,t)$– amplifies the error in a high frequency component by the factor

$$\exp[(\sqrt{(w_1^2+w_2^2)^2+w_3^2}+(w_1^2+w_2^2))]^{1/2},$$

$$-\infty < \omega_1, \omega_2, \omega_3 < +\infty.$$

Thus, the three-dimensional inverse problem is severely ill-posed in the high frequency components.

Remark:

Similar analysis shows that for the one-dimensional IHCP – solving for $f(t)$ and $q(t)$ from $F(t)$ and $Q(t)$ – the high frequency components of the error are amplified by the factor $\exp(\sqrt{w_3})$; for the two-dimensional IHCP – solving for $f(y,t)$ and $q(y,t)$ from $F(y,t)$ and $Q(y,t)$ – the high frequency components of the error are amplified by the factor $\exp(\sqrt{w_1^4+w_3^2}+w_1^2)^{1/2}$. Hence, the three-dimensional IHCP is much more ill-posed (in the high frequency components) than the lower dimensional problems.

2.1. Stabilized problem. Introducing the three-dimensional Gaussian kernel

$$\rho\left(y,z,t,\delta_1,\delta_2,\delta_3\right)=\frac{1}{\delta_1\delta_2\delta_3\pi^{3/2}}\exp\left[-\left(\frac{y^2}{\delta_1^2}+\frac{z^2}{\delta_2^2}+\frac{t^2}{\delta_3^2}\right)\right]$$

and denoting $\rho\left(y,z,t,\delta_1,\delta_2,\delta_3\right)$ by $\rho_\delta\left(y,z,t\right)$ if $\delta_1=\delta_2=\delta_3=\delta$, the three-dimensional convolution of any locally integrable function $g\left(y,z,t\right)$ with $\rho_\delta\left(y,z,t\right)$ – the mollification of the function $g(y,z,t)$– is defined by

$$J_\delta g\left(y,z,t\right)=\left(\rho_\delta * g\right)\left(y,z,t\right)$$

$$(2)\qquad =\int_{-\infty}^{\infty}\int_{-\infty}^{\infty}\int_{-\infty}^{\infty}\rho_\delta\left(y',z',t'\right)g\left(y-y',z-z',t-t'\right)dy'dz'dt'.$$

We notice that the Gaussian kernel $\rho_\delta\left(y,z,t\right)$ is always positive and nearly vanishes outside the ball centered at the origin and radius approximately 3δ. Moreover, $J_\delta g\left(y,z,t\right)$ is infinitely differentiable and $\|J_\delta g-g\|\to 0$ as $\delta\to 0$, the convergence being uniform on any compact set where g is continuous.

Mollifying system (1), we obtain the following associated problem: For some $\delta>0$, find $J_\delta f_m\left(y,z,t\right)=J_\delta v\left(x_1,y,z,t\right)$ and $J_\delta f_{1m}\left(y,z,t\right)=-J_\delta v_x\left(x_1,y,z,t\right)$ at the (y,z,t) points of interest, given that the mollified temperature function $J_\delta v\left(x,y,z,t\right)$ verifies

$$
(3) \quad
\begin{array}{ll}
(J_\delta u)_t = (J_\delta u)_{xx} + (J_\delta u)_{yy} + (J_\delta u)_{zz}, & 0 < x < x_1, y > 0, z > 0, t > 0, \\
J_\delta u(0, y, z, t) = J_\delta F_m(y, z, t), & y > 0, z > 0, t > 0, \\
-J_\delta u_x(0, y, z, t) = J_\delta Q_m(y, z, t), & y > 0, z > 0, t > 0, \\
J_\delta u(x, 0, z, t) = 0, & 0 < x < x_1, z > 0, t > 0, \\
J_\delta u(x, y, 0, t) = 0, & 0 < x < x_1, y > 0, t > 0, \\
J_\delta u(x, y, z, 0) = 0, & 0 \le x \le x_1, y \ge 0, z \ge 0.
\end{array}
$$

This problem and its solutions satisfy the following theorem.

Theorem 1. *Suppose that* $\| F - F_m \| \le \epsilon$ *and* $\| Q - Q_m \| \le \epsilon$. *Then,*

1. Problem (3) is a formally stable problem with respect to perturbations in the data.

2. If the exact boundary temperature f *and boundary heat flux* q *have compact support* Ω *(*Ω *convex with radius* r, $r > 3\delta$*) and* $f, q \in C^2(\Omega)$, *then there exist* $M > 0$ *such that*

$$\|f - J_\delta f_m\|_\Omega \le |\Omega|^{1/2} \tfrac{3M}{\pi^{1/2}} \delta + 2\epsilon \exp(11\delta^{-2})$$

and

$$\|f_1 - J_\delta f_{1m}\|_\Omega \le |\Omega|^{1/2} \tfrac{3M}{\pi^{1/2}} \delta + 2\epsilon \exp(11\delta^{-2}).$$

The proof of these statements can be found in D. A. Murio and H. Zheng [9].

2.2. Numerical method and error analysis. Let $v = J_\delta u$ and $w = -\frac{\partial v}{\partial x}$. Then system (3) is equivalent to

$$
(4) \quad
\begin{array}{lll}
\frac{\partial v}{\partial t} = -\frac{\partial w}{\partial x} + \frac{\partial^2 v}{\partial y^2} + \frac{\partial^2 v}{\partial z^2}, & 0 < x < x_1,\ y > 0,\ z > 0,\ t > 0, & (4a) \\
-\frac{\partial v}{\partial x} = w, & 0 < x < x_1,\ y > 0,\ z > 0,\ t > 0, & (4b) \\
v(0, y, z, t) = J_\delta F_m(y, z, t), & y > 0,\ z > 0,\ t > 0 & (4c) \\
w(0, y, z, t) = J_\delta Q_m(y, z, t), & y > 0,\ z > 0,\ t > 0 & (4d) \\
v(x, 0, z, t) = 0, & 0 < x < x_1,\ z > 0, t > 0 & (4e) \\
v(x, y, 0, t) = 0, & 0 < x < x_1,\ y > 0,\ t > 0, & (4f) \\
v(x, y, z, 0) = 0, & 0 \le x \le x_1,\ y \ge 0,\ z \ge 0, & (4g) \\
v(x_1, y, z, t) = J_\delta f_m(y, z, t), & y > 0,\ z > 0,\ t > 0,\ \text{unknown}, & (4h) \\
w(x_1, y, z, t) = J_\delta q_m(y, z, t), & y > 0,\ z > 0,\ t > 0,\ \text{unknown}, & (4i)
\end{array}
$$

Without loss of generality, suppose we seek to reconstruct the unknown mollified boundary temperature function $J_\delta f_m$ and the mollified boundary heat flux function $J_\delta q_m$ in the unit cube $D_{x_1} = [0,1] \times [0,1] \times [0,1]$ of the (y, z, t) space at $x = x_1$. To this effect, consider a uniform grid in the (x, y, z, t) space:

$$\{(\xi_i = ih,\ y_j = js,\ z_l = lp,\ t_n = nk),\ i = 0, 1, \cdots, N_{x_1},\ N_{x_1} h = x_1;\ j = 0, 1, 2, \cdots, N_y,\ N_y s = 1;\ l = 0, 1, 2, \cdots, N_z,\ N_z p = 1;\ n = 0, 1, 2, \cdots, N_t,\ N_t k = 1\},$$

and let the grid functions V and W be defined by

$$V_{i,j,l}^{n}=v\left(\xi_{i},y_{j},z_{l},t_{n}\right),\quad W_{i,j,l}^{n}=w\left(\xi_{i},y_{j},z_{l},t_{n}\right),$$

$i=0,1,\cdots,N_{x_1};\ j=0,1,2,\cdots,N_y;\ l=0,1,2,\cdots,N_z;\ n=0,1,2,\cdots,N_t.$

The first two differential equations in system (4) can be approximated with the consistent finite difference schemes

$$\frac{V_{i,j,l}^{n+1}-V_{i,j,l}^{n-1}}{2k}=-\frac{W_{i+1,j,l}^{n}-W_{i,j,l}^{n}}{h}+\frac{V_{i,j+1,l}^{n}-2V_{i,j,l}^{n}+V_{i,j-1,l}^{n}}{s^2}$$

$$+\frac{V_{i,j,l+1}^{n}-2V_{i,j,l}^{n}+V_{i,j,l-1}^{n}}{p^2},$$

$$\frac{V_{i+1,j,l}^{n}-V_{i,j,l}^{n}}{h}=-W_{i+1,j,l}^{n}.$$

Reordering these last two equations and introducing the initial and boundary conditions for the marching scheme, we obtained the discrete system of equations

$$(5)\quad \begin{aligned}
&W_{i+1,j,l}^{n}=W_{i,j,l}^{n}+\tfrac{h}{s^2}\left(V_{i,j+1,l}^{n}-2V_{i,j,l}^{n}+V_{i,j-1,l}^{n}\right)\\
&\qquad+\tfrac{h}{p^2}\left(V_{i,j,l+1}^{n}-2V_{i,j,l}^{n}+V_{i,j,l-1}^{n}\right)-\tfrac{h}{2k}\left(V_{i,j,l}^{n+1}-V_{i,j,l}^{n-1}\right), &&(5a)\\
&V_{i+1,j,l}^{n}=V_{i,j,l}^{n}-hW_{i+1,j,l}^{n}, &&(5b)\\
&i=0,1,\cdots,N_{x_1}-1;\ j=1,\cdots,N_y-1;\\
&l=1,\cdots,N_z-1;\ n=1,2,\cdots,N_t-1,\\
&V_{0,j,l}^{n}=J_\delta F_m\left(y_j,z_l,t_n\right), &&(5c)\\
&j=0,1,2,\cdots,N_y;\ l=0,1,2,\cdots,N_z;\ n=0,1,2,\cdots,N_t,\\
&W_{0,j,l}^{n}=J_\delta Q_m\left(y_j,z_l,t_n\right), &&(5d)\\
&j=0,1,2,\cdots,N_y;\ l=0,1,2,\cdots,N_z;\ n=0,1,2,\cdots,N_t\\
&V_{i,0,l}^{n}=0,\quad i=0,1,2,\cdots,N_{x_1};\ l=0,1,2,\cdots,N_z;\ n=0,1,2,\cdots,N_t. &&(5e)\\
&V_{i,j,0}^{n}=0,\quad i=0,1,2,\cdots,N_{x_1};\ j=0,1,2,\cdots,N_y;\ n=0,1,2,\cdots,N_t. &&(5f)\\
&V_{i,j,l}^{0}=0,\quad i=0,1,2,\cdots,N_{x_1};\ j=0,1,2,\cdots,N_y;\ l=0,1,2,\cdots,N_z. &&(5g)
\end{aligned}$$

The stability of the finite difference scheme (5) and the approximation of the numerical solution to the exact mollified solution of problem (4), when restricted to the grid points, are discussed in D. A. Murio and H. Zheng [9], where the following theorem is proved.

Theorem 2. *If* $\delta\geq\max\left(\sqrt{2sx_1/\pi},\sqrt{2px_1/\pi},\sqrt{2kx_1/\pi}\right)$, *the finite difference scheme (5) is numerically stable and approximates the solution of the mollified problem (4) for small and fixed* h, s, p *and* k. *More precisely,*

$$\left\|V_{N_{x_1}}\right\|_2\leq\exp\left(3x_1^2\delta^{-2}\right)\left\{\|F_m\|_2+\|Q_m\|_2\right\}$$

and

$$\left\|W_{N_{x_1}}\right\|_2 \leq \exp\left(3x_1^2\delta^{-2}\right)\{\|F_m\|_2 + \|Q_m\|_2\},$$

where

$$\left\|V_{N_{\xi_i}}\right\|_2 = \left(spk \sum_{j=-\infty}^{j=\infty} \sum_{l=-\infty}^{l=\infty} \sum_{n=-\infty}^{n=\infty} \mid V(\xi_i, y_j, z_l, t_n) \mid^2\right)^{1/2}$$

indicates the l_2 norm of the discrete function $V(\xi_i, y_j, z_l, t_n)$ suitably extended to the whole three-dimensional discrete space (y, z, t) at $x = \xi_i$.

Remarks:

1. Similar results for the two-dimensional IHCP can be found in L. Guo and D. A. Murio [4].

2. As we march in the x-direction, we must drop the estimation of the interior temperature and heat flux functions from the highest point in time and the right-most points in the y and z- directions. Since we are interested in the evaluation of $\{U^n_{N_{x_1},j,l}\}$ and $\{W^n_{N_{x_1},j,l}\}$ at the grid points of the unit cube $C_{x_1} = [0,1] \times [0,1] \times [0,1]$ – in the (y, z, t) space at $x = x_1$– after N_{x_1} iterations, the *available* initial data functions should be read, in principle, in a three-dimensional cube – in the (y, z, t) space at $x = x_0$– of minimal dimensions $[0, 1 - j + j/h] \times [0, 1 - l + l/h] \times [0, 1 - n + n/h]$.

3. The actual mollification of the initial data functions with a Gaussian kernel of blurring radius δ requires these functions to be defined in a slightly "larger" domain. Namely, the data cube $D_{data} = [-3\delta, 1-j+j/h+3\delta] \times [-3\delta, 1-l+l/h+3\delta] \times [-3\delta, 1-n+n/h+3\delta]$, at $x = x_0$, is the smallest domain necessary for the successful mollification of the noisy data functions F_m and Q_m if the reconstruction of the discrete functions f_m and q_m is needed over the domain C_{x_1}. Thus, we extend the original domain of the data functions F_m and Q_m by setting them equal to zero whenever $y < 0$ or $z < 0$ or $t < 0$.

It is precisely the feasibility of this type of extensions, for nonhomogeneous boundary data, that will play an important role in the numerical implementation of the method for the approximate solution of the three-dimensional IHCP on a bounded space domain as explained in section 5.

4. The radius of mollification, δ, can be selected automatically as a function of the level of noise in the data. In fact, for a given $\epsilon > 0$, there is a unique $\delta > 0$, such that

$$\|J_\delta F_m - F_m\|_D = \epsilon. \tag{6}$$

(See D. A. Murio and H. Zheng [9].)

5. It is also possible to replace the discrete three-dimensional mollification of the data functions by three successive one-dimensional mollifications in time, y and z-space. In this manner, the data filtering task can be executed as a parallel process.

6. If $F(y, z, t)$ and $Q(y, z, t)$ denote the exact temperature and heat flux functions, respectively, the noisy data is obtained by adding random errors to $F(y_j, z_l, t_n)$ and $Q(y_j, z_l, t_n)$ at each grid point. That is,

$$F_m(y_j, z_l, t_n) = F(y_j, z_l, t_n) + \epsilon^F_{j,n,l},$$

and

$$Q_m(y_j, z_l, t_n) = Q(y_j, z_l, t_n) + \epsilon^Q_{j,n,l},$$

where $\epsilon^F_{j,n,l}$ and $\epsilon^Q_{j,n,l}$ are Gaussian random variables with variance $\sigma^2 = \epsilon^2$.

3. Two-D IHCP in a bounded domain. In this section we follow the presentation in Y. Liu and D. A. Murio [6]. We investigate the two-dimensional IHCP when the space domain in the (x, y) plane is restricted to the bounded prototype rectangle $R = [0, x_1] \times [0, 1]$. Mathematically, the new inverse problem can be stated as follows.

The temperature function $u(x, y, t)$ satisfies:

$$(7)\quad \begin{array}{ll}
u_t(x,y,t) = u_{xx}(x,y,t) + u_{yy}(x,y,t), & (7a) \\
0 < x < x_1,\ 0 < y < 1,\ t > 0, & \\
u(0,y,t) = F(y,t),\ 0 < y \le 1,\ t > 0, & (7b) \\
\text{with corresponding approximate data function } F_m(y,t), & \\
u_x(0,y,t) = Q(y,t),\ 0 < y \le 1,\ t > 0, & (7c) \\
\text{with corresponding approximate data function } Q_m(y,t), & \\
u(x,y,0) = \beta(x,y),\ \text{unknown},\ 0 \le x \le x_1,\ 0 \le y \le 1, & (7d) \\
u(x,0,t) = \gamma(x,t),\ \text{unknown},\ 0 < x < x_1,\ t > 0, & (7e) \\
u(x,1,t) = \alpha(x,t),\ \text{unknown},\ 0 < x < x_1,\ t > 0, & (7f) \\
u(x_1,y,t) = f(y,t),\ \text{unknown},\ 0 < y \le 1,\ t > 0, & (7g) \\
u(x_1,y,t) = q(y,t),\ \text{unknown},\ 0 < y \le 1,\ t > 0, & (7h)
\end{array}$$

For a discussion of the uniqueness of the solution of this problem, the reader is referred to D. A. Murio [8].

3.1. "Interior" cube. We begin the numerical testing of the two-dimensional space marching implementation of the mollification method. The numerical process, as well as the results of the recovery, depends on a set of parameters that includes the sample step-sizes, the radii of mollification and some others. We would like to show practical stability and determine an optimal set of parameters for the fully explicit algorithm. Accordingly, we test the method on five different model examples. The overall performance of the algorithm is "problem independent".

3.1.1. The list of examples. Here we list the five examples that are used to illustrate the results of the numerical experiments. As mentioned before, the general conclusions do not depend on any particular example and, most of the time, we will present the numerical results associated with the prototype **example 1**.

Example 1:

$$u(x,y,t) = e^{5-2t} \sin x \sin y$$

$$u_x(x,y,t) = e^{5-2t} \cos x \sin y$$

Example 2:

$$u(x,y,t) = e^{5-4.25t} \sin \frac{1}{2} x \sin 2y$$

$$u_x(x,y,t) = \frac{1}{2} e^{5-4.25t} \cos \frac{1}{2} x \sin 2y$$

Example 3:

$$u(x,y,t) = e^{t+\frac{1}{\sqrt{2}}(x+y)}$$

$$u_x(x,y,t) = \frac{1}{\sqrt{2}} e^{t+\frac{1}{\sqrt{2}}(x+y)}$$

Example 4:

$$u(x,y,t) = \frac{1}{3} x^3 y + 3txy + \frac{1}{6} xy^3$$

$$u_x(x,y,t) = x^2 y + 3ty + \frac{1}{6} y^3$$

Example 5:

$$u(x,y,t) = 50x - 100x^2 + 50y^2 - 100t$$

$$u_x(x,y,t) = 50 - 200x$$

All these functions satisfy the heat-conduction equation $(1a)$.

We study now the two-dimensional IHCP when the space domain in the (x,y) plane is the rectangle $R = [0,x_1] \times [0,1]$ with initial data temperature and heat flux functions $F_m(y,t)$ and $Q_m(y,t)$ measured at the discrete grid points of the domain $D_{data} = [0,1] \times [0,1]$ in the (y,t) space at $x = 0$, with sample step sizes s and k in the y and t directions, respectively. Once the radii of mollification δ_F and δ_Q have been obtained – after solving the discrete version of equation (6) – setting $\delta = \max(\delta_F, \delta_Q)$, δ a multiple of s and k, the discrete mollified data functions $J_\delta F_m$ and $J_\delta Q_m$ – computed using a rectangular quadrature rule in the original grid – are evaluated on a coarser grid with step sizes $\Delta y = n_y s$ and $\Delta t = n_t k$ $(n_y, n_t > 1)$, in the domain $D_\delta = [3\delta, 1-3\delta] \times [3\delta, 1-3\delta]$ of the (y,t) plane at $x = 0$. The numbers n_y and n_t are the refine constants of the algorithm.

Our new immediate task is to recover the temperature and heat flux functions at the grid points of the three-dimensional "interior" cube $\Omega_\delta = [0,x_1] \times [3\delta, 1-3\delta] \times [3\delta, 1-3\delta]$ of the (x,y,t) space with the knowledge of the discrete data functions F_m and Q_m at the grid points of the domain D_δ.

Remark:

In section 5, for the three-dimensional IHCP, we will follow a somewhat different strategy. The domain of the original data functions will be initially extended to allow the recovery on the whole domain while marching in the x-direction.

3.1.2. Numerical procedure. We consider the uniform grid in the (x,y,t) space:

$$\{(\xi_i = ih,\ y_j = 3\delta + j\Delta y,\ t_n = 3\delta + n\Delta t),\ i = 0,1,\dots,N_x,\ N_x h = x_1;$$

$$j = 0,1,\dots,N_y,\ N_y\Delta y = 1-6\delta;\ n = 0,1,\dots,N_t,\ N_t\Delta t = 1-6\delta\}.$$

We define the grid functions

$$V_{i,j}^n = v(\xi_i, y_j, t_n),\quad W_{i,j}^n = (\xi_i, y_j, t_n),$$

$$0 \le i \le N_x,\quad 0 \le j \le N_y,\quad 0 \le n \le N_t,$$

and with $v = J_\delta u$ and $w = v_x$, we introduce the consistent finite difference scheme

$$(8)\qquad \begin{aligned} W_{i+1,j}^n &= W_{i,j}^n + \tfrac{h}{2k}(V_{i,j}^{n+1} - V_{i,j}^{n-1}) \\ &\qquad -\tfrac{h}{s^2}(V_{i,j-1}^n - 2V_{i,j}^n + V_{i,j+1}^n) \\ V_{i+1,j}^n &= V_{i,j}^n + hW_{i-1,j}^n, \end{aligned}$$

$$\begin{array}{lll} i = 0,1,\dots,N_x-1; & j = 1,2,\dots,N_y-1; & n = 1,2,\dots,N_t-1, \\ V_{0,j}^n = (J_\delta F_m)_j^n, & j = 0,1,\dots,N_y; & n = 1,2,\dots,N_t, \\ W_{0,j}^n = (J_\delta Q_m)_j^n, & j = 1,2,\dots,N_y; & n = 1,2,\dots,N_t, \end{array}$$

As we march forward in the x-direction in space, we must drop the estimation of $V_{i,j}^n$ associated with the boundary values of j and n. We extrapolate to recover the boundary conditions at $y = 3\delta$, $y = 1-3\delta$, $t = 3\delta$ and $t = 1-3\delta$. For example, at each step of the marching scheme, the boundary value for $j = N_y$ is recovered by extrapolation from the corresponding values of the discrete heat flux function at $j = N_y - 1$, $N_y - 2$, $N_y - 3$ and $N_y - 4$, introducing a $O(\Delta y^4)$ error in the process. We recover the discrete temperatures at $j = 0$, $n = N_t$ and $n = 0$ in a similar way. The algorithm is continued for $i = 1,2,\cdots,N_{x_1}$ until all the required quantities $\{W_{N_{x_1},j}^n\}$ and $\{V_{N_{x_1},j}^n\}$ are computed. These values are then taken as the accepted approximations for the heat flux and temperature functions, respectively, at each point of the discrete domain $[3\delta, 1-3\delta] \times [3\delta, 1-3\delta]$ of the (y,t) space at $x = x_1$.

It is desirable to make the grids denser in order to get more detailed information about the computed temperature and heat-flux functions. However, there is a trade off between the step-size and the accuracy of the recovery since the IHCP is an ill-posed problem. Keeping the accuracy at a certain level, we try to make the sample step-sizes

$\triangle y, \triangle t$ and space-marching step-size h as small as possible. The performance of the algorithm is mostly justified by the consistency and the magnitude of these parameters.

We must test the stability of the algorithm under moderately high levels of noise. However, higher noise level means that a larger mollification constant is required to stabilize the numerical process, and, therefore, we have a smaller inner cube to recover.

The accuracy of a recovery is measured by the l_2 error norm of the recovered function **over the whole discrete interior cube** Ω_δ.

The stability of the algorithm is indirectly measured by the size of the working grid, i.e., by the sample step-size and the space-marching step-size of the recovery. For the same level of accuracy, an algorithm is considered more stable if it allows for a denser working grid, i.e., smaller sample step-size and space marching step-size.

To put ourselves in a comparable contention, we consider (most of the time) the noise level constant, ε, up to 0.005 in our experiments. The corresponding mollification constant δ is 0.0667. A recovery is considered "accurate" (acceptable) if the l_2 error norm of the recovered temperature function is less than 0.01 and that of the heat flux function is less than 1. An algorithm is **"quasi-stable"** if it allows an accurate recovery with $h, \triangle y, \triangle t \leq 0.1$.

No specific requirements are imposed on the grid refine constants. We take the smallest integers that stabilize the numerical process. In most cases, $n_y = n_t = 10$. We also fix $x_1 = 0.5$.

3.1.3. The mollification results. For the mollification method, with $\varepsilon = 0.005$ and $\delta = 0.0667$, the optimal combination of the parameter values, according to our numerical experience, are $N_x = 10$, $N_y = N_t = 40$, $n_y = n_t = 10$.

For this specific choice of the parameter set, the l_2 error norms of the recovered temperature and heat-flux functions on the whole cube Ω_δ are shown in table 1.

Example 1	Temperature	Heat-flux
l_2 Error	0.0079	0.4490
Relative l_2 Error	0.0009	0.0153

Table 1: The l_2 error norms. Example 1.
$N_x = 10,\ N_y = N_t = 40,\ \varepsilon = 0.005,\ \delta = 0.0667,\ n_y = n_t = 10.$

Here and in what follows,

$$\text{Relative } l_2 \text{ error norm} = \frac{l_2 \text{ error norm}}{l_2 \text{ norm of the function}}.$$

Table 2 contains the l_2 error norms for a list of 2-surfaces for fixed values of x.

x	l_2 for $u(x,y,t)$	R. l_2 for $u(x,y,t)$	l_2 for $u_x(x,y,t)$	R. l_2 for $u_x(x,y,t)$
0.1	0.0153	0.0050	0.1776	0.0058
0.2	0.0573	0.0094	0.2514	0.0084
0.3	0.1295	0.0143	0.3246	0.0111
0.4	0.2310	0.0193	0.4367	0.0154
0.5	0.3612	0.0245	1.0794	0.0400

Table 2: The l_2 error norms for 2-surfaces for fixed value of x . Example 1.
$N_x = 10, \ N_y = N_t = 40, \ \varepsilon = 0.005, \ \delta = 0.0667, \ n_y = n_t = 10.$

Next we take the previous parameter set values as a reference to test the performance of the algorithm under a different choice of parameters. At each experience, only one of the parameters is changed.

Table 3 lists the dependency between the two parameter ε and δ. All the error norms are for the temperature functions.

ε	δ	l_2 error	R. l_2 error
0.0005	0.0417	0.0061	0.0006
0.0010	0.0417	0.0061	0.0006
0.0020	0.0500	0.0069	0.0007
0.0030	0.0583	0.0071	0.0008
0.0040	0.0583	0.0071	0.0008
0.0050	0.0667	0.0079	0.0009

Table 3: The optimal δ for different ε . Example 1.
$N_x = 10, \ N_y = N_t = 40, \ n_y = n_t = 10.$

Table 4 shows the performance of the algorithm when the space marching step-size h changes. Table 5 is for the change of the sample step-sizes $\triangle y$ and $\triangle t$.

N_x	l_2 for $u(x,y,t)$	R. l_2 for $u(x,y,t)$	l_2 for $u_x(x,y,t)$	R. l_2 for $u_x(x,y,t)$
5	0.0138	0.0015	0.5242	0.0179
10	0.0079	0.0009	0.4490	0.0153
15	0.0058	0.0007	1.7076	0.0580
20	0.0060	0.0007	7.2921	0.2477
25	0.0093	0.0011	26.6089	0.9035

Table 4: The change of the space marching step-size. Example 1.
$N_y = N_t = 40, \ \varepsilon = 0.005, \ \delta = 0.0667, \ n_y = n_t = 10.$

$N_y = N_t$	l_2 for $u(x,y,t)$	R. l_2 for $u(x,y,t)$	l_2 for $u_x(x,y,t)$	R. l_2 for $u_x(x,y,t)$
10	0.0301	0.0033	0.3188	0.0105
20	0.0151	0.0017	0.4476	0.0150
40	0.0079	0.0009	0.4490	0.0153
60	0.0053	0.0010	0.5186	0.0177
80	0.0040	0.0005	0.3826	0.0131
100	0.0032	0.0004	0.5165	0.0177

Table 5: The change of the initial sample step-sizes. Example 1.
$N_x = 10,\ \varepsilon = 0.005,\ \delta = 0.0667,\ n_y = n_t = 10.$

We recall that all the tables listed above are for example 1. The performance of the algorithm is the same when applied to other examples. The l_2 error norms of the temperature and heat flux functions for the other examples are in table 6. In all cases, we set $\varepsilon = 0.005,\ \delta = 0.0667,\ h = 0.05,\ \Delta y = \Delta t = 0.025,\ n_y = n_t = 10$.

Example	l_2 for $u(x,y,t)$	R. l_2 for $u(x,y,t)$	l_2 for $u_x(x,y,t)$	R. l_2 for $u_x(x,y,t)$
1	0.0079	0.0009	0.4490	0.0153
2	0.0042	0.0012	0.3857	0.0332
3	0.0028	0.0009	0.3307	0.1571
4	0.0017	0.0063	0.3306	0.3569
5	0.0917	0.0024	0.4842	0.0153

Table 6: The l_2 error norms, all the examples.
$N_x = 10,\ N_y = N_t = 40,\ \varepsilon = 0.005,\ \delta = 0.0667,\ n_y = n_t = 10.$

Remarks:

1. As we mentioned earlier, one would like to make the number of steps of the space marching, N_x, as big as possible to get more detailed information of the temperature and heat-flux functions on the cube Ω_δ. However, due to the ill-posedness of the problem, the accuracy of the recovery is very sensitive to N_x. As illustrated in table 4, the accuracy drops dramatically on every increment of five units in N_x. According to our experience, N_x is one of the most important stability measurements. To keep the accuracy at the same level, allowing an increment of ten in N_x should be regarded as a substantial improvement on the stability of a fully explicit algorithm.

The major objective for a quasi-stable algorithm is to make $\mathbf{N}_x$, as well as $\mathbf{N}_y$ and $\mathbf{N}_t$, as large as possible while keeping the same level of accuracy.

2. For our example 1, the exact temperature function on the initial surface is identically zero. Nevertheless, we do add randomly generated noise to the **zero data.**

The perturbed "zero data" causes specific trouble on the stability and accuracy of the recovery. As illustrated in table 6, the accuracy is only related to the magnitude of the noise added into the initial data sample, but not to the individual example model. The performance of the mollification method, when applied to different problems, is consistent.

3.2. The Stable Algorithm. In this section we describe different ways to improve the stability of the mollification algorithm. In particular, we will combine the mollification algorithm with a singular perturbation method. The addition of a suitably chosen term to the heat conduction equation will greatly improve the stability of the mollification algorithm.

3.2.1. The singular perturbation scheme. The discussion in this section is motivated by the realization that it is possible to modify the previous numerical algorithm in such a way as to allow for some extra filtering of the high frequency components of the noise. In particular, we would like to improve the stability related with the approximation of the second order partial derivative of the solution temperature with respect to the y-variable as we march in the x-direction. What we have in mind is a potential perturbation of the differential operator itself, so that the finite-difference implementation of the algorithm is not substantially changed.

We recall that in the frequency space, the relationship

$$\widehat{u_{xx}}(x, w_1, w_3) = (iw_3 + w_1^2)\widehat{u}(x, w_1, w_3), \tag{9}$$

where w_1 and w_3 are the frequency variables corresponding to the real variables y and t respectively, leads to the general solution

$$\widehat{u}(x, w_1, w_3) = A(w_1, w_3)e^{\sqrt{iw_3+w_1^2}x} + B(w_1, w_3)e^{-\sqrt{iw_3+w_1^2}x}.$$

We propose to replace the solution of equation (9) by the function

$$\widehat{v}(x, w_1, w_3) = (A(w_1, w_3)e^{\sqrt{iw_3+w_1^2}x} + B(w_1, w_3)e^{-\sqrt{iw_3+w_1^2}x})e^{-\lambda w_1^2 x}, \tag{10}$$

where λ is a small positive constant. We observe that in real space, *the function v is simply the convolution of the function u with a Gaussian kernel that acts only in the y-variable* and, consequently, the analysis of the high frequency components of section 2 applies directly to this situation. The fundamental task is now reduced to finding the governing differential equation for v.

To this end, we proceed as follows. Introducing $\alpha = \sqrt{iw_3 + w_1^2}$, taking partial derivatives with respect to x and using (10), we have

$$\widehat{v_x}(x, w_1, w_3) = (\alpha - \lambda w_1^2)A(w_1, w_3)^{(\alpha-\lambda w_1^2)x} - (\alpha + \lambda w_1^2)B(w_1, w_3)e^{-(\alpha+\lambda w_1^2)x}$$

$$= \alpha(A(w_1,w_3)e^{\alpha x} - B(w_1,w_3)e^{-\alpha x})e^{-\lambda w_1^2 x} - \lambda w_1^2 \hat{v}(x,w_1,w_3). \tag{11}$$

Similarly, utilizing (10) and (11),

$$\widehat{v_{xx}}(x,w_1,w_3) = \alpha^2 \hat{v}(x,w_1,w_3) - \alpha\lambda w_1^2 (A(w_1,w_3)e^{\alpha x}$$

$$-B(w_1,w_3)e^{-\alpha x})e^{-\lambda w_1^2 x} - \lambda w_1^2 \widehat{v_x}(x,w_1,w_3)$$

$$= \hat{v}_t(x,w_1,w_3) - \widehat{v_{yy}}(x,w_1,w_3) - 2\lambda w_1^2 \widehat{v_x}(x,w_1,w_3) - \lambda^2 w_1^4 \hat{v}(x,w_1,w_3).$$

By the linearity of the Fourier transformations, the function $v(x,y,t)$ satisfies the linear partial differential equation

$$v_{xx}(x,y,t) = v_t(x,y,t) - v_{yy}(x,y,t) + 2\lambda v_{xyy}(x,y,t) - \lambda^2 v_{yyyy}(x,y,t),$$

a singular perturbation of the original diffusion equation, that depends on the small positive parameter λ.

In actual implementations, assuming the solution is sufficiently smooth, we only retain the first order term in λ , i.e., numerically we use the approximation

$$v_{xx}(x,y,t) \cong v_t(x,y,t) - v_{yy}(x,y,t) + 2\lambda v_{xyy}(x,y,t).$$

Thus, instead of solving the equation

$$u_{xx} = u_t - u_{yy},$$

we will work with the perturbed equation

$$u_{xx} = u_t - u_{yy} + 2\lambda u_{xyy}.$$

For the numerical implementation, we use the following consistent finite difference schemes to obtain the necessary formulas for the space-marching algorithm:

$$W_{i+1,j}^n = W_{i,j}^n + \frac{h}{2\Delta t}(V_{i,j}^{n+1} - V_{i,j}^{n-1}) - \frac{h}{\Delta y^2}(V_{i,j-1}^n - 2V_{i,j}^n + V_{i,j+1}^n)$$

$$+ 2\lambda \frac{h}{\Delta y^2}(W_{i,j-1}^n - 2W_{i,j}^n + W_{i,j+1}^n), \tag{12}$$

$$V_{i+1,j}^n = V_{i,j}^n + hW_{i-1,j}^n.$$

3.2.2. Combined mollification-singular perturbation method. To add the singular perturbation to the mollification algorithm, we only change the corresponding space-marching formulas as indicated in (12); everything else is kept the same. We again consider $\varepsilon = 0.005$ and $\delta = 0.0667$.

To see the improvements, we set $N_y = N_t = 40$ and $\lambda = 0.010$. One can compare Table 7 with Table 4 to appreciate the differences made by the singular perturbation term.

N_x	l_2 for $u(x, y, t)$	R. l_2 for $u(x, y, t)$	l_2 for $u_x(x, y, t)$	R. l_2 for $u_x(x, y, t)$
5	0.0047	0.0013	0.1602	0.0138
10	0.0023	0.0007	0.5075	0.0437
15	0.0015	0.0004	0.1927	0.0166
20	0.0015	0.0004	0.2544	0.0219
25	0.0017	0.0005	0.3531	0.0304

Table 7: The l_2 error norms, with singular perturbation. Example1.
$N_y = N_t = 40,\ \varepsilon = 0.005,\ \delta = 0.0667,\ n_y = n_t = 10,\ \lambda = 0.010$.

Table 7 illustrates a significant improvement in the stability of the algorithm. Much smaller space marching step size is now allowed, and all the results satisfy our accuracy criterion.

We provide one more comparison. Table 8 contains the results without singular perturbation and table 9, those with singular perturbation. The parameter values are $\varepsilon = 0.005$, $\delta = 0.0667$, $N_y = N_t = 60$ and $\lambda = 0.004$.

N_x	l_2 for $u(x, y, t)$	R. l_2 for $u(x, y, t)$	l_2 for $u_x(x, y, t)$	R. l_2 for $u_x(x, y, t)$
5	0.0093	0.0010	0.5186	0.0177
10	0.0053	0.0006	0.3436	0.0117
15	0.0036	0.0004	1.3039	0.0445
20	0.0037	0.0004	9.9876	0.3405
25	0.0134	0.0015	78.3677	2.6714

Table 8: The l_2 error norms, without singular perturbation. Example 1.
$N_y = N_t = 60,\ \varepsilon = 0.005,\ \delta = 0.0667,\ n_y = n_t = 10,\ \lambda = 0.000$.

N_x	l_2 for $u(x,y,t)$	R. l_2 for $u(x,y,t)$	l_2 for $u_x(x,y,t)$	R. l_2 for $u_x(x,y,t)$
5	0.0091	0.0010	0.4455	0.0125
10	0.0051	0.0006	0.2363	0.0081
15	0.0035	0.0004	0.2495	0.0085
20	0.0027	0.0003	0.5511	0.0188
25	0.0024	0.0003	1.3109	0.0447

Table 9: The l_2 error norms, with singular perturbation. Example 1.
$N_y = N_t = 60,\ \varepsilon = 0.005,\ \delta = 0.0667,\ n_y = n_t = 10,\ \lambda = 0.004.$

The graphs in this section, illustrate the qualitative behavior of the reconstructed temperatures and heat fluxes at $x = 0.5$, corresponding to Example 1 and the set of parameters $N_x = 20$, $N_y = N_t = 40$, $\varepsilon = 0.005$, $\delta = 0.0667$, $n_y = n_t = 10$ and $\lambda = 0.01$.

Figures 1 and 2, show the computed temperature and heat flux functions, respectively, for y and t between 0.2 and 0.8.

The discrete relative error functions associated with the temperature and heat flux functions $\left(\frac{V_{i,j}^n - u(ih,j\Delta y,n\Delta t)}{u(ih,j\Delta y,n\Delta t)} \text{ and } \frac{W_{i,j}^n - u_x(ih,j\Delta y,n\Delta t)}{u_x(ih,j\Delta y,n\Delta t)}\right)$, respectively, are displayed in Figures 3 and 4 for y and t between 0.2 and 0.8.

A general conclusion is that the algorithm presented in this section remains a viable procedure for the numerical solution of the 2-D IHCP in "interior" bounded subdomains.

4. Extension to the "whole" domain. The first step of the numerical procedure for the mollification scheme is to filter the high frequency components of the initial sample data collected on the 2-surface $x = x_0 = 0$. To compute the value of the mollified function at a given point P at $x = 0$, all the initial sample data in the interior of the square centered at P and width 6δ are required. Since the initial sample data are collected on the unit square $[0,1] \times [0,1]$, the mollified function can only be computed on the set $(y,t) \in [3\delta, 1-3\delta] \times [3\delta, 1-3\delta]$. Therefore, the domain of the recovered temperature and heat-flux functions is restricted to the "inner cube" $[0,0.5] \times [3\delta, 1-3\delta] \times [3\delta, 1-3\delta]$.

We try to extend the domain of the recovered functions from the "inner cube" to the "whole cube" to resolve this "lost boundary" problem. Based on the fact that the t-direction is less ill-posed than the y-direction – the parabolic operator involves first partial derivatives with respect to time and second partials derivatives with respect to y – we propose completely different extension techniques for the individual directions. We first introduce the "dynamical radius of mollification" in the mollification of the initial data to extend the domain in the t-direction and, later, the space marching scheme is again called in to extend the computed solutions in the y-direction.

4.1. The differences between y and t-directions. In all the discussions so far, we have only consider the case $N_y = N_t$. However, since only the first derivative for the variable t is involved in the heat conduction equation, one could expect that the

t-direction is "less ill-posed" than the y-direction. Hence, the stability of the algorithm should be more sensitive to the magnitude of N_y than that of N_t.

The numerical results for $N_y = 40$ and $N_t = 100$ are illustrated in table 10 with $\lambda = 0.01$. As expected, the accuracy remains almost the same when N_t is increased, but drops rapidly when N_y is increased.

N_x	l_2 for $u(x,y,t)$	R. l_2 for $u(x,y,t)$	l_2 for $u_x(x,y,t)$	R. l_2 for $u_x(x,y,t)$
5	0.0085	0.0009	0.3381	0.0116
10	0.0046	0.0005	0.2921	0.0100
15	0.0030	0.0003	0.0951	0.0033
20	0.0023	0.0003	0.1180	0.0040
25	0.0018	0.0002	0.1757	0.0060

Table 10: The l_2 error norms. Example 1.
$N_y = 40,\ N_t = 100,\ \varepsilon = 0.005,\ \delta = 0.0667,\ n_y = 10,\ n_t = 6,\ \lambda = 0.010.$

For the same reason, the data mollification in the y and t-directions can also be treated differently. For instance, instead of using $\rho_\delta(y,t) = \frac{1}{\pi\delta^2}\exp\left(-\frac{t^2+y^2}{\delta^2}\right)$, we could use the more general Gaussian kernel

$$\rho(y,t,\delta_y,\delta_t) = \frac{1}{\pi\delta_y\delta_t}\exp\left[-\left(\frac{y^2}{\delta_y^2}+\frac{t^2}{\delta_t^2}\right)\right]$$

to mollify the two-dimensional initial data. The radius of mollification δ_t can be made smaller than its counterpart δ_y to obtain larger recovered regions for the solution functions.

Table 11 illustrates some results involving these ideas. Here $\delta_y = 0.0667$ and $\delta_t = 0.050$.

N_x	l_2 for $u(x,y,t)$	R. l_2 for $u(x,y,t)$	l_2 for $u_x(x,y,t)$	R. l_2 for $u_x(x,y,t)$
5	0.0119	0.0013	0.3114	0.0102
10	0.0063	0.0007	0.5655	0.0185
15	0.0041	0.0004	0.1422	0.0046
20	0.0030	0.0003	0.2303	0.0075
25	0.0025	0.0003	0.3552	0.0116

Table 11: The l_2 error norms for the general Gaussian Kernel. Example 1.
$N_y = N_t = 40,\ \varepsilon = 0.005,\ \delta_y = 0.0667,\ \delta_t = 0.050,\ n_y = n_t = 10,\ \lambda = 0.010.$

4.2. The extension in t-direction. Up to now, in all the theoretical and numerical studies, the mollification radii δ_y and δ_t have always been regarded as constants. One observes that this assumption is, indeed, not essential for the mollification method. The data mollification is in fact a point by point process. For a given point (y_0, t_0) on the surface $x = 0$, one applies the Gaussian kernel to calculate $J(\delta_y, \delta_t, y_0, t_0)$ by using all the sample data in the square $[y_0 - 3\delta_y, y_0 + 3\delta_y] \times [t_0 - 3\delta_t, t_0 + 3\delta_t]$. Practically, one can regard δ_y and δ_t as functions of (y_0, t_0). We will call a mollification of this kind (with non-constant mollification radii) "dynamical mollification". The former method with constant mollification radii will be referred as the "uniform mollification".

The change from the uniform mollification to the dynamical mollification makes the "lost boundary" problem somehow disappear. For a point (y_0, t_0) closer to the boundary of the square $[0,1] \times [0,1]$, smaller radii of mollification will be assigned to this point to perform the required integration to mollify the data functions there.

Let δ be the radius of mollification corresponding to uniform mollification. We define the dynamical mollification radius δ_y and δ_t on the set $(y_0, t_0) \in [3\delta, 1-3\delta] \times [\,0,1]$ by

$$\delta_y(y_0, t_0) = \delta$$

and

$$\delta_t(y_0, t_0) = \begin{cases} \delta, & (y_0, t_0) \in [3\delta, 1-3\delta] \times [3\delta, 1-3\delta], \\ \frac{1}{3}(1 - t_0), & (y_0, t_0) \in [3\delta, 1-3\delta] \times [1-3\delta, 1], \\ \frac{1}{3}t_0, & (y_0, t_0) \in [3\delta, 1-3\delta] \times [0, 3\delta]. \end{cases}$$

Dynamical mollification with the given mollification radii $\delta_y(y_0, t_0), \delta_t(y_0, t_0)$ produces the mollified initial data on the subset $(y, t) \in [3\delta, 1-3\delta] \times [0,1]$ of the surface $x = 0$.

At this point, we merely apply the space marching scheme in the x-direction to recover the temperature and heat-flux functions in the domain $(x, y, t) = [0, 0.5] \times [3\delta, 1-3\delta] \times [0,1]$. By following this simple procedure, part of the lost boundary in the t-direction is actually recovered.

We recall that for those points that are very close to the lines $t = 0$ and $t = 1$, the dynamical mollification approach does not filter the original sample data. Consequently, for the last three-points next to the line $t = 0$ and $t = 1$, a 4-th order extrapolation polynomial is introduced to create the filtered initial data.

Table 12 illustrates the l_2 error of the recovery in the t-direction by dynamical mollification for example 1. We again set $N_y = N_t = 40$. The mollification radius is $\delta = 0.0667$ and the noise level constant is set as $\varepsilon = 0.005$.

N_x	l_2 for $u(x,y,t)$	R. l_2 for $u(x,y,t)$	l_2 for $u_x(x,y,t)$	R. l_2 for $u_x(x,y,t)$
5	0.0137	0.0010	1.9262	0.0425
10	0.0080	0.0006	1.1449	0.0252
15	0.0048	0.0004	0.7141	0.0157
20	0.0042	0.0003	0.7852	0.0173
25	0.0066	0.0005	1.2742	0.0280

Table 12: The recovery in the t-direction. Example 1.
$N_t = N_y = 40,\ \varepsilon = 0.005,\ \delta = 0.0667,\ n_t = n_y = 10,\ \lambda = 0.01.$

4.3. Recovery for the whole domain. We have extended the domain of the recovered functions from the inner set $[0, 0.5] \times [3\delta, 1 - 3\delta] \times [3\delta, 1 - 3\delta]$ to the set $[0, 0.5] \times [3\delta, 1 - 3\delta] \times [0, 1]$. Dynamical mollification can not be used in the y-direction because this direction is more ill-posed than the t-direction. However, starting from the 2-surface $(y, x, t) = \{3\delta\} \times [0, 0.5] \times [0, 1]$, we can now solve the inverse heat conduction problem by marching in the y-direction. We only need to march a short distance, 3δ, up to the surface $y = 0$ to recover one side of the lost boundary.

The formula for the implementation of the space marching scheme now is a little different from what was presented in section 3. We have to compute $u_y(x, y, t)$ together with $u_x(x, y, t)$. Setting $v = J_\delta u$, $w = v_x(x, y, t)$ and $r = v_y(x, y, t)$, the grid functions are now defined by

$$V_{i,j}^n = v(\xi_i, y_j, t_n), \qquad W_{i,j}^n = w(\xi_i, y_j, t_n), \qquad R_{i,j}^n = r(\xi_i, y_j, t_n).$$

The corresponding formulas for the space marching scheme read

$$R_{i,j}^n = \frac{V_{i,j}^n - V_{i,j-2}^n}{2\Delta y},$$

$$R_{i,j+1}^n = R_{i,j}^n + \frac{\Delta y}{2\Delta t}\left(V_{i,j}^{n+1} - V_{i,j}^{n-1}\right) - \frac{\Delta y}{2h}\left(W_{i+1,j}^n - W_{i-1,j}^n\right),$$

$$V_{i,j+1}^n = V_{i,j}^n + R_{i,j+1}^n \Delta y,$$

$$W_{i,j+1}^n = \frac{V_{i+1,j+1}^n - V_{i-1,j+1}^n}{2h}.$$

The discrete error norms of the recoveries for example 1 are presented in table 13. We again set $N_t = N_y = 40$, $\varepsilon = 0.005$ and $\delta = 0.667$. The perturbation constant is $\lambda = 0.01$.

N_x	l_2 for $u(x,y,t)$	R. l_2 for $u(x,y,t)$	l_2 for $u_x(x,y,t)$	R. l_2 for $u_x(x,y,t)$
5	0.0120	0.0006	2.8108	0.0459
10	0.0249	0.0014	2.7965	0.0456
15	0.0023	0.0001	1.8666	0.0304
20	0.0177	0.0010	1.4889	0.0243
25	0.0420	0.0023	2.9487	0.0480

Table 13: The recovery of the whole cube. Example 1.
$N_t = N_y = 40,\ \varepsilon = 0.005,\ \delta = 0.0667,\ n_t = n_y = 10,\ \lambda = 0.01.$

Figures 5 and 6 illustrate the qualitative behavior of the relative grid error functions for the temperature and the heat flux functions respectively for example 1, at the 2-surface $x = 0.5$ for $0 \le t, y \le 1$.

5. Three-dimensional IHCP on a finite cube. We investigate now the three-dimensional IHCP when the temperature and heat flux domains, in the (x, y, z) space, are restricted to the bounded prototype cube $[0, x_1] \times [0,1] \times [0,1]$. The new inverse problem can be stated as follows:

$$
(13)\quad \begin{array}{ll}
u_t(x,y,z,t) = u_{xx}(x,y,z,t) + u_{yy}(x,y,z,t) + u_{zz}(x,y,z,t), & (13a)\\
0 < x < x_1, \quad 0 < y,z < 1,\ t > 0, & \\
u(0,y,z,t) = F(y,z,t), \qquad 0 < y,z \le 1,\ t > 0, & (13b)\\
\text{with corresponding approximate data function } F_m(y,z,t), & \\
-u_x(0,y,z,t) = Q(y,z,t), \qquad 0 < y,z \le 1,\ t > 0 & (13c)\\
\text{with corresponding approximate data function } Q_m(y,z,t), & \\
u(x,y,z,0) = 0, \qquad 0 \le x \le x_1,\ 0 \le y,z \le 1, & (13d)\\
u(x,0,z,t) = \alpha_0(x,z,t),\ \text{unknown},\ 0 < x < x_1, 0 < z < 1, t > 0, & (13e)\\
u(x,1,z,t) = \alpha_1(x,z,t),\ \text{unknown},\ 0 < x < x_1, 0 < z < 1, t > 0, & (13f)\\
u(x,y,0,t) = \beta_0(x,y,t),\ \text{unknown},\ 0 < x < x_1, 0 < y < 1, t > 0, & (13g)\\
u(x,y,1,t) = \beta_1(x,y,t),\ \text{unknown},\ 0 < x < x_1, 0 < y < 1, t > 0, & (13h)\\
u(x_1,y,z,t) = f(y,z,t),\ \text{unknown},\ 0 < y,z \le 1,\ t > 0, & (13i)\\
-u_x(x_1,y,z,t) = q(y,z,t),\ \text{unknown},\ 0 < y,z \le 1,\ t > 0, & (13j)
\end{array}
$$

Proceeding as in section 2, mollifying system (13) and setting $v = J_\delta u$ and $w = -\frac{\partial v}{\partial x}$, we obtain

$$
(14)\quad \begin{array}{lll}
\frac{\partial v}{\partial t} = -\frac{\partial w}{\partial x} + \frac{\partial^2 v}{\partial y^2} + \frac{\partial^2 v}{\partial z^2}, & 0 < x < x_1,\ 0 < y,z < 1,\ t > 0, & (14a)\\
-\frac{\partial v}{\partial x} = w, & 0 < x < x_1,\ 0 < y,z < 1,\ t > 0, & (14b)\\
v(0,y,z,t) = J_\delta F_m(y,z,t), & 0 < y,z \le 1,\ t > 0, & (14c)\\
w(0,y,z,t) = J_\delta Q_m(y,z,t), & 0 < y,z \le 1,\ t > 0, & (14d)\\
v(x,0,z,t) = \alpha_0(x,z,t),\ \text{unknown}, & 0 < x < x_1, 0 < z < 1, t > 0, & (14e)\\
v(x,1,z,t) = \alpha_1(x,z,t),\ \text{unknown}, & 0 < x < x_1, 0 < z < 1, t > 0, & (14f)\\
v(x,y,0,t) = \beta_0(x,y,t),\ \text{unknown}, & 0 < x < x_1, 0 < y < 1, t > 0, & (14g)\\
v(x,y,1,t) = \beta_1(x,y,t),\ \text{unknown}, & 0 < x < x_1, 0 < y < 1, t > 0, & (14h)\\
v(x,y,z,0) = 0, & 0 \le x \le x_1,\ 0 \le y,z \le 1, & (14i)\\
v(x_1,y,z,t) = J_\delta f_m(y,z,t),\ \text{unknown}, & 0 < y,z \le 1,\ t > 0, & (14j)\\
w(x_1,y,z,t) = J_\delta q_m(y,z,t),\ \text{unknown}, & 0 < y,z \le 1,\ t > 0, & (14k)
\end{array}
$$

5.1. Extension of initial data functions . The most striking difference between system (14) and the ones discussed in section 2 is that, now, the data functions F_m and Q_m are only available – at $x = x_0 = 0$ – at the grid points of the cube $D_T = [0,1] \times [0,1] \times [0,T]$ of the (y,z,t) space, $T > 0$, arbitrary. We recall that we are interested in the evaluation of $\{V^n_{N_{x_1},j,l}\}$ and $\{W^n_{N_{x_1},j,l}\}$ at the grid points of the unit cube $C_{x_1} = [0,1] \times [0,1] \times [0,1]$ – in the (y,z,t) space at $x = x_1$ – after N_{x_1} iterations. It is then necessary, in order to mollify the data functions F_m and Q_m in the domain D_T, we have to extend F_m and Q_m to the grid points of the domain $D_{T,\delta} = [-3\delta, 1+3\delta] \times [-3\delta, 1+3\delta] \times [-3\delta, T+3\delta]$ in the (y,z,t) space, at $x = x_0 = 0$, and with $T = 1-n+n/h$. We notice that the extension in the t-axis is obtained by setting F_m and Q_m equal to zero for $t < 0$ and simply reading the data up to time $T + 3\delta$.

It is sufficient to explain the extension of the data functions F_m and Q_m to the y interval $(1, 1+3\delta]$; we proceed similarly for the y-intervals $[-3\delta, 0)$ and the corresponding z-intervals. We must consider two cases: 1) For each z_l and $t_n, 0 \le z_l \le 1$, $0 \le t_n \le T$, we divide the y-interval $[1, 1+3\delta]$ in three equal parts, introducing the points $y_\delta = y(1+\delta)$, $y_{2\delta} = y(1+2\delta)$ and $y_{3\delta} = y(1+3\delta)$. The step size $\delta = n's$ $(n' \ge 1)$ generates, in general a much coarser grid than the original one. The function extension is now obtained by a one-dimensional polynomial extrapolation procedure in the coarse grid, as follows:

Each of the second degree polynomials determined by the triplet of pairs

$$\begin{aligned}
&\{\{(1,z_l,t_n),\ F_m(1,z_l,t_n)\},\ \{(1-\delta,z_l,t_n),\ F_m(1-\delta,z_l,t_n)\},\\
&\{(1-2\delta,z_l,t_n),\ F_m(1-2\delta,z_l,t_n)\}\};\\
&\{\{(1,z_l,t_n),\ F_m(1,z_l,t_n)\},\ \{(1-2\delta,z_l,t_n),\ F_m(1-2\delta,z_l,t_n)\},\\
&\{(1-4\delta,z_l,t_n),\ F_m(1-4\delta,z_l,t_n)\}\};\\
&\{\{(1,z_l,t_n),\ F_m(1,z_l,t_n)\},\ \{(1-3\delta,z_l,t_n),\ F_m(1-3\delta,z_l,t_n)\},\\
&\{(1-6\delta,z_l,t_n),\ F_m(1-6\delta,z_l,t_n)\}\};
\end{aligned}$$

is used to uniquely evaluate $\overline{F_m}(1+\delta, z_l, t_n)$, $\overline{F_m}(1+2\delta, z_l, t_n)$, $\overline{F_m}(1+3\delta, z_l, t_n)$, respectively. 2) For each z_l and t_n, $1 < z_l \le 1+3\delta$, $0 \le t_n \le T$, we proceed as before but only for the grid points in the "diagonal" $y_l = t_l$ for simplicity we consider $s = p$). Linear interpolation among these new values provides the desired extension of F_m to the finer original grid, in the y-interval $[1, 1+3\delta]$.

The extrapolation error, at each grid point, is the sum of two terms. The first one is due to the extrapolation error underlying the exact function F, assumed sufficiently smooth. It is well known that this error is proportional to $(3\delta)^3$ and $(3\sqrt{2}\delta)^3$ in cases 1 and 2, respectively. The second term involves the extrapolation error of the measured data function F_m. If the pointwise error at the grid points satisfies $| F_m(y_j, z_l, t_n) - F(y_j, z_l, t_n) | \le \eta$, $(y_j, z_l, t_n) \in D_T$, the Lagrange interpolating polynomial immediately gives the uniform estimate $| \overline{F_m}(y_j, z_l, t_n) - \overline{F}(y_j, z_l, t_n) | \le 7\eta$, $(y_j, z_l, t_n) \in D_{T,\delta} - D_T$.

In what follows we consider the noisy data functions F_m and Q_m to be discrete functions measured at equally spaced points in the domain $D_{T,\delta} = [-3\delta, 1+3\delta] \times [-3\delta, 1+3\delta] \times [-3\delta, T+3\delta]$ with sample step sizes s, p and k for the y, z and t-intervals, respectively.

5.2. Numerical scheme. Once the radius of mollification has been decided, after extending the data functions to the domain $D_{T,\delta}$, as indicated in the last section, the mollified data functions $J_\delta F_m$ and $J_\delta Q_m$ are computed at the grid points of the domain $D_T = [0,1] \times [0,1] \times [0,T]$ in the (y,z,t) space at $x = x_0 = 0$. We use a rectangular quadrature formula to approximate the exact convolution with step sizes s, p and k in the y, z and t variables respectively, but we evaluate $J_\delta F_m$ and $J_\delta Q_m$ at a coarser grid with step sizes $\Delta y = n_y s$, $\Delta z = n_z p$ and $\Delta t = n_t k$, $\min(n_y, n_z, n_t) \geq 1$.

We approximate system (14) with the consistent finite difference schemes

$$(15)\quad \begin{aligned} W^n_{i+1,j,k} &= W^n_{i,j,k} + \tfrac{h}{s^2}\left(V^n_{i,j+1,l} - 2V^n_{i,j,l} + V^n_{i,j-1,l}\right) \\ &\quad + \tfrac{h}{p^2}\left(V^n_{i,j,l+1} - 2V^n_{i,j,l} + V^n_{i,j,l-1}\right) - \tfrac{h}{2k}\left(V^{n+1}_{i,j,l} - V^{n-1}_{i,j,l}\right), \qquad (15a) \\ V^n_{i+1,j,l} &= V^n_{i,j,l} - hW^n_{i+1,j,l}, \qquad (15b) \\ i &= 0,1,\cdots,N_{x_1}-1;\ j = 1,\cdots,N_y-1; \\ l &= 1,\cdots,N_z-1;\ \ n = 1,2,\cdots,N_t-1, \end{aligned}$$

$$U^0_{i,j,l} = 0, \quad i = 0,1,2,\cdots,N_{x_1};\ j = 0,1,2,\cdots,N_y;\ l = 0,1,2,\cdots,N_z. \quad (15c)$$

Using these schemes to evaluate $W^n_{i+1,j,l}$ and $V^n_{i+1,j,l}$, the computations are ordered as

$$W^n_{1,j,l} \to V^n_{1,j,l} \to W^n_{2,j,l} \to V^n_{2,j,l} \to W^n_{3,j,l} \to V^n_{3,j,l} \to \cdots,$$

and for each fixed i_0, the time evaluations are performed according to

$$W^1_{i_0,j,l} \to W^2_{i_0,j,l} \to W^3_{i_0,j,l} \to \cdots,$$

and

$$V^1_{i_0,j,l} \to V^2_{i_0,j,l} \to V^3_{i_0,j,l} \to \cdots.$$

As we march forward in the x-direction in space, we must drop the estimation of $V^n_{i,j,l}$ associated with the boundary values of j and l and the largest value of n. We have already anticipated this behavior and have collected "enough" data in the t-direction, but we need to extrapolate to recover the boundary values in the y and z-directions. For example, at each step of the marching scheme, the boundary value for $j = N_y$ is recovered by extrapolation from the corresponding values of the discrete temperature function at $j = N_y - 1$, $N_y - 2$ and $N_y - 3$, introducing a $O(\Delta y^3)$ error in the process. The algorithm is continued for $i = 1, 2, \ldots, N_{x_1}$ until the values $V^n_{N_{x_1},j,l}$ and $W^n_{N_{x_1},j,l}$, $0 \leq j \leq N_y$, $0 \leq l \leq N_z$, $0 \leq n \leq N_t$, are computed at all the grid points of the cube C_{x_1} in the hyperplane $x = x_1$. These values are then taken as the accepted approximations for the temperature and heat flux histories, respectively.

5.3. Numerical example. In order to test the numerical algorithm described in the previous sections, following H. Zheng and D. A. Murio [12], we consider its application to one model problem of interest.

The three-dimensional IHCP is described by the following system of equations.

$$v_t = v_{xx} + v_{yy} + v_{zz}\ ,\ 0 \le x, y, z \le 1,\ t > 0,$$

$$v(0, y, z, t) = F(y, z, t) = 0,\ 0 \le y, z \le 1,\ t > 0,$$

$$-v_x(0, y, z, t) = Q(y, z, t) = -e^{-3t} \sin y \sin z,\ 0 \le y, z \le 1,\ t > 0,$$

$$v(x, y, z, 0) = l(x, y, z) = \sin x \sin y \sin z,\quad 0 \le x, y, z \le 1,$$

$$v\left(x, 0, z, t\right) = \alpha_0\left(x, z, t\right),\ \text{unknown},\ 0 < x < x_1,\ 0 < z < 1,\ t > 0,$$

$$v\left(x, 1, z, t\right) = \alpha_1\left(x, z, t\right),\ \text{unknown},\ 0 < x < x_1,\ 0 < z < 1,\ t > 0,$$

$$v(x, y, 0, t) = \beta_0(x, y, t),\ \text{unknown},\ 0 < x, y < 1,\ t > 0,$$

$$v(x, y, 1, t) = \beta_1(x, y, t),\ \text{unknown},\ 0 < x < x_1,\ 0 < y < 1,\ t > 0,$$

$$v(x_1, y, z, t) = f(y, z, t),\ \text{unknown},\ 0 \le y, z \le 1,\ t > 0,$$

$$-v_x(x_1, y, z, t) = q(y, z, t),\ \text{unknown},\ 0 \le y, z \le 1,\ t > 0.$$

We attempt to reconstruct the mollified surface temperature function $J_\delta f(y, z, t)$ and the mollified surface heat flux function $J_\delta q(y, z, t)$ at $x = x_1 = 0.4$ for $0 \le y, z \le 1$ and $0 \le t \le 1$.

The exact boundary temperature and boundary heat flux solutions for this problem are

$$v(0.4, y, z, t) = f(y, z, t) = e^{-3t} \sin(0.4) \sin y \sin z, \ 0 \le y, z \le 1, \ 0 \le t \le 1,$$

and

$$-v_x(0.4, y, z, t) = q(y, z, t) = -e^{-3t} \cos(0.4) \sin y \sin z, \ 0 \le y, z \le 1, \ 0 \le t \le 1.$$

We also reconstruct the mollified unknown boundary functions

$$J_\delta \alpha_0(x, z, t), \ J_\delta \alpha_1(x, z, t), \ J_\delta \beta_0(x, y, t) \text{ and } J_\delta \beta_1(x, y, t).$$

For example, the exact solution for the temperature function at $y = 1$ is given by

$$\alpha_1(x, z, t) = v(x, 1, z, t) = e^{-3t} \sin x \sin 1 \sin z, \ 0 < x < x_1, \ 0 \le z, t \le 1$$

and Figure 7 displays the computed temperature history at the surface of coordinates $y = 1$, $x = 0.4$, $z = 0.4$ corresponding to an average noise level $\epsilon = 0.005$.

We set $h = \Delta x = 0.1$, and consider the data functions F_m and Q_m to be defined at the grid points of the domain

$$D_T = [0, 1] \times [0, 1] \times [0, T], \ T = 1.5,$$

of the (y, z, t) space, $(x = x_0 = 0)$, with grid parameters $s = p = k = 0.01$ and $\delta = 0.05$ on the fine grid and $\Delta y = 10s$, $\Delta z = 10p$ and $\Delta t = 10k$ on the coarse grid. Before convolving the data, as indicated in section 5.1, we apply the extrapolation method to extend the functions F_m and Q_m to the grid points of the domain

$$D_{T,\delta} = [-3\delta, 1 + 3\delta] \times [-3\delta, 1 + 3\delta] \times [-3\delta, T], \ T = 1.5.$$

During the computations, when marching forward in the x-direction, at each step we implement the one-dimensional extrapolation procedure described in section 5.2 to compute all the boundary values in the y and z-directions.

In order to verify the numerical stability of the algorithm, we test it against several different average perturbations for $\epsilon = 0.000$, 0.001, 0.002, 0.003, 0.004 and 0.005.

Table 14 summarizes the behavior of the l_2 norms of the errors for the reconstructed temperature and surface heat flux histories in the unit time interval and for one typical surface point of coordinates $x = 0.4$, $y = z = 0.6$, as functions of the amount of noise in the data.

	Temperature		Heat flux	
ϵ	δ_g	Error norm	δ_{g_1}	Error norm
.000	.05	.00028388	.05	.04424880
.001	.05	.00022937	.05	.04424573
.002	.05	.00018646	.05	.04424345
.003	.05	.00016452	.05	.04424122
.004	.05	.00017178	.05	.04423904
.005	.05	.00020515	.05	.04423690

Table 14 : *Error Norms as functions of noise in the data for $J_\delta f$ and $J_\delta q$ at $x = 0.4$, $y = z = 0.6$, $0 \leq t \leq 1$.*

Table 15 summarizes the behavior of the l_2 norms of the errors for the reconstructed boundary temperature in the unit time interval and for one typical surface point of coordinates $x = z = 0.4$, $y = 1$, as functions of the amount of noise in the data.

	Temperature	
ϵ	δ_g	Error norm
.000	.05	.01868815
.001	.05	.01869499
.002	.05	.01870215
.003	.05	.01870963
.004	.05	.01871742
.005	.05	.01872552

Table 15 : *Error Norms as functions of noise in the data for $J_\delta \alpha_1$ at $x = 0.4$, $y = 1$, $z = 0.4$, $0 \leq t \leq 1$.*

Figures 8 and 9 illustrate the temperature and heat flux histories, respectively, at the interior point of coordinates $x = 0.4$, $y = z = 0.4$, for an average perturbation $\epsilon = 0.005$ of the discrete data functions F_m and Q_m.

Finally, figure 10 shows the qualitative behavior of the difference between the computed and the exact temperatures at the grid points of the surface $x = 0.4$ for the final time $t = 1$.

REFERENCES

[1] B. R. Bass and L. J. Ott, "A finite element formulation of the two-dimensional inverse heat conduction problem", Adv. Comput. Technol., Vol. 2, 1980, pp. 238-248.

[2] J. Baumeister and H. J. Reinhardt, "On the approximate solution of a two-dimensional inverse heat conduction problem", Inverse and Ill-posed Problems, H. G. Engl and C. W. Groetsch, (Eds.), Academic Press, Orlando, Florida, 1987, pp. 325-344.

[3] J. V. Beck, "Nonlinear estimation applied to the nonlinear inverse heat conduction problem", Int. J. Heat Mass Transfer, Vol. 13, 1970, pp. 703-716.

[4] L. Guo and D. A. Murio, "A mollified space-marching finite difference algorithm for the two-dimensional inverse heat conduction problem with slab symmetry", Inverse Problems, Vol. 7, 1991, pp. 247-259.

[5] M. Imber, "Temperature extrapolation mechanism for two-dimensional heat flow", AIAA J., Vol. 12, 1974, pp. 1089-1093.

[6] Y. Liu and D. A. Murio, "Numerical experiments in 2-D IHCP on bounded domains. Part I: The "interior" cube problem", to appear.

[7] Y. Liu and D. A. Murio, "Numerical experiments in 2-D IHCP on bounded domains. Part II: The "lost boundary" problem", submitted for publication.

[8] D. A. Murio, *The Mollification Method and the Numerical Solution of Ill-Posed Problems,* John Wiley & Sons, Interscience, New York, New York, 1993.

[9] D. A. Murio and H. Zheng, "A stable algorithm for 3D-IHCP", to appear.

[10] T. Yoshimura and K. Ituka, "Inverse heat conduction problem by finite element formulation", Int. J. Systems Sci., Vol. 16, 1985, pp. 1365-1376.

[11] N. Zabaras and J. C. Liu, "An analysis of two-dimensional linear inverse heat transfer problems using an integral method", Numer. Heat Transfer, Vol. 13, 1988, pp. 527-533.

[12] H. Zheng and D. A. Murio, "3D-IHCP on a finite cube", submitted for publication.

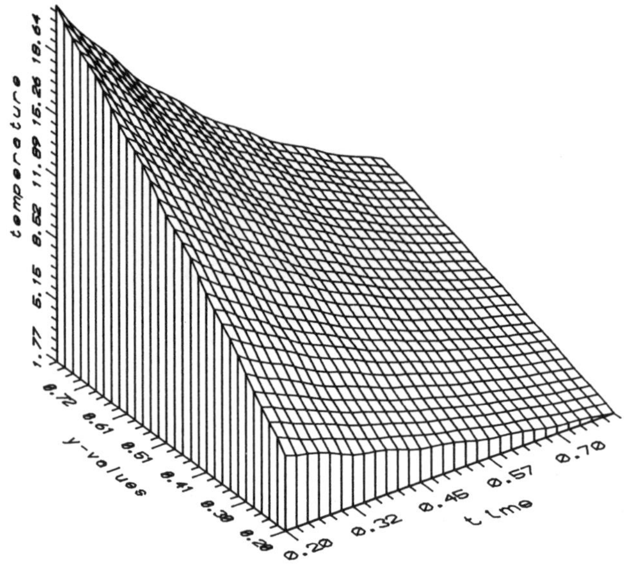

Figure 1: Reconstructed temperature function at $x = 0.5$. Example 1.
$N_x = 20,\ N_y = N_t = 40,\ \epsilon = 0.005,\ \delta = 0.0667,\ n_y = n_t = 10,\ \lambda = 0.010.$

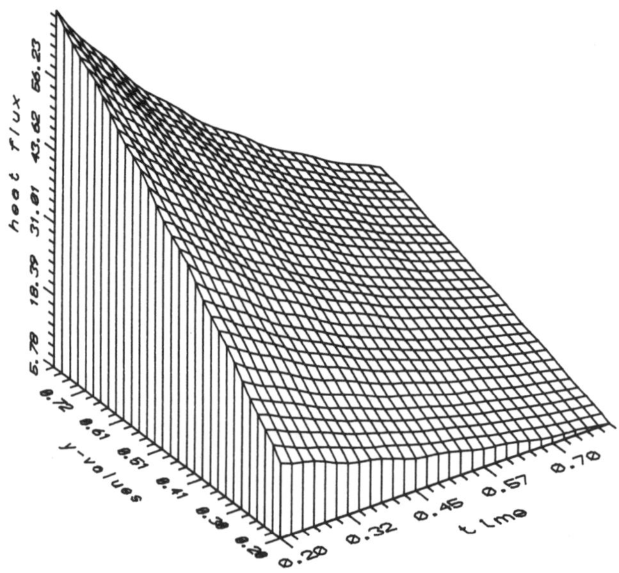

Figure 2: Reconstructed heat flux function at $x = 0.5$. Example 1.
$N_x = 20,\ N_y = N_t = 40,\ \epsilon = 0.005,\ \delta = 0.0667,\ n_y = n_t = 10,\ \lambda = 0.010.$

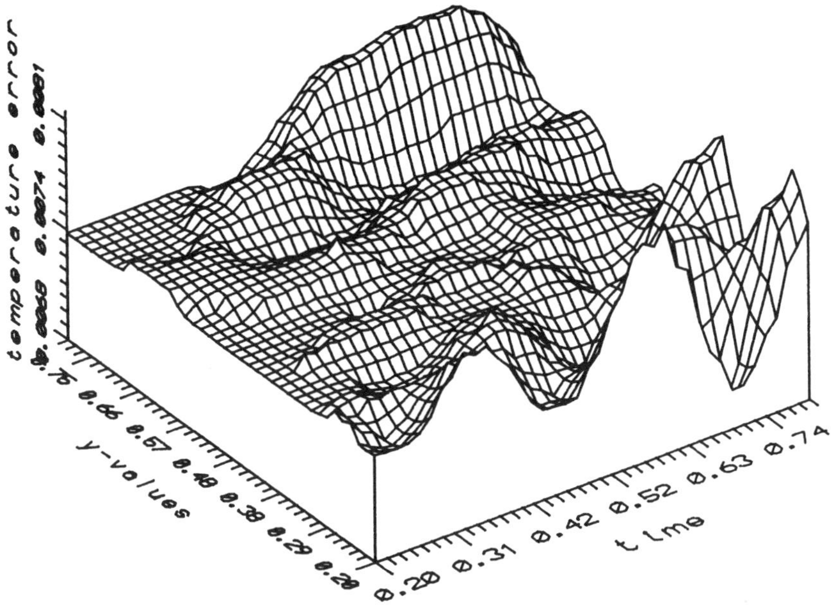

Figure 3: Relative errors for the temperature function at $x = 0.5$. Example 1.
$N_x = 20$, $N_y = N_t = 40$, $\epsilon = 0.005$, $\delta = 0.0667$, $n_y = n_t = 10$, $\lambda = 0.010$.

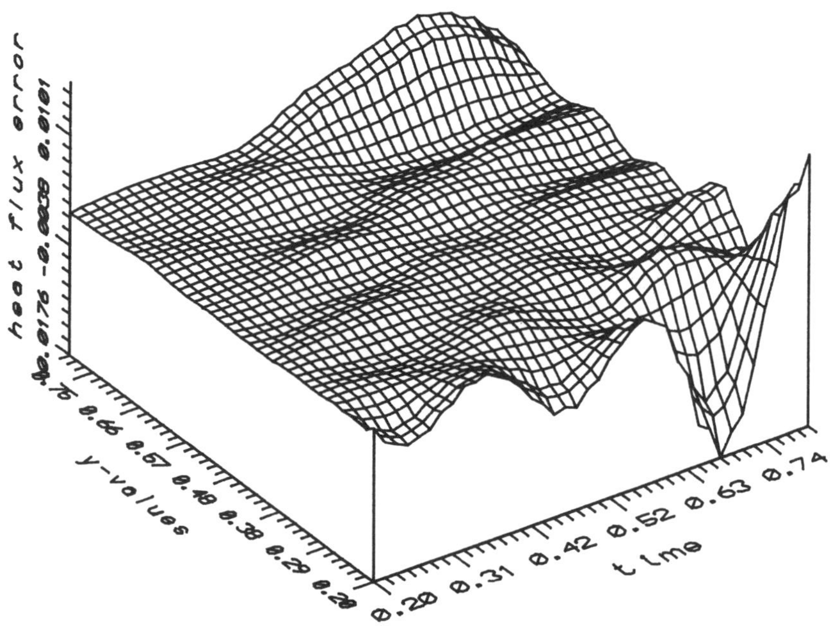

Figure 4: Relative errors for the heat flux function at $x = 0.5$. Example 1.
$N_x = 20$, $N_y = N_t = 40$, $\epsilon = 0.005$, $\delta = 0.0667$, $n_y = n_t = 10$, $\lambda = 0.010$.

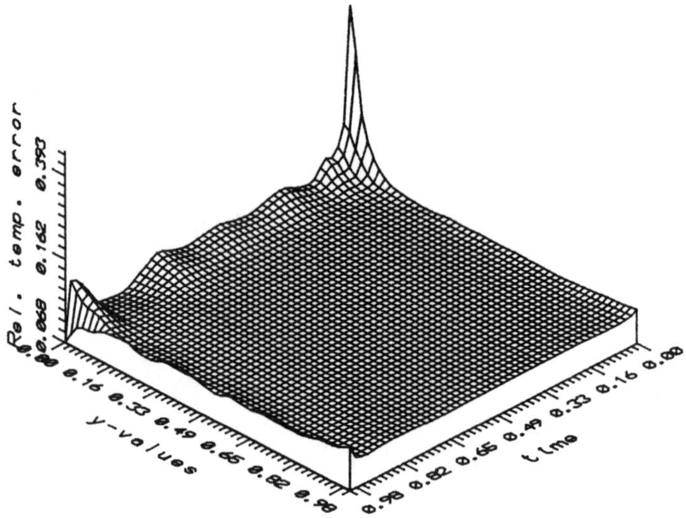

Figure 5: Relative errors for the temperature function at $x = 0.5$. Whole cube, including the boundary. Example 1.
$N_x = 20, N_y = N_t = 40, \epsilon = 0.005, \delta = 0.0667, n_y = n_t = 10, \lambda = 0.010.$

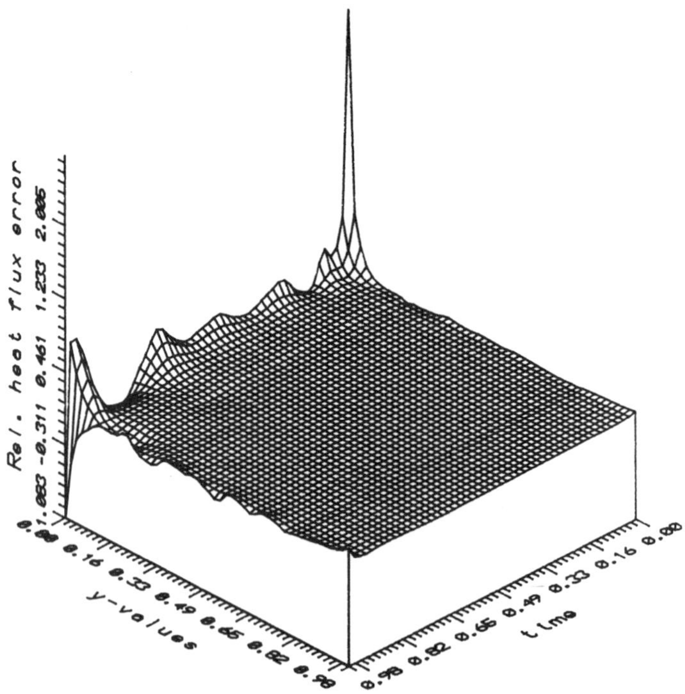

Figure 6: Relative errors for the heat flux function at $x = 0.5$. Whole cube, including the boundary. Example 1.
$N_x = 20, N_y = N_t = 40, \epsilon = 0.005, \delta = 0.0667, n_y = n_t = 10, \lambda = 0.010.$

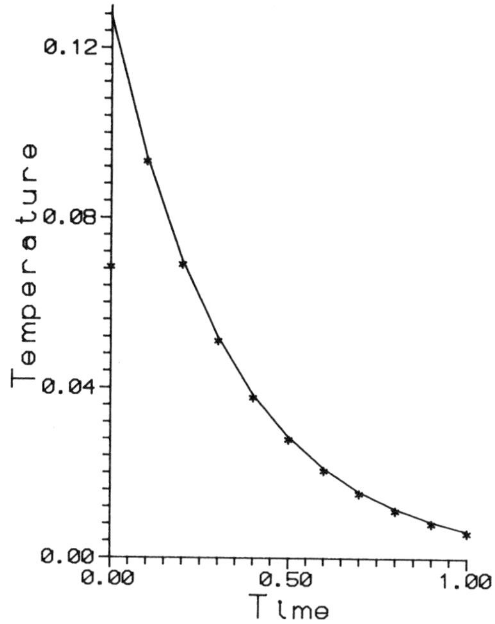

Figure 7: Reconstructed (*) and exact boundary temperature histories at $x = 0.4$, $y = 1$, $z = 0.4$. Example 1.
$\Delta x = 0.1,\ s = p = k = 0.01,\ n_y = n_x = n_t = 10,\ \epsilon = 0.05,\ \delta = 0.05.$

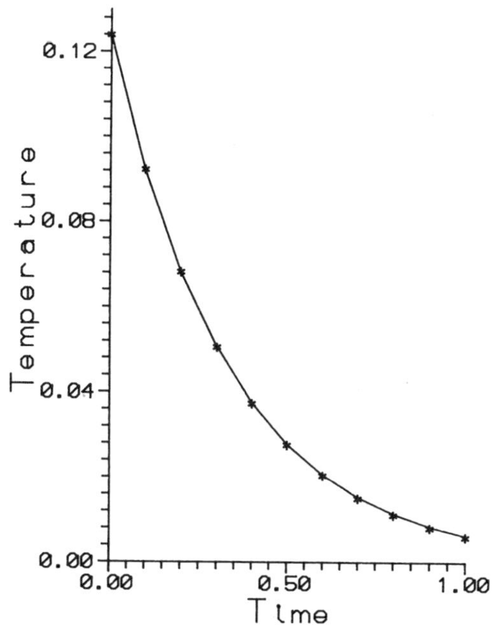

Figure 8: Reconstructed (*) and exact temperature histories at $x = 0.4$, $y = z = 0.4$. Example 1.
$\Delta x = 0.1,\ s = p = k = 0.01,\ n_y = n_x = n_t = 10,\ \epsilon = 0.05,\ \delta = 0.05.$

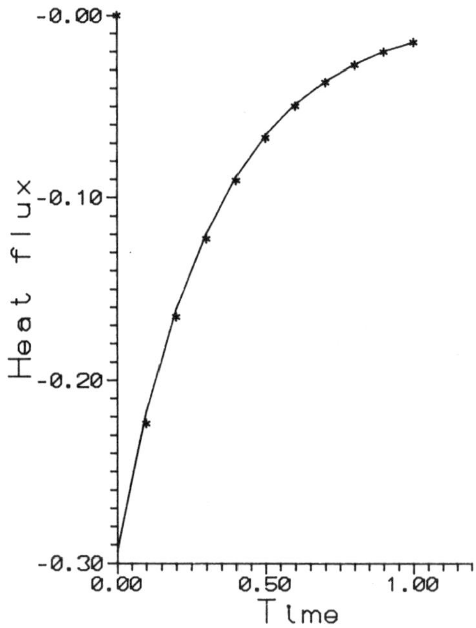

Figure 9: Reconstructed (*) and exact heat flux histories at $x = 0.4$, $y = z = 0.4$. Example 1.
$\Delta x = 0.1,\ s = p = k = 0.01,\ n_y = n_x = n_t = 10,\ \epsilon = 0.05,\ \delta = 0.05.$

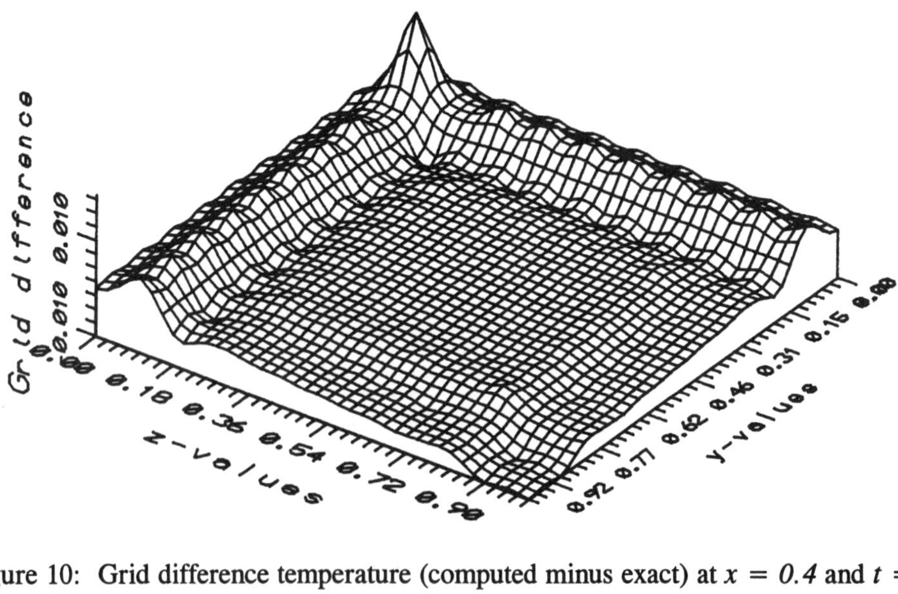

Figure 10: Grid difference temperature (computed minus exact) at $x = 0.4$ and $t = 1$. Example 1.
$\Delta x = 0.1,\ s = p = k = 0.01,\ n_y = n_x = n_t = 10,\ \epsilon = 0.05,\ \delta = 0.05.$

Recent and New Results on Determination of Unknown Coefficients in Parabolic Partial Differential Equations with Over-specified Conditions

Hong-Ming Yin[1]

Abstract In this presentation, we shall review various results on determination of unknown coefficient in parabolic equations. Special attention will focus on the case where a unknown function is time-dependent only. Some recent and new results are presented.

AMS(MOS) Subject Classification: 35R30, 35R25, 35K55.

1. Introduction

In real world, many physical problems can be modeled as inverse problems in partial differential equations. Studying these problems has a great deal of importance in engineering and other applied sciences. In the present paper, we shall discuss a class of inverse problems associated with parabolic equations. More precisely, we would like to find functions $u(x,t)$ and $p = p(t), p(x)$ or $p(u)$ such that

$$(1.1) \qquad u_t = F(x,t,u,u_x,u_{xx},p) \qquad \textit{in } Q_T = R^n \times (0,T] \text{ or } \Omega \times (0,T]$$

along with initial or initial-boundary condition, where $F(x,t,u,q,r,p)$ defined on $Q_T \times R \times R^n \times R^{n^2} \times R$ satisfies the ellipticity condition:

$$\lambda|\xi|^2 \leq F_{r_{ij}}\xi_i\xi_j \leq \Lambda|\xi|^2, \qquad \xi \in R^n, \ (\lambda \leq \Lambda).$$

When p is given, there is a rather satisfactory theory for this parabolic problem regarding to well-posedness and other related properties. When p is unknown, in order to find a solution as well as p one needs some additional condition(s), say,

$$(1.2) \qquad Au = 0,$$

[1] Department of Mathematics, University of Notre Dame, Notre Dame, IN 46556, USA.

where A is an operator acting on the function space $C^{2+\alpha,1+\frac{\alpha}{2}}(\bar{Q}_T)$ such as

$$Au = \int_{\Omega} f(x,t,u,u_x)dx, \int_{\partial\Omega} \nabla_{\nu} u ds, u(x_0,t), u(x,T), x_0 \in \Omega, etc.$$

Here and throughout this paper, ν denotes the outward normal on $\partial\Omega$. The actual form of this additional condition depends on the concrete physical model. It is well-known that inverse problems are, in general, not well-posed. An important question is what conditions can ensure the problem to be well-posed? How much additional information is needed in order to recover the unknown parameter uniquely? These are very important questions both theoretically and practically. In this paper, we shall discuss these questions and shall pay special attention to the case where $p = p(t)$. The following general observation which has been known for a long time is useful: when a direct problem is well-posed for a given $p = p(t)$ or $p = p(x)$, then the over-specified condition(s) Au should have the same variable as p, namely, when $p = p(t)$ (or $p = p(x)$) is unknown, the over-data shall be given in the direction of t (or x). Counterexamples can be easily constructed (cf.[34]) if this general rule fails. On the other hand, even the over-specified condition(s) satisfies the general rule, the problem may not be well posed. For example, consider the following simple inverse problem of finding $u(x,t)$ and $a(t) > 0$ such that

$$\begin{aligned} &u_t - a(t)u_{xx} = 1, \qquad (x,t) \in R \times (0,\infty); \\ &u(x,0) = 0, \qquad x \in R \end{aligned}$$

and an additional condition:

$$u(0,t) = t, \qquad t \geq 0.$$

It is clear that $u(x,t) = t$ and an arbitrary function $a(t) > 0$ will be a solution of the inverse problem. The task we face is to find reasonable conditions (physically acceptable) which will guarantee the well-posedness of the problem.

In Section 2, we shall give a review about determining $p = p(t)$ in parabolic equations. In Section 3, we will outline some general results on parabolic equations with nonlocal functionals and illustrate its application to some inverse problems. In Section 4, some remarks are presented on the case of determination of $p = p(x)$ and $p = p(u)$.

2. A Review on the Case $p = p(t)$

In the mathematical literature, the study of the problem of determining a function $p(t)$ in a parabolic equation began in 1960's. In the paper [29], B.F. Jones studied the following problem of finding $u(x,t)$ and $a(t)$ such that

$$u_t = a(t)u_{xx}, \qquad x > 0, t > 0,$$
$$u(x,0) = 0,\ x \geq 0, \qquad u(0,t) = f(t),\ t \geq 0$$

and an extra condition

$$-a(t)u_x(0,t) = g(t),\ t \geq 0.$$

By introducing a suitable new variable, the author transformed the problem into a nonlinear integral equation of $a(t)$. The existence of a unique solution was obtained by means of Schauder's fixed point theorem. The numerical analysis was treated in [22]. Later, J. R. Cannon [6] as well as Jones [30], considered the same problem in different regions and proved the well-posedness of the problem. The reader can find a clear and neat treatment of the problem in [5]. Since the argument in [6] and [29]-[30] heavily depends on the representation of the solution for the heat equation, there was not much progress on this topic afterwards. In 1967, B.M. Budak and A.D. Iskenderov [1] considered the following nonlinear problem: Find $(a(t), c(t), u(x,t))$ such that

$$u_t = a(t)u_{xx} - c(t)u + H(x,t,a(t),c(t)),\ x \in (0,1),\ t \in (0,T],$$
$$u(0,t) = f_1(t),\ u(1,t) = f_2(t), 0 \leq t \leq T,$$
$$u(x,0) = \psi(x),\ x \in [0,1],$$

and additional conditions:

$$-a(t)u_x(0,t) = g(t),\ -a(t)u_x(1,t) = c(t)\phi(t), \qquad t \in [0,T].$$

Under various conditions on the known data and H, the existence of a unique solution (where $a(t), c(t) \in C[0,T]$ and $u(x,t) \in C^{2,1}(Q_T)$) was established via certain successive iteration procedure and compactness argument. Moreover, the continuous dependence upon data was obtained. A slight generalization was given in [2]. In 1975,

A.D. Iskenderov [28] considered the following problem of finding $(u(x,y,t), a(x,t))$ such that

$$u_t = a(x,t)\Delta u + h(x,y,t), \qquad (x,y,t) \in Q_T = \Omega \times (0,T],$$
$$u(x,y,0) = \psi(x,y), \qquad (x,y) \in \Omega,$$
$$u(x,y,t) = f(x,y,t), \qquad (x,y,t) \in S_T = \partial\Omega \times [0,T]$$

and an additional restriction:

$$a(x,t)u_\nu = g(x,y,t), \qquad (x,y,t) \in S_T,$$

where ν is the outward normal on $\partial\Omega$. Under certain conditions, the existence of a unique stable solution locally in time was proved. A similar result to the above problem was obtained in [3] by N. Ya. Beznoshchenko and sufficient conditions for the global existence were given in [4]. In 1978, Uzlov [45] studied the following problem of finding $(u(x,t), f(t))$ which satisfy:

$$u_t = a(x,t)u_{xx} + b(x,t)u_x + c(x,t)u + f(t)g_1(x,t) + g_2(x,t),$$

along with initial or initial-boundary conditions on the parabolic boundary and an over-specified constraint: $u(d,t) = h(t)$, where d is an interior point of the interval. By using the solution representation, a unique solution was obtained. In 1980, Rundell [38] studied the problem of determining u and $p(t)$ such that $u_t = Au + p(t)g(x)$ along with initial-boundary conditions: $u(x,0) = u_0(x)$ on Ω; $u(x,t) = 0$ on S_T and the additional condition $u_\nu(x_0,t) = h(t)$ (where ν is the outward normal at $x = x_0 \in \partial\Omega$). An explicit solution was obtained. In 1981, Lorenzi [33] studied the following inverse problem: find $(p(t), u(x,t))$ such that

$$u_t = p(t)c(u)u_{xx}, (x,t) \in (0,\infty) \times (0,T]$$

associated with initial-boundary conditions: $u(x,0) = 0, u(0,t) = h_0(t)$ and an additional condition: $u_x(0,t) = h_1(t)$. A stability result was established. In 1982, Cannon and Zachman [17] considered the following equation:

$$u_t = u_{xx} + f(u, u_x, p(t)), \quad \text{where } f(u,u_x,p(t)) = p(t), p(t)u \text{ or } p(t)u_x.$$

Again based on the solution representation of the heat equation, they proved the existence of a unique stable solution. In 1984-1986, Prilepko and Orlovskii considered the following problem of determining u and p such that $u_t = Au + f(t, u, p(t))$, where A is a linear elliptic operator. Under some sufficient conditions, some existence results were established. In [35] part I, a local existence theorem was proved for $f = p(t)u$. In part II of [35], they dealt with the following equation:

$$u_t = (a_{ij}(x)u_{x_j})_{x_i} + c(x)u + f(x, t, u, p(t))$$

associated with initial-boundary conditions, where

$$f(x, t, u, p(t)) = p(t)g(x, t) \text{ or } p(t)u(x, t) + g(x, t).$$

The additional condition is given by

$$\int_\Omega u(x, t)\psi(x)dx = \phi(t).$$

Again the existence of a solution was established. In part III of [35], when the additional condition is given by $u(x_0, t)$ and $f(x, u, p) = \psi(x, t)p(t) + F(t)$, the existence of a unique solution was obtained as an application of an abstract result. For this same problem in one space dimension with $f = p(t)\psi(x, t) + g(x, t)$, a numerical procedure was presented by Prilepko and Solov'ev [36] in 1987 and the convergence of the numerical approximation was proved. During 1988 to 1991, Cannon and Lin [9]-[10] considered the equation

$$u_t = (a_{ij}(x, t)u_{x_i} + b_j(x, t, u))_{x_j} + f(x, t, u, p)$$

along with initial and conormal boundary conditions and an additional constraint:

$$\int_\Omega \psi(x, t)u(x, t)dx = E(t).$$

A weak solution first and then a classical solution were obtained for $f = G(x, t, u) + h(x, t)p(t)$ and $f = G(x, t, u) + u(x, t)p(t)$ and the solution is local in the latter case. In [11] and [32], the authors considered the following equations, respectively,

$$u_t = u_{xx} + p(t)u + f(x, t, u, u_x, p(t)),$$
$$u_t = a(x, t, u, u_x)_x + p(t)u + f(x, t, u, p(t)).$$

The additional condition is as follows:

$$\int_0^1 \psi(x, t)udx = E(t), \text{or} \int_0^{s(t)} \psi(x, t)udx = E(t),$$

where $\psi(x,t), E(t)$ and $s(t)$ are known functions. Under some growth conditions on f, a classical solution was obtained. In 1991, Cannon and Rundell [12] considered the following problem: Find $a(t)$ and $u(x,t)$ such that

$$u_t = a(t)\Delta u, \qquad (x,t) \in Q_T = \Omega \times (0,T]$$

along with initial and Dirichlet boundary conditions and an over-specified condition: $a(t)u_\nu(x_0,t) = h(t)$, where $x_0 \in \partial\Omega$. The existence of a solution as well as the continuous dependence was established. Since 1987, this author along with his colleagues took a rather different approach to consider these various inverse problems. By studying parabolic equations with certain nonlocal functionals, they were able to use a unified and systematic method to handle a large class of parabolic inverse problems. Many previous results were recovered via this unified method. A number of new results were obtained (cf. [13]-[16], [18]-[19], [26], [43]-[44], etc.). The argument characterizes the well-posedness of a large class of parabolic inverse problems. The basic idea will be described in next Section. The reader may find further details in [13]-[16], [43]-[44], etc. More recently, motivated by nuclear reaction processes, the authors of [26] considered the following inverse problem of identifying $p(t)$ and $u(x,t)$ such that:

$$(2.1) \qquad u_t = \Delta u + f(u,p(t)), (x,t) \in Q_T = \Omega \times (0,T],$$

$$(2.2) \qquad u_\nu(x.t) = 0, \qquad (x,t) \in S_T,$$

$$(2.3) \qquad u(x,0) = u_0(x), \qquad x \in \Omega$$

and an extra-datum:

$$\int_\Omega u(x,t)dx = h(t),$$

where $f(u,p(t)) = |u|^q - p(t)$, or $p(t)|u|^q$, and $q \geq 1$. There is an essential difference between the current problem and the previous ones. The solution of the direct problem that is, the solution of (2.1)-(2.3) when $p(t)$ is given, may not have a global solution because of the strong reaction term u^q (recall $q \geq 1$) in the system and the integral $\int_\Omega u dx$ blows up in finite time if the initial value is suitably large. On the other hand, for the inverse problem the finite energy condition stabilizes the reaction. It is not clear whether the problem has a solution for all time. This problem was solved recently in [26]. The problem has many interesting phenomena which are quite different from the classical theory. The interested reader is referred to [26] for further details.

There are other works related to the above type of parabolic inverse problems; examples are Isakov [27], Suzuki [42], etc.

3. Parabolic Equations with nonlocal functionals

Before discussing an inverse problem, we always assume that the corresponding direct problem has a unique classical solution (at least locally in time) when $p = p(t)$ is given in a suitable function space. There are many excellent books (cf. [23], [31] for examples) in which one can find conditions such that this direct problem has a unique stable solution. As we mentioned in Section 1, we shall give an additional condition in the t-direction in order to possibly recover $p = p(t)$. This additional condition depends on various concrete physical problems. We shall first state the general idea and then give a number of examples.

To find a solution $(u(x,t), p(t))$, we first use the additional condition along with the equation to eliminate the unknown function $p(t)$. When $p(t)$ occurs in (1.2), it is often an algebraic equation for $p(t)$ and then is easy to solve for $p(t)$ in terms of solution and its derivatives. On the other hand, if the condition (1.2) does not contain $p(t)$, then we formally differentiate the equation (1.2) with respect to t to have

$$A_t u + Au_t = 0, \tag{3.1}$$

where we have assumed that the operator A is differentiable w.r.t. t. From the equation (3.1), we use the equation (1.1) to obtain:

$$A_t u + AF(x, t, u, u_x, u_{xx}, p(t)) = 0, \ t \geq 0. \tag{3.2}$$

Therefore, we need the following fundamental assumption
(H): the function $p(t)$ is solvable from the equation (1.2) or (3.2).
As $p(t)$ is the coefficient of the equation (1.1), the above assumption is often necessary in order to have a unique solution to the inverse problem. With the above fundamental assumption (H), we know that there exists a functional S such that

$$p(t) = S(x, t, u, u_x, u_{xx}) \equiv Su. \tag{3.3}$$

The actual form of the functional S varies and it depends on the concrete physical model. Now the original inverse problem is transformed into the following parabolic problem with a nonlocal functional:

$$u_t = F(x, t, u, u_x, u_{xx}, Su) \qquad in\, Q_T = \Omega \times (0, T] \tag{3.4}$$

along with initial condition (if $\Omega = R^n$)

$$u(x,0) = u_0(x), \qquad x \in R^n \tag{3.5}$$

or initial-boundary condition (if Ω is a bounded domain in R^n)

$$u(x,0) = u_0(x), \qquad x \in \Omega, \tag{3.6}$$

$$Bu(x,t) = 0, \qquad (x,t) \in S_T = \partial\Omega \times (0,T], \tag{3.7}$$

where B is a boundary operator.

With the above assumptions, the solvability of the original inverse problem is equivalent to that of the problem (3.4)-(3.5) (Cauchy problem) or (3.4), (3.6)-(3.7) (initial-boundary value problem). Here we give an example.

Example 1: Find $(u(x,t), p(t))$ with $p(t) > 0$ such that

$$u_t = \nabla[p(t)a(u)\nabla u] + b(x,t,u,u_x), \qquad (x,t) \in Q_T, \tag{3.8}$$

$$u(x,t) = 0, \qquad (x,t) \in S_T, \tag{3.9}$$

$$u(x,0) = u_0(x), \qquad x \in \Omega. \tag{3.10}$$

and an additional condition: the heat flux is given at a fixed point, say, x_0 on the boundary $\partial\Omega$, i.e.

$$p(t)a(u)\nabla_\nu u|_{x=x_0} = h(t). \tag{3.11}$$

For this inverse problem, the fundamental assumption (H) is that $a(0)\nabla_\nu u(x_0,t) \neq 0$ for all $t \geq 0$. This is actually a necessary condition in order to recover $p(t)$ uniquely (otherwise, one can take $p(t)$ to be arbitrary for those t with $a(0)\nabla_\nu u(x_0,t) = 0$). It follows that

$$p(t) = \frac{h(t)}{a(0)\nabla_\nu u|_{x=x_0}} \equiv S_1 u$$

and the problem (3.8)-(3.11) is now equivalent to

$$u_t = \nabla\left[\frac{h(t)}{a(0)\nabla_\nu u|_{x=x_0}} a(u)\nabla u\right] + b(x,t,u,u_x) \tag{3.12}$$

along with (3.9)-(3.10).

If we replace the condition (3.11) by prescribing the weighted energy:

$$\int_\Omega \psi(x,t)u(x,t)dx = h(t),$$

then we have, after differentiating with respect to t and using the equation, that

$$\int_\Omega \psi_t u dx + \int_\Omega \psi\{\nabla[p(t)a(u)\nabla u] + b\}dx = h'(t).$$

It follows that

$$p(t) = \frac{h'(t) - \int_\Omega[\psi_t u + \psi b]dx}{\int_{\partial\Omega}\psi\{a(u)\nabla_\nu u\}ds - \int_\Omega a(u)\nabla\psi\nabla u dx} \equiv S_2 u,$$

where we need the assumption $\int_{\partial\Omega}\psi\{a(u)\nabla_\nu u\}ds - \int_\Omega a(u)\nabla\psi\nabla u dx \neq 0$.
If we assume that the temperature at an interior point $x_0 \in \Omega$ can be measured:

$$u(x_0, t) = h(t), \qquad\qquad t \geq 0,$$

then we can differentiate the above equation with respect to t and use the equation to obtain:

$$p(t) = \frac{h'(t) - b(x_0, t, h(t), u_x(x_0, t))}{\nabla[a(u)\nabla u]|_{x=x_0}} \equiv S_3 u,$$

where again we need the fundamental assumption $\nabla[a(u)\nabla u]|_{x=x_0} \neq 0$.

One can give many more examples when $p(t)$ appears as the coefficient of lower order terms in a parabolic equation(cf.[13]). It does not reduce the difficulties by transforming an inverse problem to a parabolic problem with nonlocal functionals. However, it does provide a systematic way to handle various this type of inverse problems. It is worthy pointing out that this type of equations is also encountered in other fields (cf. [13], [24]). We will give an overview of the progress on this topic. As we have seen from the above example, the functional S often maps a function in $C^{1+\alpha,\frac{1+\alpha}{2}}(\bar{Q}_T)$ or $C^{2+\alpha,1+\frac{\alpha}{2}}(\bar{Q}_T)$ into a function in $C^{\frac{1+\alpha}{2}}[0,T]$ or $C^{\frac{\alpha}{2}}[0,T]$, although we can not give an explicit expression of the functional. The degree of difficulty of the problem depends on the order of the highest derivative involved in the expression of the functional S. For example, it is harder to deal with the equation (3.4) with the functional $S_3 u$ than that with the functional $S_1 u$ or $S_2 u$ since the functional $S_1 u$ or $S_2 u$ contains only the first order derivative u_x while $S_3 u$ contains the second order derivative. Here we shall discuss the well-posedness of this type of problem.

(I) Existence.

It is difficult, in general, to prove the existence of a solution for a parabolic problem with nonlocal functionals. This can be seen from the following observation: for convenience, we take the functional $Su = u_x(x_0, t)$ where $x_0 \in \Omega$ is an arbitrary fixed point. With this nonlocal term in a parabolic equation, a priori estimates are very hard to derive since the maximum principle is no longer applicable in most cases. For example, consider the following simple equation:

$$u_t = u_{xx} + u_x(0,t)^2, \qquad (x,t) \in (0,1) \times (0,T], (\text{ here } Su = u_x(0,t)^2)$$

along with initial and Dirichlet boundary conditions: $u(x,0) = u_0(x)$, $x \in [0,1]$ and $u(0,t) = f_1(t)$, $u(1,t) = f_2(t)$. It is not clear whether or not $u(x,t)$ is uniformly bounded (See an answer in the recent paper [16]). As x_0 can be any point in Ω, $u_x(x_0,t)$ is equivalent in some sense to the term $\max_{x\in\Omega} |u_x(x,t)|$. In deriving a priori estimates, one often needs "good" terms with higher order derivatives in order to dominate this "bad" term. When $Su = u_{xx}(x_0,t)$, the problem becomes even more difficult. On the other hand, since S has certain special form, we do get some general results. We shall consider two different cases.

Case 1: The functional S is only acting on x, t, u and $u_{x_i}, i = 1, 2, \cdots, n$.

Theorem 1: Assume that the equation (3.4) along with appropriate initial-boundary conditions has a unique classical solution for a given $p(t) = Su = S(x,t,u,u_x)$, then this problem with a functional also has a classical solution locally in time.

The proof can be done via a fixed point argument. The outline of the proof is as follows: For a given $p(t) \in K = \{p(t) \in C^{\frac{\alpha}{2}}[0,T],\ p(0) = Su_0,\ ||p||_{C^{\frac{\alpha}{2}}[0,T]} \le K_0\}$, where $K_0 > 0$ is to be determined later, by the assumption and the classical theory, the direct problem where Su is replaced by $p(t)$ has a unique solution $u(x,t) \in C^{2+\alpha,1+\frac{\alpha}{2}}(\bar{Q}_T)$ and

$$||u||_{C^{2+\alpha,1+\frac{\alpha}{2}}(\bar{Q}_T)} \le C(K_0).$$

It follows that $u_x \in C^{1+\alpha,\frac{1+\alpha}{2}}(\bar{Q}_T)$ which implies, by the form of Su,

$$p^*(t) \equiv Su \in C^{\frac{1+\alpha}{2}}[0,T] \text{ and } ||Su||_{C^{\frac{1+\alpha}{2}}[0,T]} \le C'(K_0).$$

As the embedding from $C^{\frac{1+\alpha}{2}}[0,T]$ into $C^{\frac{\alpha}{2}}[0,T]$ is compact, the mapping from $p(t) \in K$ into $p^*(t) = Su \in C^{\frac{1+\alpha}{2}}[0,T]$ is precompact. It is easy to verify the mapping is also continuous. Moreover, it can be shown that the mapping is from K into K if we take K_0 to be large and T to be sufficiently small. Schauder's fixed point theorem yields the existence of a local solution.

Case 2: The functional S is acting on x, t, u, u_{x_i} and $u_{x_ix_j}, i, j = 1, 2, \cdots, n$.

When the functional S contains the second order derivative with respect to the space variables, the problem becomes much more difficult. Indeed, one can use the same procedure as above to define a mapping. However, since Su contains the second derivative w.r.t. the space variables, $p^*(t) = Su \in C^{\frac{\alpha}{2}}[0,T]$. This means that the mapping is from the space $C^{\frac{\alpha}{2}}[0,T]$ to $C^{\frac{\alpha}{2}}[0,T]$ and no compactness can be obtained for the mapping in general. To obtain the compactness, one needs to use the special structure of the equation. For example, consider the following problem:

$$\begin{aligned}
&u_t = u_{xx} + f(u, u_{xx}(x_0,t)), && (x,t) \in Q_T = (0,1) \times (0,T], \\
&u(0,t) = u(1,t) = 0, && t \in [0,T], \\
&u(x,0) = u_0(x), && x \in [0,1].
\end{aligned}$$

where $x_0 \in (0,1)$ is a fixed point.

If we introduce $v(x,t) = u_x(x,t)$, then the above problem is equivalent to the following one:

$$\begin{aligned}
&v_t = v_{xx} + f_u\left(\int_0^x v dx, v_x(x_0,t)\right)v, && (x,t) \in Q_T = (0,1) \times (0,T], \\
&v_x(0,t) = v_x(1,t) = -f(0, v_x(x_0,t)), && \in [0,T], \\
&v(x,0) = u_0'(x), && x \in [0,1].
\end{aligned}$$

For this problem, one can use the same argument as that in case 1 to prove the existence of a local solution. We shall not give the details here.

Global existence is extremely difficult to prove in both cases. In one space dimension, there are some general results (cf.[13], [16], [44]). In higher space dimensions, global existence strongly relies on the special structure of the problem (cf. [14] etc.).

(II) Uniqueness and Continuous Dependence upon data.

Theorem 2: If there exists a constant C such that the functional Su satisfies the following condition: for any $u_1, u_2 \in C^{2+\alpha,1+\frac{\alpha}{2}}(\bar{Q}_T)$,

$$||Su_1 - Su_2||_{L^\infty(0,T)} \leq C||u_1 - u_2||_{L^\infty(0,T;W^2_\infty(\Omega))},$$

then the solution of the equation (3.4) along with initial and Dirichlet boundary conditions is unique. Moreover, it continuously depends on the known data.

This result gives a rather complete answer to the question of the uniqueness and continuous dependence upon data for the equation with nonlocal functional. It covers a large class of parabolic inverse problems where $p = p(t)$ is unknown. The proof of the above theorem is quite complicated and it relies on a crucial estimate on Green's function of a linear parabolic operator associated with a Dirichlet boundary condition. The interested reader may find more details in [43]. With some simple modification, the argument in [43] can be carried over. When the boundary condition is replaced by a Neumann type one, the question remains open.

Here we give an example to illustrate the above general approach to a specific parabolic inverse problem. The result is new in the author's opinion (a local existence was obtained in [12]).

Example 1: Find $(u(x,t), p(t))$ with $p(t) > 0$ such that

$$u_t = p(t)\Delta u, \qquad (x,t) \in Q_T = \Omega \times (0,T], \tag{3.13}$$

$$u(x,t) = 0, \qquad (x,t) \in S_T = \partial\Omega \times [0,T], \tag{3.14}$$

$$u(x,t) = u_0(x) \geq 0, \qquad x \in \Omega \tag{3.15}$$

and an additional condition:

$$\int_\Omega u(x,t)dx = h(t).$$

We shall assume that the known data $h(t), u_0(x)$ are suitable smooth, say, $C^{1+\alpha}[0,T]$ and $C^m(\bar{\Omega})$ with a large $m > 0$, and the compatibility conditions hold on $\partial\Omega$ and

$$p(0) = \frac{h'(0)}{\int_{\partial\Omega} \nabla_\nu u_0 ds} > 0.$$

By using the additional condition, one can easily find that

$$p(t) = \frac{h'(t)}{\int_{\partial\Omega} \nabla_\nu u ds},$$

where ν is the outward normal on $\partial\Omega$ and ds is the element of surface area.

Now we consider the following equivalent problem:

$$u_t = \frac{h'(t)}{\int_{\partial\Omega} \nabla_\nu u ds}\Delta u, \qquad (x,t) \in Q_T, \tag{3.16}$$

$$u(x,t) = 0, \qquad (x,t) \in S_T, \tag{3.17}$$

$$u(x,t) = u_0(x) \geq 0, \qquad x \in \Omega. \tag{3.18}$$

Assume that $h'(t) < 0$ on $[0,T]$. As $u_0(x) \geq 0$, Hopf's lemma implies that $\nabla_\nu u < 0$ on S_T. Consequently, the local existence follows from Theorem 1. Moreover, uniqueness and continuous dependence upon data can be obtained as a special case of Theorem 2. To show the global existence, we need to derive some a priori estimates.

Let Δ^m denote the m-th Laplace operator and $w_m(x,t) = \Delta^m u$.

Lemma 1: There exists a constant C_{2m} such that

$$max_{Q_T}|w_n| \leq max_\Omega|\Delta^m u_0|,$$
$$\int_\Omega w_{mx}^2 dx + \int\int_{Q_T}\left[\frac{h'(t)}{\int_{\partial\Omega}\nabla_\nu uds}\right]|\Delta w_m|^2 dxdt \leq C_{2m},\ m = 1,2,\cdots,$$

where C_{2m} depends only on $u_0(x), h(t),\ \Omega,\ T$ and m.

Proof: It is clear that $w_m(x,t)$ solves the following problem:

$$w_{mt} = \frac{h'(t)}{\int_{\partial\Omega}\nabla_\nu uds}\Delta w_m, \qquad (x,t) \in Q_T,$$
$$w_m(x,t) = 0, \qquad (x,t) \in S_T.$$
$$w_m(x,t) = \Delta^m u_0(x), \qquad x \in \Omega.$$

The first estimate directly follows from the maximum principle. Now we derive the second estimate. Multiplying the equation by Δw_m and taking the integration over Q_T, we can derive the desired estimate. By the norm equivalence (cf. [40]) and Sobolev's embedding, when m is large enough, we have

$$||u_x||_{L^\infty(Q_T)} \leq C_1,$$

where C_1 depends only on C_{2m} and Ω.

Lemma 2: There exists a positive constant C_2 such that

$$\nabla_\nu u \leq -C_2 < 0, \qquad (x,t) \in S_T,$$

where C_2 depends only on $h(t)$, u_0 and Ω.

The proof follows from the same argument as that in [44], since $||u_x||_{L^\infty(Q_T)}$ is bounded. Combining the above estimate and the result in Lemma 1, we see that there exist two constants C_3 and C_4 depending only on known data such that

$$0 < C_3 \leq \frac{h'(t)}{\int_{\partial\Omega}\nabla_\nu uds} \leq C_4 < \infty.$$

Using the well-known DeGiorgi-Nash-Moser result (cf. [31]), we have a constant C_5 depending only on known data such that

$$||u||_{C^{\alpha,\frac{\alpha}{2}}(\bar{Q}_T)} \leq C_5.$$

By considering the equation for w_m, we obtain

$$||u||_{C^{1+\alpha,\frac{1+\alpha}{2}}(\bar{Q}_T)} \leq C_6,$$

where C_6 depends only on the known data. Finally, Schauder's theory implies

$$||u||_{C^{2+\alpha,1+\frac{\alpha}{2}}(\bar{Q}_T)} \leq C_7,$$

where C_7 depends only on known data. With the above a priori estimate in hand, we apply the result of [14] to obtain the following:

Theorem 3: The parabolic inverse problem (3.13)-(3.15) has a unique solution globally, provided that $u_0(x)$ and $h(t)$ are smooth with $h'(t) < 0$ on $[0, T]$ and $p(0) > 0$.

4. Some Remarks on the cases $p = p(x)$ and $p = p(u)$

It has a long history to the problem of determination of $p(x)$ along with solution in a partial differential equation (cf. [27], [41], [21], etc.). On the other hand, much attention has paid recently to the study of inverse problems where $p = p(u)$ is unknown. A review on this part will need much longer space and it is beyond the scope of this presentation. In this section we shall give a few remarks on inverse problems where a unknown function depends on space variables or the solution itself.

When $p = p(x)$, as we mentioned in Section 1, in order to find $p(x)$ in a parabolic equation, an extra-condition must be given in the space variables. A natural choice is to give the distribution at a specific time, say, T. Along the idea of Section 3, we can transform this type of inverse problems into the following equation with a functional $Su = u_t(x, T)$

$$u_t = F(x, t, u, u_x, u_{xx}, u_t(x, T))$$

subject to an initial or initial-boundary conditions. One can see the functional $u_t(x, T)$ is much like the second order derivative with respect to the space variables. The existence problem is very difficult to deal with. Nevertheless, when the equation has certain specific structure, some results can be accomplished. The interested reader can find some recent results in [20], [25], [27], etc. and the references therein. In general, uniqueness is open.

We shall mention that when the over-data is not given in space direction, one often need a lot of more information in order to recover $p(x)$ uniquely. One of these is to use the Dirichlet to Neumann mapping. The reader can find the review paper [39] and other related work in [21].

When $p = p(u)$ is unknown in a parabolic equation, the problem has many similarities to ones with the case $p = p(t)$. For example, consider the following problem of finding $(u(x,t), p(u))$ such that

$$u_t = \Delta u + p(u)$$

along with appropriate initial and boundary data. An extra condition is given by

$$u(x_0, t) = h(t),$$

where x_0 is an interior fixed point in the region. By evaluating the equation at x_0, we see

$$p(h(t)) = h'(t) - \Delta u(x_0, t).$$

We assume that $h(t)$ is invertible. Then

$$p(s) = h'(h^{-1}(s)) - \Delta u(x_0, h^{-1}(s)).$$

Consequently, the original inverse problem reduces into a parabolic equation with functional $Su = \Delta u(x_0, g(u(x,t)))$ for some known function $g(s)$. There are quite a number of results on existence, uniqueness and stability about this type of problem. The reader may find recent reviews given by Pilant-Rundell [34] and Cannon-DuChateau-Steube [7]. Many more references can be found in the bibliography of [7], [15], [34], etc.

Remark: The references listed below are mainly for parabolic inverse problems where $p = p(t)$ is unknown. Most of references on $p = p(x)$ or $p = p(u)$ are not listed here.

Acknowledgment: The author would like to thank Professor John R. Cannon and

Professor W. Rundell for many valuable comments. Many thanks also go to the referee for some helpful suggestions.

References

1. B.M. Budak and A.D. Iskenderov, On a class of boundary value problems with unknown coefficients, Soviet Math. Dokl., 8(1967), 786-789.
2. B.M. Budak and A.D. Iskenderov, A class of inverse boundary value problems with unknown coefficients, Soviet Math. Dokl., 8(1967), 1026-1029.
3. N Ya. Beznoshchenko, Determination of the coefficients of the highest derivatives in parabolic equations, Diff. Eqs., 11(1975), 19-26.
4. N. Ya. Beznoshchenko, Sufficient conditions for the existence of a solution to problems of determining the coefficients multiplying the highest derivatives of a parabolic equation, Diff. Eqs. 19(1983), 1908-1915.
5. J. R. Cannon, The one-dimensional heat equation, Addison-Wesley, 1984.
6. J.R. Cannon, Determination of certain parameters in heat conduction problems, J. Math. Anal. Appl., 8(1964), 188-201.
7. J.R. Cannon, P. DuChateau and K. Steube, Unknown ingredient Inverse Problems and Trace-type functional differential equations, in [21], page 185-200.
8. J. R. Cannon, P. DuChateau, K. Steube, Identifying a time-dependent unknown coefficient in a nonlinear heat equation, in *Nonlinear Diffusion Equations and their Equilibrium States, 3*, ed., N. G, Lloyd, W. M. Ni, L. P. Peletier, J. Serrin, Page 153-170, Birkhauser, Boston, 1992.
9. J.R. Cannon and Y. Lin, Determination of a parameter $p(t)$ in some quasilinear parabolic differential equations, Inverse Problems, 4(1988), 35-45.
10. J.R. Cannon and Y. Lin, Determination of a parameter $p(t)$ in Holder class for some semilinear parabolic differential equations, Inverse problems 4(1988), 596-605.
11. J.R. Cannon and Y. Lin, An inverse problem of finding a parameter in a semilinear heat equation, J. Math. Anal. Appl. 145(1990), 470-484.
12. J.R. Cannon and William Rundell, Recovering a time dependent coefficient in a parabolic differential equations, J. Math. Anal. Appl., 160(1991), 572-582.
13. J.R. Cannon and H.M. Yin, A class of nonlinear nonclassical parabolic problems, J. Diff. Eqs., 79(1989), 266-288.
14. J. R. Cannon and H.M. Yin, A class of multidimensional nonclassical parabolic equations, in Proceeding of International Conference on " Theory and Applications of Differential Equations" (Ed. A. R. Aftabizadeh), Vol.I, 122-127, Ohio University Press, 1989.

15. J. R. Cannon and H. M. Yin, On nonlinear parabolic equations with nonlinear trace type functionals, Inverse Problems, 7(1991), 149-161.

16. J. R. Cannon and H. M. Yin, Blowup of solutions for parabolic equations with nonlocal functionals, preprint, 1994.

17. J.R. Cannon and D. Zachman, Parameter determination in parabolic differential equations from over-specified boundary data, Int. J. Engng. Sci., 20(1982), 779-788.

18. J. M. Chadam, A. Perice and H. M. Yin, The blowup property of solutions to a chemical diffusion equation with localized reactions, J. Math. Anal. Appl., 169(1992), 313-328.

19. J. R. Chadam and H. M. Yin, An iteration procedure for a class of integrodifferential equations of parabolic type, J. of Integral Equations and Applications, 2(1989), 31-47.

20. J. R. Chadam and H. M. Yin, The determination of an unknown function in a parabolic equation with an over-specified condition, Mathematical Methods in Applied Sciences, 13(1990), 421-430.

21. D. Colton, R. Ewing and W. Rundell, Inverse problems in partial differential equations, SIAM Press, Philadelphia, 1990.

22. J. Douglas, Jr. and B.F. Jones, Jr., The determination of a coefficient in a parabolic differential equation, Part II, Numerical Approximation, J. Math. Mech. 11(1962), 919-926.

23. A. Friedman, Partial Differential Equation of Parabolic Type, Prentice Hall, Englewood Cliffs, N.J., 1964.

24. A. Friedman, Analyticity of the free boundary for the Stefan problem, Arch. Rat. Mech. Anal., 61(1976), 97-125.

25. Bei Hu and H.M. Yin, Determination of the leading coefficient $a(x)$ in the heat equation $u_t = a(x)\Delta u$, Quarterly of Applied Mathematics, LI(1993), 577-583.

26. Bei Hu and H. M. Yin, On a class of semilinear equations with prescribed energy, to appear in Rendiconti di Palermo.

27. V. Isakov, Inverse Source Problems, Mathematical Surveys and Monographs, 34, AMS, Providence, Rhode Island, 1990.

28. A.D. Iskenderov, Multidimensional inverse problems for linear and quasilinear parabolic equations, Soviet Math. Dokl., 16(1975), 1564-1568.

29. B.F. Jones, Jr., The determination of a coefficient in a parabolic differential equation, Part I, Existence and Uniqueness, J. Math. Mech., 11(1962),907-918.

30. B.F. Jones, Jr., Various methods for finding unknown coefficients in parabolic

differential equations, Comm. Pure and Appl. Math., 16(1963),33-44.

31. O. A. Ladyzenskaja, V. A. Solonnikov and N.N. Ural'ceva, Linear and Quasilinear Equations of Parabolic type, Trans. Math. Monog., vol. 23, AMS, Providence, R.I., 1968.
32. Y. Lin, An inverse problem for a class of quasilinear parabolic equations, SIAM J. Math. Anal., 22(1991), 146-156.
33. A. Lorenzi, Determination of a time-dependent coefficient in a quasilinear parabolic equation, 1981.
34. M.S. Pilant and W. Rundell, Undetermined coefficient problems for quasilinear parabolic equations, in [21], page 165-185.
35. A. I. Prilepko and D.G. Orlovskii, Determination of evolution parameter of an equation and inverse problems of mathematical physics I, Diff. Eqs., 21(1985), 119-129, II, 21(1985), 694-701; III, 23(1987), 910-918.
36. A. I. Prilepko and V.V. Solov'ev, Solvability of the inverse boundary-value problem of finding a coefficient of a lower-order derivative in a parabolic equation, Diff. Eqs., 23(1987), 136-143.
37. M. H. Protter and H. F. Weinberger, Maximum Principle in Differential Equations, Prentice-Hall, Englewood Cliffs, NJ., 1967.
38. W. Rundell, Determination of an unknown non-homogeneous term in a linear partial differential equation from overspecified boundary data, Appl. Anal., 10(1980), 231-242.
39. J. Sylvester and G. Uhlmann, The Dirichlet to Neumann map and applications, in [21], page 99-138.
40. R. Temam, Infinite Dimensional Dynamical Systems in Mechanics and Physics, Springer-Verlag, New York, 1988.
41. T. Suzuki, Gel'fand-Levitan's theory, deformation formulas and inverse problems, J. Fac. Sci. Univ. Tokyo, Sect. IA, Math., 32(1985), 223-271.
42. T. Suzuki, On a multidimensional inverse parabolic problem, Proceedings of the Japan Acdemy, 62(1986), 83-86.
43. H.M. Yin, A uniqueness theorem for a class of nonclassical parabolic equations, Applicable Analysis, 34(1989), 67-78.
44. H. M. Yin, Global Solvability for some parabolic inverse problems, J. of Math. Anal. Appl., 162(1991), 392-403.
45. A. E. Uzlov, Existence and Uniqueness theorems for solutions of inverse problems for second-order parabolic equations, Diff. Eqs., 15(1979), 261-263.

Numerical Methods for Parameter Estimation Problems*

Karl Kunisch†

Abstract

A survey on some of the advances in the area of parameter estimation problems for partial differential equations during the last five years is given. Parameter estimation problems give rise to a class of nonlinear illposed inverse problems with a special structure. Among other topics first and second order augmented Lagrangian methods, estimation of state-dependent coefficients, and optimal input to maximize the robustness of the parameter-to-observation mapping are treated.

1 Introduction

The objective of this note is to survey some of our recent contributions in the area of numerical methods for parameter estimation problems for elliptic equations. For a survey of methods developed up to 1989 we refer to Chapter VI of [1]. For the most part of this paper we concentrate on the identification of the diffusion coefficient $a = a(x)$ in

$$(1) \qquad \begin{cases} -\mathrm{div}\,(a \,\mathrm{grad}\, u) + b\,u &= f \quad \text{in} \quad \Omega \\ u &= 0 \quad \text{on} \quad \partial\Omega, \end{cases}$$

where $f \in L^2, b \in L^\infty, b \geq 0$, and Ω is a bounded domain with Lipschitzian boundary $\partial\Omega$. Unless otherwise specified all function spaces are considered over the domain Ω. To identify the unknown diffusion coefficient a we assume the availability of data z. The problem then consists in determining a in the set of admissible parameters denoted by Q_{ad} in such a way that $u(a)$, the solution to (1), best approximates z. Most frequently parameter estimation problems are formulated as least squares problems, i. e. one considers

$$(2) \qquad \min \frac{1}{2}\, |\, u\,(a) - z\,|_H^2 \quad \text{over} \quad a \in Q_{ad},$$

where H is a Hilbert space with the property that H_0^1 embeds continuously into H and $z \in H$. Further $|\cdot|_H$ denotes the norm in H. Together with (2) we also consider the regularized least squares formulation given by

$$(3) \qquad \min \frac{1}{2}|\, u\,(a) - z\,|_H^2 + \frac{\beta}{2}\,|\, a\,|_V^2 \quad \text{over} \quad Q_{ad},$$

where $\beta > 0$ is the regularization parameter and V is a Hilbert space which embeds continuously into L^∞. Parameter estimation problems of the type considered here arise

*Research partially supported by the Christian Doppler Laboratory for Parameter Estimation and Inverse Problems.

†Technische Universität Berlin, Fachbereich Mathematik, MA 6-2, Straße des 17. Juni 136, D-10623 Berlin

in ground-water management, where a represents the porosity of the rock and u stands for pressure or hydraulic head [21].

In (2) and (3) we assumed the availability of a distributed observation z defined on all of Ω. In this situation identifiability, i. e. injectivity of $a \to u(a)$ is well-studied; we refer to [9], and the references given there. In practice the availability of distributed observations is, of course, unrealistic and z will arise from interpolation of point measurements in the interior of Ω. While several of the methods that we shall present in the following sections can be used in the case of point measurements without interpolation to a distributed measurement z or in the case of boundary measurements along $\partial\Omega$ by an obvious modification of the least squares functionals in (2) and (3), one cannot expect identifiability and hence uniqueness of solutions to (2) in these cases. Further comments on partial observations will follow below. In (3) we introduced the regularization term involving the norm in V. In practice one will replace this norm by an appropriate and easily implementable seminorm regularization term. Moreover, if a-priori information about the "true" parameter is available then the regularization term in (3) can be replaced by $\beta \,|\, a - a_{apriori} \,|_V^2$.

The numerical methods that we shall concentrate on here are regularized least squares methods with the partial differential equation treated as constraint. Thus – differently from what the notation in (2) and (3) suggests – for the numerical solution the state variable u of the differential equation is not considered as a function of the unknown coefficient, here a, but rather the parameter variable a and the state variable u are both treated as independent variables. They are linked by equation (1). We arrive at a constraint optimization problem of the form

$$
(4) \qquad \begin{cases} \min \; F(a,u) \\ e(a,u) = 0 \\ a \in Q_{ad}. \end{cases}
$$

In the context of (2) and (3) we choose

$$
(5) \qquad F(a,u) = \frac{1}{2} \,|\, u - z \,|_H^2 + \frac{\beta}{2} \,|\, a \,|_V^2,
$$

and

$$
(6) \qquad e(a,u) = (-\Delta)^{-1} \Big[\operatorname{div} (a \operatorname{grad} u) - b\, u + f \Big],
$$

where $F : X \to \mathbb{R}, e : X \to H_0^1, X = V \times H_0^1$, and $\Delta : H_0^1 \to H^{-1}$ denotes the Laplacian. Clearly (3) and (4) are equivalent as optimization problems. Algorithms to solve (3) and (4) differ significantly, however. Let us point out that the cost functional in (3) is not convex and unstructured, while in (4) with (5), (6) holding, the cost is quadratic and the equation constraint is bilinear. To solve (4) – (6) numerically we had previously proposed the first order augmented Lagrangian (AL-) method [8], [11], which proved to be numerically very robust. The initialization phase of the first order AL-method contains an equation error step, i. e. the initial guess for a is constructed from

$$
(7) \qquad \min \; |\nabla (a \nabla z) - b\, z + f \,|_{H^{-1}}^2 + \beta \,|\, a \,|_V^2 \quad \text{over} \quad a \in Q_{ad}.
$$

Note that the cost in (7) is quadratic in a. We shall use the structure of least squares problems to obtain a good initial guess for the Lagrange multiplier associated with the equality constraint as well.

The shortcoming of the AL-method is its linear convergence rate. We therefore extended the framework of AL-methods to second order techniques. This led to the AL-SQP algorithm which we shall explain in detail in Section 2. As the nomenclature indicates, this algorithm combines features of the AL approach and of the sequential quadratic programming (SQP-) technique. It has second order convergence rate and the practically attractive feature that globalization is built into the cost functional with no additional globalization strategy (like a line search) required. Convergence is obtained for a wide range of initial values. In Section 2 we also compare the AL-SQP-method to the reduced SQP-method developed in [18].

In Section 3 we return to the first order AL-method. The first order AL method contains a step which requires minimization in the variables a and u simultaneously. The question arises whether this minimization can be carried out iteratively (sequentially) or even in parallel. We shall give an affirmation answer in the case of small residue problems.

In Section 4 we divert from the linear equation (1) and consider the nonlinear differential inclusion

$$
\begin{aligned}
-\mathrm{div}\left(\partial j(u)\right) + b\,u \;&\ni\; f \quad \text{in} \quad \Omega \\
u \;&=\; 0 \quad \text{on} \quad \partial\Omega,
\end{aligned}
\tag{8}
$$

where j is now the unknown "parameter" contained in a set of convex functions and ∂j denotes the subdifferential of j. We shall describe a general approach, based on Fenchel duality, to develop numerical schemes for the estimation of j and we describe specific results for (8) from [2].

Within this short survey we do not aim at completeness. Concerning numerical methods which are not covered in this survey nor in [1] we mention an equation error type method by Vainikko [20], (see also his monograph). The lack of completeness not only concerns numerical methods for the actual computation of the unknown parameter, but also certain method-independent topics. For example, we do not cover identifiability or details of regularization. For (3) these topics have been treated in the literature. There are additional topics on which the literature is less complete. Among them are the optimal choice of input (in (1) this would be f), the optimal choice of cost functional [7], optimal choice of measurement location in the case of point measurements, and the statistical interpretation of numerical results [6]. In the final Section 5 we summarize our findings on the problem of optimal input for stable reconstruction of the parameter a.

2 Augmented Lagrangian Methods

In this section we describe augmented Lagrangian methods developed in [8], [10], [11], for the solution of parameter estimation problems. We focus on (4) – (6) with $H = L^2$ or $H = H^1$, V a Hilbert space that embeds continuously into L^∞, and

$$Q_{ad} = \left\{a \in V : a(x) \geq \nu > 0 \quad \text{a. e. on} \quad \Omega\right\}$$

The essential conditions for applicability of augmented Lagrangian methods are

(i) existence of a solution (a^*, u^*)

(ii) existence of a Lagrange multiplier associated with the constraints in (4) – (6)

(iii) second order sufficient optimality.

Let us discuss (i) – (iii). It is simple to argue that (4) – (6) has a solution (a^*, u^*). If only seminorm regularization is used and the seminorm is expressed by $|\,Na\,|_W$, with N a bounded linear operator from V to the Hilbert space W, then an appropriate condition on the behavior of the solution $u(a)$ of (1) for $a \in \ker N$ must hold to guarantee existence of a solution. Such a condition is given by

$$\inf_{a \in \bar{Q}_{ad}} |u(a) - z|_H < |z|_H,$$

where $\bar{Q}_{ad} = \left\{a \in (\ker N)^{\perp} \cap V : a(x) \geq \nu \text{ a. e. on } \Omega\right\}$. If $\beta = 0$, then (4) – (6) may have no solution. Existence is restored, if an a-priori bound is imposed on the set of admissible parameters.

The Lagrangian $L : V \times H_0^1 \times H_0^1 \times V \to \mathbb{R}$ associated with (4) – (6) is given by

$$L(a, u, \lambda, \eta) = F(a, u) + \left\langle \lambda, e(a, u) \right\rangle_{H_0^1} + \left\langle \eta, a - \nu \right\rangle_V.$$

It can be shown that there exists a Lagrange multiplier $(\lambda^*, \eta^*) \in H_0^1 \times V$ satisfying

$$\text{(9)} \qquad \nabla_{(a,u)} L(a^*, u^*, \lambda^*, \eta^*)(h, v) = 0 \quad \text{for all} \quad (h, v) \in V \times H_0^1,$$

where $\nabla_{(a,u)}$ denotes the Fréchet derivative with respect to (a, u).

Together with the Lagrangian we also consider the augmented Lagrangian functional which is defined by

$$L_c(a, u, \lambda, \eta) = L(a, u, \lambda, \eta) + \frac{c}{2} |e(a, u)|^2_{H_0^1},$$

where $c > 0$ is the penalty parameter. We require the following second order sufficient optimality condition:

$$\text{(10)} \qquad \begin{cases} \text{there exist constants} \quad \sigma > 0 \quad \text{and} \quad c_0 \geq 0 \quad \text{such} \\ \text{that the second Fréchet derivative of} \quad L_c \quad \text{satisfies} \\ \nabla^2_{(a,u)} L_{c_0}(a^*, u^*, \lambda^*, \eta^*)(h, v)^2 \geq \sigma |(h, v)|^2_{V \times H_0^1} \\ \text{for all} \quad (h, v) \in V \times H_0^1. \end{cases}$$

With (10) holding it can be shown that (a^*, u^*) is a local solution to (4) – (6) if and only if it is a solution to

$$\text{(11)} \quad \min \; F(a, u) + \left\langle \lambda^*, e(a, u) \right\rangle_{H_0^1} + \frac{c}{2} |e(a, u)|^2_{H_0^1} \quad \text{over} \quad (a, u) \in V \times H_0^1, \quad a \geq \nu,$$

for $c \geq c_0$. In particular $e(a, u) = 0$ is eliminated from the explicit constraints in (4) – (6).

The second order sufficient optimality condition (10) was analyzed in [11], [12], [19] for different choices of H and for regularization with full as well as with seminorm. In the case that the data z are attainable, i. e. that $u(a) = z$ for some $a \in Q_{ad}$, it can be shown that there exists $\bar{\beta} > 0$ such that (10) holds for every $\beta \in (0, \bar{\beta}]$. Of course, σ depends on β and $\sigma(\beta) \to 0$ as $\beta \to 0$. In the nonattainable case existence of a nontrivial interval of regularization parameters for which (10) holds is guaranteed provided that $\inf \left\{ |u(a^*) - z| : a \in Q_{ad} \right\}$ is sufficiently small. — Let us also mention that (10) is implied by the (standard) second order sufficient optimality condition requiring the existence of $\sigma > 0$ such that

$$\text{(12)} \qquad \nabla^2_{(a,u)} L(a^*, u^*, \lambda^*, \eta^*)(h, v)^2 \geq \sigma |(h, v)|^2_{V \times H_0^1},$$

for all $(h,v) \in \ker\left(\nabla_{(a,u)}\, e(a^*,u^*)\right)$, [10], [12]. In the special case that $H = H_0^1$, and with regularization by full norm, (12) holds for all $(h,v) \in V \times H_0^1$, provided that $\inf\left\{|\,u(a^*) - z\,|_{H_0^1} : a \in Q_{ad}\right\}$ is sufficiently small and β is chosen appropriately. In this case the solution to (the regularized problem) (4) – (6) is even globally unique, [19].

Augmented Lagrangian methods for parameter estimation problems focus on solving (11) numerically. The iteration is carried out with respect to the Lagrange variable λ. From the beginning let us stress that the penalty term is essential for the numerical success of the method. Choosing $c = 0$ is numerically unfeasible.

We define

$$\mathcal{L}_c(a,u,\lambda) = F(a,u) + \left\langle \lambda, e(a,u)\right\rangle_{H_0^1} + \frac{c}{2}\,|\,e(a,u)\,|_{H_0^1}^2.$$

Algorithm 1 (First order).

(i) Choose $c > c_0$, *and set* $\lambda_1 = 0, k = 1$,

(ii) determine (a_k, u_k) *as a (local) solution to*

(P_{aux}) $\qquad$ $\min \mathcal{L}_c(a,u,\lambda_{k-1})$ *over* $(a,u) \in Q_{ad} \times H_0^1$,

(iii) set $\lambda_{k+1} = \lambda_k + (c - c_0)\, e(a_k, u_k)$,

(iv) set $k = k+1$, *goto (ii).*

Several comments are necessary.

Remark 2.1.

(a) *The constant value c for the penalty parameter in (iii) of Algorithm 1 can be replaced by a monotonically increasing sequence* $\left\{c_k\right\}_{k=1}^{\infty}$ *with* $c_1 > c_0$ *[8], [11]. In numerical experiments increasing c during the iteration had only little effect on the results.*

(b) *The choice* $\lambda_1 = 0$ *in (ii) is suggested by the first order optimality condition (9), which implies that*

$$\lambda^* = \left(\nabla \cdot (a^* \nabla u^*)\right)^{-1} \Delta(u^* - z), \tag{13}$$

if $H = H_0^1$, *and* Δ *is replaced by* -1 *if* $H = L^2$.

(c) *Calculation of the gradient of the cost functional in (ii) is straightforward. For (2), on the contrary, a convenient representation of the gradient requires the solution of the adjoint equation [4]. We note from (13) that at the minimizer the adjoint state and* λ^* *coincide.*

(d) *Since in the auxiliary problem* (P_{aux}) *of Algorithm 1 the equality constraint* $e(a,u) = 0$ *is not enforced, existence of a minimizer to* (P_{aux}) *may not hold. Let us observe, however, that*

$$\mathcal{L}_c(a,u,\lambda) = F(a,u) + \frac{c}{2}\,|\,e(a,u) + \frac{\lambda}{c}\,|_{H_0^1}^2 - \frac{1}{2c}\,|\,\lambda\,|_{H_0^1}^2,$$

and hence existence can easily be argued in the case that $H = H_0^1$ *and regularization with the* $|\cdot|_V$*-norm is carried out. Otherwise – at least for the sake of the theory – we need to replace (ii) of Algorithm 1 by*

$$(14) \qquad (a_k, u_k) = \mathit{argmin}\left\{\mathcal{L}_c(a, u, \lambda_{k-1}) : (a, u) \in (Q_{ad} \times H_0^1) \cap B(a^*, u^*; r)\right\},$$

where $B(a^*, u^*; r)$ *is a closed ball with center* (a^*, u^*) *and radius* r *in* $V \times H_0^1$*. From Theorem 2.1 below it will follow that with (10) holding* (a_k, u_k) *can be found in the interior of* $B(a^*, u^*, r)$ *if* $\frac{1}{c-c_0}|\lambda_1 - \lambda^*|$ *is sufficiently small. — Theorem 2.2 remains correct with* (a_k, u_k) *in step (ii) replaced by*

$$\mathcal{L}_c(a_k, u_k, \lambda_{k-1}) \leq \mathcal{L}_c(a^*, u^*, \lambda_{k-1}), \quad (a_k, u_k) \in Q_{ad} \times H_0^1,$$

i. e. if (a_k, u_k) *is a suboptimal solution to* (P_{aux}).

(e) *In the numerical experiments that we carried out we did not enforce the constraint* $a \geq \nu$*. Apparently this had no effect on the numerical results.*

(f) *The cost functional in* (P_{aux}) *is quadratic in* a *and* u *separately. This suggests splitting the cost functional* L_c *and minimizing with respect to* a *and* u *either sequentially or in parallel. This will be analyzed in Section 3.*

(g) *If (ii) is solved sequentially and minimization is started with respect to the* a*-variable, then for the first iteration with* $k = 1$ *it is suggested to initialize the* u*-variable with* $u_0 = z$*. This results in*

$$\min |e(a, z)|_{H_0^1}^2 + \frac{\beta}{2}|a|_V^2 \quad \text{over} \quad a \in Q_{ad}$$

or equivalently

$$\min |\nabla(a\nabla z) - bz + f|_{H^{-1}}^2 + \frac{\beta}{2}|a|_V^2 \quad \text{over} \quad a \in Q_{ad}.$$

Thus the first step of the first order augmented Lagrangian method is a regularized equation error step.

(h) *In [17] one can find the description of a MATLAB code for the estimation of* a *as well as of* b *in (1) which is based on Algorithm 1. It can be used as a tool to experiment with the effect of noise in the data, as well as with the choice of the regularization parameter* β *and the penalty parameter* c *on the numerical solution. It also allows for different discretizations of the parameter variables* a *respectively* b *and the state variable* u*, and to presmooth noisy data* z*.*

Concerning the convergence of Algorithm 1 we have the following result.

THEOREM 2.1. *Assume that (10) holds and replace (ii) of Algorithm 1 by (14). Then there exists* $\bar{\sigma} = \bar{\sigma}(\sigma)$ *such that*

$$(15) \quad \bar{\sigma}\left|(a_k, u_k) - (a^*, u^*)\right|_{V \times H_0^1}^2 + \frac{1}{2(c - c_0)}|\lambda_{k+1} - \lambda^*|_{H_0^1}^2 \leq \frac{1}{2(c - c_0)}|\lambda_k - \lambda^*|_{H_0^1}^2,$$

for all $k = 1, 2, \ldots$. *This implies*

$$\sum_{k=1}^{\infty} \left|(a_k, u_k) - (a^*, u^*)\right|^2 \le \frac{1}{2\bar{\sigma}(c - c_0)} |\lambda_1 - \lambda^*|^2.$$

Moreover $(a_k, u_k) \in$ interior $\left(B(a^*, u^*; r)\right)$ *provided that* $\frac{1}{c-c_0}|\lambda_1 - \lambda^*|^2$ *is sufficiently small. In this case there exists a constant* K, *independent of* c *and* k, *such that*

$$|\lambda_k - \lambda^*|^2_{H_0^1} + |\eta_{k-1} - \eta^*|^2_V \le \left(\frac{K}{\bar{\sigma}(c - c_0)}\right)^{k-1} |\lambda_1 - \lambda^*|^2_{H_0^1}$$

for all $k = 2, 3, \ldots$. *Here* η_k *denotes the Lagrange multiplier associated with the constraint* $a \ge \nu$ *in* (P_{aux}).

In numerical testing, Algorithm 1 proved to be a very reliable algorithm. The first order convergence rate is a disadvantage, however. The key for an algorithm with higher convergence rate is a modification of the Lagrange multiplier update, which we shall describe next.

It will be convenient to introduce $x = (a, u) \in X := V \times H_0^1$. The dual function associated with $\mathcal{L}_c$ is given by

$$d_c(\lambda) := \min \mathcal{L}_c(x, \lambda) \quad \text{subject to } x \in B_\varepsilon = \left\{x : |x - x^*| \le \varepsilon\right\}, \tag{16}$$

for some sufficiently small $\varepsilon > 0$. The first and second derivatives of d_c are given by

$$\nabla_\lambda d_c(\lambda) = e(x(\lambda))$$

and

$$\nabla_\lambda^2 d(\lambda) = -B(x(\lambda), \lambda)$$

where

$$B(x, \lambda) = \nabla_x e(x)\left(\nabla_x^2 \mathcal{L}_c(x, \lambda)\right)^{-1}\left(\nabla_x e(x)\right)^*.$$

Here $x(\lambda)$ stands for argmin $\{\mathcal{L}_c(x, \lambda) : x \in B_\varepsilon\}$. Step (iii) of Algorithm 1 can therefore be interpreted as a first order steepest ascent for the dual problem of maximizing d_c. Replacing the first order step by a Newton step results in

$$\lambda_{k+1} = \lambda_k + B(x_k, \lambda_k)^{-1} e(x_k).$$

Since computation of $B(x_k, \lambda_k)^{-1}$ would be rather burdensome, some reformulations are carried out which lead to attractive algorithms [12]. To describe them we introduce

$$M(x, \lambda) = \begin{pmatrix} \mathcal{L}_0''(x, \lambda) & e'(x)^* \\ e'(x) & 0 \end{pmatrix},$$

where for notational convenience derivatives with respect to x are denoted by primes. Due to lack of convexity of $\mathcal{L}_c(a, u, \lambda)$ in the variable $x = (a, u)$, the algorithms below require choice of local solutions during the iteration. Thus let $\varepsilon > 0$ denote a sufficiently small constant that must guarantee among other properties that (10) holds for all $x \in B(x^*, \varepsilon)$, where $x^* = (a^*, u^*)$. In practice we did not enforce the constraint that results from localization. For small residue problems this was theoretically justified in [19].

ALGORITHM 2 (SECOND ORDER WITH EXACT FIRST ORDER STEP).

(i) Choose $c > c_0$, *and set* $\lambda_1 = 0, k = 1$,

(ii) determine $\tilde{x}$ *as solution to*

$$(P_{aux}) \qquad \min \mathcal{L}_c(x, \lambda_k) \quad \text{over } (Q_{ad} \times H_0^1) \cap B(x^*, \varepsilon),$$

(iii) set $\tilde{\lambda} = \lambda_k + (c - c_0)e(\tilde{x})$,

(iv) solve

$$M(\tilde{x}, \tilde{\lambda}) \begin{pmatrix} x_k - \tilde{x} \\ \lambda_{k+1} - \tilde{\lambda} \end{pmatrix} = -\begin{pmatrix} \mathcal{L}_0'(\tilde{x}, \tilde{\lambda}) \\ e(\tilde{x}) \end{pmatrix} \tag{17}$$

for (x_k, λ_{k+1}),

(v) set $k = k + 1$ *and goto (ii)* .

Variants of Algorithm 2 differ with respect to the first order step (ii) of Algorithm 2.

ALGORITHM 3 (SECOND ORDER WITH SUBOPTIMAL FIRST ORDER STEP).

This coincides with Algorithm 2, except that (ii) is replaced by

(ii') determine $\tilde{x} \in (Q_{ad} \times H_0^1) \cap B(x^*, \varepsilon)$ *such that*

$$\mathcal{L}_c(\tilde{x}, \lambda_k) \leq \mathcal{L}_c(x^*, \lambda_k) = F(x^*).$$

ALGORITHM 4 (SECOND ORDER WITH FIRST ORDER STEP DELETED).

Here step (ii) of Algorithm 2 is deleted and (i) , (iii) are replaced by

(i') Choose $c > c_0, x_0$, *and set* $\lambda_1 = 0, k = 1$,

(iii') set $\tilde{\lambda} = \lambda_k + (c - c_0)e(x_{k-1})$.

In Algorithm 4 the constraint $a(x) \geq \nu$ is not enforced. — Algorithms 2 – 4 are similar to the SQP method applied to

$$\min \mathcal{L}_0(x, \lambda) \quad \text{subject to } e(x) = 0.$$

The SQP method consists essentially in the iteration (17) with $(\tilde{x}, \tilde{\lambda})$ replaced by (x_{k-1}, λ_k). We refer to Algorithms 2 – 4 as augmented Lagrangian – SQP methods. They differ from the SQP method in the use of the penalty term and the intermediate first order Lagrangian update.

Convergence properties of Algorithms 2 – 4 are summarized in the following theorem.

THEOREM 2.2. *Assume that (10) holds.*

(i) If $\frac{1}{c-c_0}|\lambda_1 - \lambda^*|^2$ *is sufficiently small, then the iterates of Algorithm 2 satisfy*

$$\left|(x_k, \lambda_{k+1}) - (x^*, \lambda^*)\right|_{V \times H_0^1 \times H_0^1} \leq \frac{K}{c - c_0}|\lambda_k - \lambda^*|^2_{H_0^1},$$

for a constant K *independent of* $k = 1, 2, \ldots$ *and* $c > c_0$.

(ii) If $|\lambda_1 - \lambda^*|$ is sufficiently small, then the iterates of Algorithm 3 satisfy

$$\left|(x_k, \lambda_{k+1}) - (x^*, \lambda^*)\right|_{V\times H_0^1\times H_0^1} \leq K\left(1 + \frac{1}{(c-c_0)}\right) |\lambda_k - \lambda^*|^2_{H_0^1}$$

for a constant K independent of $k = 1, 2, \ldots$ and $c > c_0$.

(iii) If $\max(2, 1+c^2\gamma^2)|(x_0, \lambda_1) - (x^*, \lambda^*)|$ is sufficiently small then the iterates of Algorithm 4 satisfy

$$\left|(x_k, \lambda_{k+1}) - (x^*, \lambda^*)\right|_{V\times H_0^1\times H_0^1} \leq K\left|(x_{k-1}, \lambda_k) - (x^*, \lambda^*)\right|^2_{V\times H_0^1\times H_0^1},$$

for a constant K independent of $k = 1, 2, \ldots$ but depending on c, and γ is the Lipschitz constant of e on $B(x^*, \varepsilon)$.

Just as Algorithm 1, Algorithms 2 – 4 also contain a regularized equation error step for $k = 1$. Moreover, for zero residue problems (which is equivalent to attainability of the data z) and with $\beta = 0$, Algorithms 2 – 4 are one–step algorithms. For details we refer to [12].

The a–coordinates of the iterates x_k need not satisfy $a_k \geq \nu$, while the a–coordinates of $\tilde{x}$ do. From the proof of Theorem 2.2 convergence of $\tilde{x}$ to x^* easily follows as well.

For computations the simplicity of the structure of M is one of the advantages of augmented Lagrangian–SQP methods. Its discretization is straightforward and several specific cases are given in [12]. In [12] one can also find numerical examples which illustrate that second order convergence is achieved for a wide range of initial values.

Let us close this section with comments on SQP–methods applied to parameter estimation problems. These methods are quite popular in the context of parameter estimation for ordinary differential equations, see [3] and the references given there. In [16] SQP–methods where applied to parameter estimation for two point boundary value problems. A general framework for reduced SQP–methods applicable to parameter estimation for partial differential equations was described in [18]. There it was pointed out that the specific structure of estimation problems suggests several modifications to a straightforward use of standard SQP–methods. In particular, the nullspace representation is carried out differently for parameter estimation than for general purpose finite dimensional optimization problems, where it is frequently based on the QR–decomposition. Similarly, the Lagrange multiplier correction according to (13) is natural for parameter estimation problems, whereas the updates for finite dimensional optimization problems are based on different considerations. In [18] superlinear convergence was proved for reduced SQP–methods with BFGS update. It was also demonstrated that superlinear convergence holds numerically. The range of initial values, within which convergence can be achieved is quite small for SQP–methods, however, when compared to augmented Lagrangian–SQP methods. This difference can be reduced by combining SQP–methods with line searches.

3 Sequential and Parallel Splitting in the Auxiliary Problem.

The auxiliary problems of Algorithms 1 – 3 are separately convex – even quadratic – in the variables a and u. This suggests the use of sequential and parallel methods which carry out minimization with respect to a and u alternatingly. In this section we address the convergence of such techniques, which is complicated by the fact that in spite of the separate convexity of $\mathcal{L}_c$ with respect to a and u, $\mathcal{L}_c$ lacks convexity with respect to (a, u).

We consider the least squares problem in the form

$$(18)\qquad \begin{cases} \min \frac{1}{2}|u-z|^2_{H^1_0} + \frac{\beta}{2}|a|^2_V \\ e(a,u) = 0 \\ (a,u) \in (K_1 \cap Q_{ad}) \times H^1_0 =: K \times H^1_0, \end{cases}$$

where K_1 is a closed convex set in V. For a discussion of H^1_0 replaced by L^2 in (18) we refer to [19]. Let $x^* = (a^*, u^*)$ be a minimizer of (18) . Further let C be a constant satisfying

$$\left|\nabla\cdot(a\nabla u)\right|_{H^{-1}} \le C|a|_V|u|_{H^1_0}, \text{ for all } (a,u) \in V \times H^1_0,$$

and let $A_a u = -\nabla\cdot(a\nabla u)$, which for $a \in Q_{ad}$ is a homeomorphism from H^1_0 onto H^{-1}. We shall assume that u^* and β are related by

$$(19)\qquad C|u^* - z|_{H^1_0}\|A^{-1}_{a^*}\|_{\mathcal{L}(H^{-1},H^1_0)} < \sqrt{\beta}.$$

This small residue assumption implies uniqueness of (a^*, u^*) as global solution to (18) . The auxiliary problems of Section 2 are of the form

$$(P_{aux})\qquad \min \mathcal{L}_c(a,u,\lambda_k) \quad \text{over } (a,u) \in K \times H^1_0.$$

In the case of Algorithm 1, (P_{aux}) has a unique solution for all k, provided that $|\lambda_0 - \lambda^*|$ is sufficiently small. The same holds true for Algorithm 2, if $\frac{1}{c}|\lambda_0 - \lambda^*|^2$ is sufficiently small.

In the sequential splitting algorithm for (P_{aux}) one proceeds as follows:

ALGORITHM 5.

(i) Choose $u^0_k \in H^1_0$ *if* $k = 1$, *else set* $u^0_k = u_{k-1}$.

(ii) For $i = 1, 2, \dots$ *find* $a^i_k \in K$ *and* $u^i_k \in H^1_0$ *such that*

$$\mathcal{L}_c(a^i_k, u^{i-1}_k, \lambda_k) \le \mathcal{L}_c(a, u^{i-1}_k, \lambda_k) \quad \textit{for all } a \in K,$$

and

$$\mathcal{L}_c(a^i_k, u^i_k, \lambda_k) \le \mathcal{L}_c(a^i_k, u, \lambda_k) \quad \textit{for all } u \in H^1_0.$$

In the following result we address convergence of Algorithms 5 when applied to (P_{aux}) in Algorithms 1 or 2. The result is also correct, if u is constrained to lie in a closed convex set containing u^*.

THEOREM 3.1. *Let (19) hold and assume that* $|\lambda_{k_0} - \lambda^*|$ *for Algorithm 1, respectively* $\frac{1}{c}|\lambda_{k_0} - \lambda^*|^2$ *for Algorithm 2 is sufficiently small if* $k_0 \in \{2, 3, \dots\}$. *If* $k_0 = 1$ *in addition we assume that* $|u^0_1 - u^*|$ *is sufficiently small. Then for every* $k \ge k_0$ *the sequence* (a^i_k, u^i_k) *of Algorithm 5 converges to* (a_k, u_k) *in case of Algorithm 1 and to* $\tilde{x}$ *in case of Algorithm 2.*

We turn to parallel splitting of Jacobi type for the auxiliary problems of Algorithms 1 – 3 of Section 2 and specify

ALGORITHM 6.

(i) Choose $(a_k^0, u_k^0) \in V \times H_0^1$ *if* $k = 1$, *otherwise set* $(a_k^0, u_k^0) = (a_{k-1}, u_{k-1})$

(ii) For $i = 1, 2, \ldots$ *find* $\hat{a}_k^i \in K$ *and* $\hat{u}_k^i \in H_0^1$ *such that*

$$\mathcal{L}_c(\hat{a}_k^i, u_k^{i-1}, \lambda_k) \leq \mathcal{L}_c(a, u_k^{i-1}, \lambda_k) \quad \text{for all } a \in K \tag{20}$$

and

$$\mathcal{L}_c(a_k^{i-1}, \hat{u}_k^i, \lambda_k) \leq \mathcal{L}_c(a_k^{i-1}, u, \lambda_k) \quad \text{for all } u \in H_0^1, \tag{21}$$

and set

$$(a_k^i, u_k^i) = \frac{1}{2}\Big[(\hat{a}_k^i, u_k^{i-1}) + (a_k^{i-1}, \hat{u}_k^i)\Big].$$

Note that (20) and (21) can be solved in parallel. Convergence of Algorithm 6 is addressed in the following result.

THEOREM 3.2. *Let (19) hold and assume that* $|\lambda_{k_0} - \lambda^*|$ *for Algorithm 1 respectively* $\frac{1}{c}|\lambda_{k_0} - \lambda^*|^2$ *for Algorithm 2 is sufficiently small if* $k_0 \in \{2, \ldots\}$. *If* $k_0 = 1$ *in addition we assume that* $|(a_1^0, u_1^0) - (a^*, u^*)|$ *is sufficiently small. Then for every* $k \geq k_0$ *the sequence* (a_k^i, u_k^i) *of Algorithm 6 converges to* (a_k, u_k) *in case of Algorithm 1 and to* $\tilde{x}$ *in case of Algorithm 2.*

The theorems of this section are proved in [19].

4 Estimation of Nonlinearities

In this section we turn to the estimation of nonlinearities in elliptic equations. We consider equations of the form

$$\begin{cases} -\text{div}\,\Big(a\,(\nabla u)\Big) &= f \quad \text{in} \quad \Omega \\ \qquad\qquad u &= 0 \quad \text{on} \quad \partial\Omega, \end{cases} \tag{22}$$

where the diffusion coefficient is a function of ∇u and with $f \in L^2$ fixed. For given $z \in L^2$ the problem consists in determining the nonlinearity a such that the solution u to (22) best approximates z. This problem was studied with convex analysis techniques in [2]. We restrict our attention to differential inclusions of the form

$$\begin{cases} -\text{div}\,(\partial j\,(\nabla u)) &\ni f \quad \text{in} \quad \Omega \\ \qquad\qquad u &= 0 \quad \text{on} \quad \partial\Omega, \end{cases} \tag{23}$$

where ∂j denotes the subdifferential of j and

$$\begin{aligned} j \in \mathcal{K} = \{j : \mathbb{R}^n \to \mathbb{R} |\ r \to j(r) - \tfrac{\alpha}{2}\,|r|^2_{\mathbb{R}^n} \quad &\text{is convex}, \\ \beta_1\,|r|^2_{\mathbb{R}^n} + \gamma_1 \ \leq\ j(r) \ \leq\ \beta_2\,|r|^2_{\mathbb{R}^n} + \gamma_2\}&, \end{aligned}$$

for fixed constants $\alpha > 0$, $\beta_2 > \beta_1 > 0$ and $\gamma_1,\ \gamma_2 \in \mathbb{R}$. The least squares problem is formulated as

$$(P) \qquad \min \quad |u - z|^2_{L^2} \qquad \text{subject to } (j, u) \in \mathcal{K} \times H_0^1 \quad \text{satisfying} \quad (23).$$

In (P) a Tikhonov regularization term analogous to the one in (3) or (5) can be added without essential changes in the results of this section. Note, however, that the use of boundedness constraints in the definition of $\mathcal{K}$ also implies a certain kind of regularization. The relationship between norm constrains and Tikhonov regularization was made precise in [15] for problems of the type (4) – (6), where it was shown that the Lagrange multiplier associated with a bound on the norm of the admissible coefficients corresponds to the regularization parameter β.

To implement (23) one uses a splitting technique and expresses (23) as

$$(24) \qquad \begin{cases} w \in \partial j\,(\nabla u) & \text{in } \Omega \\ \text{div } w = -f & \text{in } \Omega \\ w = 0 & \text{on } \partial\Omega, \end{cases}$$

where $(u, w) \in H_0^1 \times L_n^2$, with $L_n^2 = \bigotimes_{i=1}^{n} L^2$. Due to the convex conjugacy formula (24) can be equivalently expressed as

$$(25) \qquad \begin{cases} I\,(j, u, w) := \int_\Omega (j(\nabla u) + j^*(w) - uf)dx = 0 \\ \text{div } w = -f. \end{cases}$$

This suggests the introduction of the following family of approximating problems for (P) :

$$(P_\varepsilon) \qquad \begin{cases} \min \frac{1}{2}|u - z|_{L^2}^2 + \frac{1}{\varepsilon} I(j, u, r) \\ \text{subject to } (j, u, w) \in \mathcal{K} \times H_0^1 \times L_n^2 \text{ and div } w = -f, \end{cases}$$

with $\varepsilon > 0$. To discuss the convergence of solution to (P_ε) as $\varepsilon \to 0^+$, the spaces H_0^1 and L_n^2 are endowed with the topology of weak convergence, denoted by $H_{0,w}^1$ and $L_{n,w}^2$ and the set $\mathcal{K}$ is endowed with the topology of uniform convergence on compact subsets of $\mathbb{R}^n$.

THEOREM 4.1. *For every $\varepsilon > 0$, (P_ε) has at least one solution $(j_\varepsilon, u_\varepsilon, w_\varepsilon) \in \mathcal{K} \times H_0^1 \times L_n^2$. The family $\{(j_\varepsilon, u_\varepsilon, w_\varepsilon)\}_{\varepsilon>0}$ has at least one cluster point $(\bar{j}, \bar{u}, \bar{w})$ in $\mathcal{K} \times H_{0,w}^1 \times L_{n,w}^2$ for $\varepsilon \to 0^+$ and every such cluster point satisfies div $\bar{z} = -f$, $(\bar{u}, \bar{w})$ is a solution to (P) and* $\liminf_{\varepsilon \to 0^+}$ (P_ε) $=$ min (P) . *Moreover, if $\tilde{u}_\varepsilon \in H_0^1$ denotes the solution to (23) with j replaced by j_ε, then* $\lim_{\varepsilon \to 0^+} |u^\varepsilon - z|^2 =$ min (P) .

Concerning the stability of solutions of (P) with respect to perturbations of the data z we have

THEOREM 4.2. *Let $\{z_n\}$ be a sequence of observations converging in L^2 to z, and let $\{(j_n, u_n)\}$ be a sequence of solutions to (P) with z replaced by z_n. Then $\{(j_n, u_n)\}$ has a cluster point in $\mathcal{K} \times H_{0,w}^1$, and every such cluster point is a solution to the unperturbed problem (P) .*

Due to the convexity constraint involved in the definition of $\mathcal{K}$ and the fact that elements of $\mathcal{K}$ are defined on all of $\mathbb{R}^n$, the problems (P_ε) are not yet amenable for computations. Next we describe how these difficulties were delt with in [2]. First let us note that j need only be defined on the range of ∇u. In fact we can only hope for identifiability of j on the range of ∇u. Using elliptic theory one derives a-priori estimates on $|\nabla u(x)|$ as j varies in the set of admissible nonlinearities. Let I_i denote appropriate intervals, such that $\{u_{x_i}(x) : x \in \Omega\} \subset I_i$, for $i = 1, ..., n$, and all $j \in \mathcal{K}$. It then suffices to restrict the domain of admissible nonlinearities to $\bigotimes_{i=1}^{n} I_i$. To arrive at an implementable form of the convexity condition, (23) is restricted to equation of the form

$$(26) \qquad \begin{cases} -\sum_{i=1}^{n} \frac{\partial}{\partial_{x_i}}(\partial j_i(u_{x_i})) = f & \text{in } \Omega \\ u = 0 \quad \text{on} \quad \partial\Omega, \end{cases}$$

with $j = (j_1, ..., j_n)$ chosen from

$$\tilde{\mathcal{K}} = \{j = (j_1, ..., j_n) \quad : \quad j_i = \varphi_i^0 + \varphi_i, \quad 0 < \alpha_i \le \varphi_i'' \le \beta_i, \\ \varphi_i \in H^2(I_i) \cap H_0^1(I_i), \ i = 1, ..., n\}$$

for fixed convex functions $\varphi_i^0 \in H^2(I_i)$ and constants α_i, β_i, $i = 1, ..., n$. The problems (P_ε) corresponding to this restricted class of nonlinearities are given by

$$(27) \qquad \begin{cases} \min \frac{1}{2}|u - z|_{L^2}^2 + \frac{1}{\varepsilon} \sum_{i=1}^{n} \int_\Omega (j_i(u_{x_i}) + j_i^*(w_i) - uf)dx \\ \text{subject to } (j, u, w) \in \mathcal{K} \times H_0^1 \times L_n^2 \text{ and } \operatorname{div} w = -f \quad \text{in } \Omega, \end{cases}$$

where it understood that the mappings j_i are extended by ∞ outside of I_i.

An alternating direction algorithm, characterization of the gradients and numerical examples for (27) can be found in [2].

The formulation (P_ε) of the estimation problem of the nonlinearity in (22) is a special case of a general method which we describe next, [5].

Consider the parameterized family of variational problems

$$(28) \qquad \min F_a(u) + G_a(A_u) \qquad \text{over} \quad u \in X,$$

where $a \in C \subset Q$, with C a closed convex set in the Banach space Q, X and Y are Banach spaces, $A \in \mathcal{L}(X, Y)$, and

$$F_a : X \to \overline{\mathbb{R}} \text{ and } G_a : Y \to \overline{\mathbb{R}}$$

are proper, convex and lower semicontinuous for every $a \in C$, with $\overline{\mathbb{R}} = \mathbb{R} \cup \{\infty\}$. We assume that X is reflexive and that $\lim_{|u| \to \infty} F_a(u) + G_a(A_u) = \infty$ for every $a \in C$.

This guarantees existence of a minimizer $\bar{u} = u(a)$ to (28).
In addition we require that

$$(29) \qquad \begin{cases} \text{for every } a \in C \text{ there exists } u \in X \text{ such that} \\ F_a(u) < \infty, \ G_a(A_u) < \infty \text{ and } G_a \text{ is continuous at } Au. \end{cases}$$

This will allow the use of the Fenchel duality theorem below. Further we denote by $B \in \mathcal{L}(X, Z)$ the observation operator, with Z a Banach space, and by $z \in Z$ the data. The least squares formulation for the varitional problems (28) is given by

$$(30) \qquad \min \frac{1}{2}|Bu(a) - z|_Z^2 , \quad \text{over } a \in C$$

where $u(a)$ is a solution to (28). In (30) the optimization problem (28) arises as a constraint, characterizing the relationship between the variables a and u. This constraint will be realized by a penalization term. If F^* denotes the convex conjugate of F,

$$F^*(x^*) = \sup_{x \in X} \Big(\langle x, x^* \rangle_{X,X^*} - F(x)\Big) \quad \text{for } x^* \in X^*,$$

then

$$(31) \qquad F_a(u) + G_a(Au) + F_a^*(A^*q^*) + G_a^*(-q^*) \ge 0$$

for all $(a, u, q^*) \in C \times X \times Y^*$. Moreover, equality holds in (31) if and only if the triple $(a, u(a), q(a))$ is such that $u(a)$ is a solution to the primal problem (28) and $q(a)$ is solution to the dual problem

$$\max_{q^* \in Y^*} -\Big\{F_a^*(A^*, q^*) + G^*(-q^*)\Big\}.$$

The term appearing on the left hand side of (31) is refered to as complementary energy. It has the property that its minimal value is zero independently of $a \in C$. This motivates the introduction of the penalized least squares formulation for the estimation of a in (28):

$$(\tilde{P}_\varepsilon) \qquad \begin{cases} \min \frac{1}{2}|Bu - z|_Z^2 + \frac{1}{\varepsilon}\{F_a(u) + G_a(Au) + F_a^*(A^*q^*) + G_a^*(-q^*)\} \\ \text{over } (a, u, q^*) \in C \times X \times Y^*, \end{cases}$$

where $\varepsilon > 0$. We refer to [5] for a detailed analysis of $(\tilde{P}_\varepsilon)$. Here we give the concrete form of $(\tilde{P}_\varepsilon)$ for two specific problems.

Example 4.3 For the estimation of a in

$$\begin{aligned} -\text{div } (a \text{ grad } u) &= f && \text{in } \Omega \\ u &= 0 && \text{on } \partial\Omega \end{aligned}$$

with $f \in H^{-1}$, one can choose

$$(32) \qquad X = H_0^1,\ Y = L_n^2, \quad A = \text{grad},$$

$$F_a(u) = -\int_\Omega f\ u\ dx, \qquad G_a(g) = \frac{1}{2}\int_\Omega a|q|_{\mathbb{R}^n}^2\ dx,$$

and

$$C = \{a \in Q\ :\ a(x) \geq \nu > 0, \text{ a.e. in } \Omega\},$$

where Q is Banach space that embeds continuously in L^∞. One argues that $A^*q = -\text{div } q$, $G_a^*(q) = \frac{1}{2}\int_\Omega \frac{1}{a}|q|_{\mathbb{R}^n}^2\ dx$, and

$$F^*(u) = \begin{cases} 0 & \text{if } \ u = -f \\ \infty & \text{if } \ u \neq -f, \end{cases}$$

and thus $(\tilde{P}_\varepsilon)$ becomes

$$(33) \qquad \begin{cases} \min \dfrac{1}{2}|Bu - z|^2 + \dfrac{1}{2\varepsilon}\displaystyle\int_\Omega \left(a|\nabla u|^2 + \frac{1}{a}|q|^2 - 2fu\right) dx \\ \text{subject to } (a, u, q) \in C \times H_0^1 \times L_n^2,\ \text{div } q = f. \end{cases}$$

We have successfully tested (33) numerically for the estimation of a in one dimensional problems. Tests for a mixed finite element formulation for two dimensional problems are presently carried out.

Example 4.4. Here we consider the estimation of j in (23) and set X, Y and A as in (31),

$$F(u) = -\int_\Omega fu\ dx, \quad G_\sigma(q) = \frac{1}{2}\int_\Omega \sigma(q)\ dx,$$

$C = \mathcal{K}$, (defined below (23)), and $Q = C(\mathbb{R}^n, \mathbb{R})$. A short calculation shows that in this case $(\tilde{P}_\varepsilon)$ becomes

$$(34) \qquad \begin{cases} \min \frac{1}{2}|Bu - z|^2 + \frac{1}{\varepsilon}I(j, u, -q^*) \\ \text{subject to } (j, u, q^*) \in C \times H_0^1 \times L_n^2 \text{ and div } q^* = f. \end{cases}$$

We observe that for $Z = L^2$ and $B = I$, (34) is equivalent to (P_ε) with $w = -q^*$.

5 Optimal Input

In this section we address the problem of optimal choice of input functions, which render a parameter estimation problem "as stable as possible". Let Φ_f denote a family of parameter-to-observation mappings from a subset of a Hilbert space X into a Hilbert space Y. In the context of (1) , these mappings are given by

$$\Phi_f(a) = u(a; f),$$

where $u(a; f)$ denotes the solution of (1) as a function of the coefficient a and the design parameter $f, X = V$, and Y could be L^2 or H^1, for example. Here the observation space coincides with the solution space. The aim is to choose f in some class of admissible functions $\mathcal{F}$ such that the inversion of Φ at the observation z can be carried out in an optimal manner. Clearly this goal needs to be made precise. It requires the introduction of a measure which

(i) quantifies stability or robustness of the estimation of the unknown coefficient with respect to the data,

(ii) is sensitive with respect to changes in the design variable f,

(iii) is computable.

Below we present one possibility for the choice of such measures and describe how to overcome inherent difficulties that arise due to the illposedness of parameter estimation problems. It will become clear that other choices are possible and probably worthy of investigation. Related questions of optimal choice of measurement locations and optimal choice of weighting functions in the least squares cost functional appear to be interesting but have not received much attention yet, see [7], however.

Concerning the optimal input (optimal experimental design problem), we propose to proceed as follows. Let a denote an a-priori guess to the true parameter, and let X_1 be a Hilbert space containing X as a subspace. Denoting by $\Phi'_f(a)$ the linearization of Φ_f at the reference parameter $a \in X$ we consider

$$\sup_{f \in \mathcal{F}} \inf_{\substack{h \in X \\ h \neq 0}} \frac{|\Phi'_f(a) h|_Y^2}{|h|_{X_1}^2}. \tag{35}$$

The focus of our attention being mappings Φ_f which are illposed in the sense of lack of continuous invertibility, one has to guard against the infimum being zero for all $f \in \mathcal{F}$, such violating (ii) above. First, this requires injectivity of $\Phi'_f(a)$ for at least one $f \in \mathcal{F}$, and secondly, for the infimum to be nonzero, an estimate of the type

$$|\Phi'_f(a) h|_Y \geq k_f |h|_{X_1} \quad \text{for all} \quad h \in X \tag{36}$$

is necessary, with k_f a constant depending on f and positive for at least some $f \in \mathcal{F}$. For illposed problems (36) is not feasible, and the infimum in (35) will be zero in general. We circumvent this difficulty by formulating "nearby" wellposed problems. These are based on

(a) restricting X to a closed subspace of X,

(b) introducing a gap in the topologies between X (natural space for parameters) and X_1,

(c) introducing a gap between the natural norm on the output space and the (finer) topology of Y,

(d) introducing regularization.

In the latter case, (35) is replaced by

$$\sup_{f\in\mathcal{F}} \inf_{\substack{h\in X\\ h\neq 0}} \frac{1}{|h|_{X_1}} \left(|\Phi'_f(a)h|_Y^2 + \beta|h|_X^2\right). \tag{37}$$

Below we give examples from [13] illustrating these points by considering the one-dimensional case of (1). For the multidimensional case we refer to [14]. — Returning to (35) we observe that we have chosen the lowest singular value of Φ'_f as a measure for distinguishing the behavior of the inverse of Φ_f as f varies. Other criteria, like the trace of the discretization of $\Phi'_f(a)^*\Phi'_f(a)$, should be of interest as well. Minimizing the lowest singular value represents a conservative measure (robustness). For parameter estimation problems such as (1) it requires the reconstruction of a to be uniformly "as stable as possible" throughout the domain Ω. Maximizing the lowest singular value as in (35) requires linearization of the nonlinear mapping $a \to \Phi_f(a)$ at a reference parameter. Therefore solving the optimal design problem (35) should be part of an iterative procedure improving the estimate of the coefficient (via least squares, for example) and of the experimental setup.

Let us turn now to the equation

$$\begin{cases} -(au_x)_x = f \quad \text{in} \quad (0,1) \\ u(0) = u(1) = 0, \end{cases} \tag{38}$$

with $f \in H^{-1}(0,1)$ and $a \in L^\infty(0,1), a(x) \geq \nu > 0$.

In this case the linearization of the parameter-to-solution mapping can be characterized explicitly by

$$\Phi'_f(a)h = -\int_0^x \frac{1}{a}(hu_x - H_a)\,ds, \quad x \in [0,1], \tag{39}$$

with

$$H_a = \frac{\int_0^1 \frac{hu_x}{a}dx}{\int_0^1 \frac{dx}{a}}.$$

If $u = u(a,f) \in W^{1,\infty}(0,1)$, then $\Phi'_f(a)h$ is welldefined on all of $L^2(0,1)$. A natural choice for X is $H^1(0,1)$. It would be desirable to choose $Y = L^2(0,1)$. But in this case $\Phi'_f(a)$ would be compact as operator from $H^1(0,1)$ or $L^2(0,1)$ to Y, and the infimum in (36) would be zero. Let us concentrate therefore on the choice $Y = H_0^1(0,1)$ and $X_1 = L^2(0,1)$. The infimum problem in (35) is then equivalent to

$$\mu^2(f) := \inf_{\substack{h\in L^2\\ h\neq 0}} \frac{|Th|_{H_0^1}^2}{|h|_{L^2}^2}, \tag{40}$$

with $T_f = \Phi'_f(a)$ given in (39). To find conditions (e. g. on f) which yield the infimum in (40) positive and sensitive with respect to f the following result on the spectrum $\sigma(T_f^*T_f)$ of $T_f^*T_f$ will be useful.

LEMMA 5.1. *If a is piecewise C^1 and $u(a;f)$ is piecewise $W^{1,\infty}$, then*

$$\sigma(T_f^*T_f) \subset \left[\inf_\xi \left|\frac{u_x(\xi)}{a(\xi)}\right|^2, \sup_\xi \left|\frac{u_x(\xi)}{a(\xi)}\right|^2\right] \cup \{0\},$$

and

$$\overline{\left\{| \frac{u_x(\xi)}{a(\xi)} |^2 : \xi \in [0,1]\right\}} \quad \subset \sigma\left(T_f^* T_f\right),$$

where the bar denotes closure.

Below we discuss two possibilities how the definition of μ^2 can be changed to eliminate 0 from the spectrum of $T_f^* T_{f'}$ so that a useful sensitivity measure can be introduced.

CASE 1. Assume that there exists a constant $\gamma > 0$ such that $|u_x(x)| \geq \gamma$ on $[0,1]$. This condition can be satisfied for example if f is a finite sum of delta distributions appropriately distributed in $(0,1)$. Then 0 is in the pointspectrum of $T_f^* T_f, 1/u_x$ is the corresponding eigenvector and $\mu^2(f) = 0$. We alter the definition of μ^2 accordingly and define the sensitivity measure

$$(41) \qquad \tilde{\mu}^2(f) = \inf \frac{|\Phi_f'(a)\, h|_{H_0^1}^2}{|h|_{L^2}^2} \quad \text{over} \quad h \in L^2(0,1) \quad \text{with} \quad \int_0^1 \frac{h}{u_x}\, dx = 0.$$

Then $\mu^2(f) > 0$ and in view of Lemma 5.1

$$\tilde{\mu}^2(f) = \inf \left| \frac{u_x(\xi)}{a(\xi)} \right|^2 \geq \gamma,$$

which is positive and easier to compute than the infimum over the Raleigh quotient in (41).

CASE 2. Assume that $u(a; f) \in C^1([0,1])$. Then there exists at least one point $x_0 \in (0,1)$ such that $u_x(x_0) = 0, \mu^2(f) = 0$ and 0 is an accumulation point of the spectrum of $T_f^* T_f$. In this case one can use the regularized version of (40) and define

$$(42) \qquad \mu_\beta^2(f) = \min |T_f\, h|_{H_0^1}^2 + \beta \int_0^1 |h_x|^2\, dx \quad \text{over} \quad h \in H^1(0,1), |h|_{L^2} = 1.$$

For a detailed study of this approach we refer to [14].

In the case of smooth u (as in CASE 2) an alternative approach to (40) is to treat T_f as closed linear operator from L^2 to $H^2 \cap H_0^1$. For details we refer to [13].

Once an appropriate measure, as for instance $\tilde{\mu}$ above, has been found, the supremum part in the saddle point problem (35) can be treated. It requires the specification of a class $\mathcal{F}$ such that $\tilde{\mu}$ is positive on a nontrivial subset of $\mathcal{F}$, and has continuity and directional differentiability properties. Again we refer to [3].

Let us close with a specific example and consider the question of optimal input, characterized by the location $\alpha \in (0,1)$ of the delta impulse in the following estimation problem:

$$\begin{cases} -(a u_x)_x = \delta_\alpha \\ u(0) = u(1) = 0, \end{cases}$$

with $\mathcal{F} = \{\delta_\alpha : 0 < \alpha < 1\}$. In this situation the measure $\tilde{\mu}^2$ is applicable. If the reference coefficient equals a constant, then it comes as no surprise that the optimal input location is characterized by $\alpha = .5$. If, on the other hand, the reference coefficient is given by $a(x) = 1 + kx$, with $k \in \mathbb{R}$ fixed, then one may wonder whether the optimal delta impulse should be located to the left or the right of $\alpha = .5$. The interested reader may wish to consult [3].

References

[1] H. T. Banks and K. Kunisch, *Estimation Techniques for Distributed Parameter Systems*, Birkhäuser, Boston, 1989.

[2] V. Barbu and K. Kunisch, *Identification of nonlinear elliptic equations*, submitted to Applied Mathematics and Optimization.

[3] H. G. Bock and J. P. Schlöder, *Recent progress in the development of algorithms and software for large-scale parameter estimation problems in chemical reaction systems, in: Kotobh (ed.) Automatic Control in Petrol, Petrochemical and Desalination Industries*, IFAC Congress, Pergamon, Oxford, (1986).

[4] G. Chavent, *New trends in identification of distributed parameter systems*, Proc. 4thIFAC Congress, Munich, (1987).

[5] G. Chavent and K. Kunisch, *Primal-dual formulation for parameter estimation problems*, manuscript.

[6] B. Fitzpatrick, *Bayesian analysis in inverse problem*, preprint.

[7] F. Guyon, Y. P. Yvon and J. Henry, *On a weighting method improving identifiability of distributed parameter systems*, preprint.

[8] K. Ito, M. Kroller, and K. Kunisch, *A numerical study of the augmented Lagrangian method for the estimation of parameters in elliptic systems*, SIAM J. on Sci. and Stat. Computing, 12 (1991), 884–910.

[9] K. Ito and K. Kunisch, *On the injectivity of the coefficient- to-solution mapping for elliptic boundary value problems and its linearization*, will be published in J. Math. Anal. and Appl.

[10] K. Ito and K. Kunisch, *The augmented Lagrangian method for equality and inequality constraints in Hilbert space*, Mathematical Programming, 46, (1990), 341–360.

[11] K. Ito and K. Kunisch, *A hybrid method combining the output least squares and the equation error approach for the estimation of parameters in elliptic systems*, SIAM J. on Control and Optimization, 28 (1990), 113–136.

[12] K. Ito and K. Kunisch, *Augmented Lagrangian SQP-methods in Hilbert spaces and application to control in the coefficients problems*, will be published in SIAM J. on Optimization.

[13] K. Ito and K. Kunisch, *Sensitivity measures for the estimation of parameters in elliptic boundary value problems*, will be published in J. of Mathematical Systems, Estimation, and Control.

[14] K. Ito and K. Kunisch, *Maximizing robustness in nonlinear ill-posed inverse problems*, to appear in SIAM J. Control and Optimization.

[15] K. Ito and K. Kunisch, *On the choice of the regularization parameter in nonlinear inverse problems*, SIAM J. on Optimization, 2 (1992) 1–29.

[16] C. T. Kelley and S. J. Wright, *Sequential quadratic programming for certain identification problems*, Nonlinear Programming (1992).

[17] M. Kroller and K. Kunisch, *MATLAB-software for parametric estimation in two point boundary value problems, in: Identification and Control of Systems Governed by Partial Differential Equations*, H. T. Banks, R. H. Fabiano, K. Ito, Eds., SIAM, Philadelphia, (1993), 59–68.

[18] K. Kunisch and E. Sachs, *Reduced SQP-methods for parameter identification problems*, SIAM J. Numerical Analysis, 29 (1992), 1793–1820.

[19] K. Kunisch and X. Tai, *Sequential and parallel splitting methods for bilinear control problems in Hilbert spaces*, submitted to SIAM J. Num. Analysis.

[20] G. Vainikko, *On the discretization and regularization of ill-posed problems with noncompact operators*, Numer. Funct. Anal. and Optimiz., 13 (1992), 381–396.

[21] W. W.-G. Yeh, *Review of parameter identification procedures in groundwater hydrology: The inverse problem*, Water Res. Rev., 22 (1986), 95–108.

Strategies for the Regularization of Nonlinear Least Squares Problems

Guy Chavent

Abstract

This paper offers a synthetic comparative presentation of various recent results on the regularization of nonlinear least squares problem : Tikhonov regularization in the general case, Tikhonov regularization of weakly nonlinear problems, state space regularization of some strongly nonlinear problems. Emphasis in the comparison is put on wether or not the regularized problems are quadratically-wellposed, *ie* have a unique stable global minimum and no parasitic local minimum.

1 Introduction

In this paper, we try to give a synthetic presentation of recent results on the regularization of nonlinear least square problems of the form

(P) $$\text{find } \hat{x} = Arg \min_{x \in C} J(x) = \frac{1}{2} \parallel z - \varphi(x) \parallel_F^2$$

where as usual

$(H_\circ - i)$ $$\begin{cases} C \subset E \text{ is a closed convex admissible set,} \\ E = \text{ Hilbert is the parameter space,} \\ F = \text{ Hilbert is the data space,} \\ z \in \varphi(C) \text{ is an error} - \text{free attainable data,} \\ \varphi \in \mathcal{C}^\circ(\mathcal{C}; \mathcal{F}) \text{ is the mapping to be inverted.} \end{cases}$$

The problem (P) may have more than one solution, so one is led to select one which is the closest to some $x_\circ$, where

$(H_\circ - ii)$ $$x_\circ \in E \text{ is a given a-priori guess.}$$

This solution, which is defined by

(MNS) $$\hat{x} = Arg\, 5 \min_{x \; sol \; of \; (P)} \frac{1}{2} \parallel x - x_\circ \parallel_E^2$$

is called the $x_\circ-$ Minimum Norm Solution ($x_\circ$ -MNS in short) of (P). Such a solution needs not to exist, nor to be unique, as the set $\varphi^{-1}(z)$ is closed but not necessarily convex because of the possible non-linearity of φ.

The classical Levenberg-Marquardt-Tikhonov regularization [9] [10] [11] for the approximation of $\hat{x}$ consists then in replacing (P) by a sequence of problems

$$(P_n) \qquad \text{find } \hat{x}_n = Arg\ \min_{x \in C} OJ_n(x) = \tfrac{1}{2} \parallel z_n - \varphi(x) \parallel_F^2 + \frac{\varepsilon_n^2}{2} \parallel x - x_\circ \parallel_E^2$$

where

$$(H_\circ - iii) \qquad \begin{cases} z_n, n \in \mathbb{N} \text{ is of sequence of noise pertubed measurements of z,} \\ \text{which is supposed to satisfy } \parallel z_n - z \parallel \leq \delta_n \rightarrow 0 \\ \varepsilon_n > 0, n \in \mathbb{N}, \text{ is the sequence of regularization parameters,} \\ \text{which is supposed to satisfy } \varepsilon_n \rightarrow 0. \end{cases}$$

We recall first in paragraph 2 the situation for the quadratic case, where φ is linear : in that case, the regularized quadratic problems (P_n) are wellposed for all n, and the sequence of $\hat{x}_n$ converges, if $\delta_n/\varepsilon_n \rightarrow 0$, to the minimum norm solution $\hat{x}$ of (P) whenever this latter exists (which is the case under hypothesis $(H_\circ - i)$ where z is supposed attainable). In the numerical application of this technique, one usually solves only one (or a few) problem(s) (P_n), and one relies on the convergence result to state that the computed $\hat{x}_n$ is an approximation to the solution $\hat{x}$ of (P). So, the quadraticity and wellposedness of (P_n) are essential in numerical applications, as they ensure that the solution $\hat{x}_n$ can actually be computed by local gradient techniques (the quadraticity of (P_n) guarantees that the optimizer will not be lured into a parasitic local minimum) in a stable way (the objective function is coercive).

The objective of this paper is to analyse under which conditions on φ and C and for what kind of regularization technique (Tikhonov or other) the regularized problems (P_n) retain, in the nonlinear case, the properties of the linear case ; the notion of well posedness carries over naturally to the nonlinear case, provided stability is required only on a neighborhood of $\varphi(C)$. Quadraticity is, of course, completely lost in the non linear case, but can be replaced by the absence of parasitic local minima, which will also ensure that local gradient methods will be effective in the search of the global minimum. These two notions are conveniently combined into the :

DEFINITION 1.1. *The least squares problem (P_n) is quadratically wellposed if there exists an (open) neighborhood ϑ of $\varphi(C)$ such that, for every $z_n \in \vartheta$,*
i) (P_n) has a unique solution $\hat{x}_n$
ii) (P_n) has no additional local minima,
iii) any minimizing sequence converges to $\hat{x}_n$,
iv) the mapping $z_n \rightarrow \hat{x}_n$ is locally lipschitz continuous from $\vartheta \subset F$ to $C \subset E$.

So we shall be looking for regularization techniques of nonlinear problems which provide a numerically effective approximation to the original problem (P), by ensuring that the regularized problems (P_n) are all (or at least for large enough n) quadratically wellposed and that $\hat{x}_n$ converges to $\hat{x}$.

We begin in paragaph 3 by recalling the situation when only regularity hypothesis are made on the nonlinear mapping φ and existence of minimizers is based on compactness [7] : as we shall see, the prospects are quite odd in that case, as the regularized problems (P_n) can have, when ε_n is small, more than one global minimum, many parasitic local minima, and convergence can be ensured only in a subsquence sense. Under this kind of hypothesis, quadratical wellposedness of (P_n) can be guaranteed only if the regularization parameter ε_n is large enough [1].

An important feature of the problem (P) to be regularized is the shape of the output set $\varphi(C)$, as solving (P) involves projecting z on $\varphi(C)$. We introduce in paragraph 4 the notion of weakly nonlinear problem ([6]) for which there exist a constant $1/R \geq 0$ such that, for all $t \in [0,1]$:

$$(1) \qquad \| A(t) \|_F \leq 1/R \| V(t) \|_F^2$$

where $V(t)$ and $A(t)$ are the velocity and accelleration along the image by φ of a segment of C. The output set $\varphi(C)$ of weakly nonlinear problems has a bounded curvature $1/R$, so that the projection on $\varphi(C)$ will behave nicely if the size of $\varphi(C)$ is not too large with respect to its curvature. This can be obtained by limiting the size of C, so that weakly nonlinear problems can be seen as a smooth perturbations of linear problems. By opposition, problems which do not satisfy equation 1 will be called strongly nonlinear, as $\varphi(C)$ may have an infinite curvature, so that the projection on $\varphi(C)$ cannot be made univocally and continuous simply by reducing the size of C.

Paragraph 5 is devoted to the Tikhonov regularization of weakly non linear problems. As it can be expected, the situation is much similar to the linear case of paragraph 2, and the problems (P_n) are all quadratically wellposed [6] , and our expectations are met.

Of course, there is no systematic way of regularizing a strongly non linear problem in such a way that the regularized problems (P_n) are all quadratically wellposed. However, this can be done under special circumstances, and we describe in paragraph 6 such a situation, which occurs when the nonlinear map φ can be seen as the composition of a nonlinear map $\tilde{\varphi} : C \to \tilde{F}$, such that the corresponding nonlinear least squares problem

$$(\tilde{P}) \qquad find\ \hat{\tilde{x}} = Arg\ \min_{x \in C} \tilde{J}(x) = \tfrac{1}{2} \| \tilde{z} - \tilde{\varphi}(x) \|_{\tilde{F}}^2$$

is quadratically well posed in $\tilde{F}$, by an injective continuous linear operator $A : \tilde{F} \to F$. Typically, in applications to identification or control of partial differential equations, $\tilde{F}$ will be the state space, equipped with a very strong norm, and A will be the observation operator, which maps the state space $\tilde{F}$ to the observation space F (usually L^2).

The state space regularization [3] [4] consists then in regularizing (P) by adding a quadratical term in the space $\tilde{F}$ (rather than in E as in the Tikhonov regularization), and by localizing the search on a subset C_n of C :

$$(P_n^*) \qquad find\ \hat{x}_n = Arg\ \min_{x \in C_n} K_n(x) = \frac{1}{2} \| z_n - \varphi(x) \|_F^2 + \frac{\varepsilon_n^2}{2} \| y_n - \tilde{\varphi}(x) \|_{\tilde{F}}^2$$

where y_n is given by the auxiliary regularization problem

$$(Q_n^*) \qquad find\ y_n = Arg\ \min_{y \in \tilde{F}}, L_n(y) = \frac{1}{2} \| z_n - Ay \|_F^2 + \frac{\varepsilon_n^2}{2} \| y - y_\circ \|_{\tilde{F}}^2,$$

where $y_\circ \in \tilde{F}$ is an a-priori guess of the state of the system. Once again, if the sequences ε_n, C_n and the a-priori guess $y_\circ$ are correctly chosen, one can proove that all the regularized problems (P_n^*) will be quadratically wellposed, so that (P_n^*) (Q_n^*) yields a constructive approximation of the strongly nonlinear problem (P).

This paper does not present original results, but organizes, along the line presented above, material which is scattered in the papers [7], [3], [4], [6], [1], [5]. No proof is given, and no attempt is made to give, for each theorem, the minimum set of hypothesis, we refer for that to the original papers.

2 Regularization of linear problems

Let hypothesis $(H_\circ)$ hold. We suppose in this paragraph that $\varphi(x)$ is linear :

$$(LIN) \qquad \varphi(x) = Ax, \qquad A \in \mathcal{L}(E;F)$$

Then $\varphi^{-1}(z)$ is closed and convex, so that the minimum norm solution $\hat{x}$ exists and is unique. Classically, convergence rates are obtained under the hypothesis that $\hat{x}$ satisfies the linear regularity condition

$$(LR) \qquad x_\circ - \hat{x} \in \text{range}A^* + T(C,\hat{x})^-$$

where $T(C,\hat{x})^-$ is the negative polar cone of the tangent cone $T(C,\hat{x})$ to C at $\hat{x}$:

$$T(C,\hat{x})^- = \{y \in E :< y,x >\leq 0 \forall x \in T(C,\hat{x})\}, \qquad T(C,\hat{x}) = \overline{\bigcup_{\lambda>@0} \frac{1}{\lambda}(C-\hat{x})}.$$

Condition (LR) is called a regularity condition because one necessarily has, for a broad class of C and A [5] :

$$x_\circ - \hat{x} \in \overline{rangeA^* + T(C,\hat{x})^-},$$

so that requiring that (LR) holds is just demanding slightly more than what is automatically granted by the definition of the $x_\circ - MNS\ \hat{x}$.

For unconstrained problems, $C = E$ and $T(C,\hat{x})^- = \{0\}$, so that (LR) reduces to the classical condition

$$x_\circ - \hat{x} \in \text{range}A^*.$$

THEOREM 2.1. *(Regularization of linear problems)*
Let hypothesis $(H_\circ)$ hold, and suppose that φ satisfies :

$$(LIN) \qquad \varphi(x) = Ax, \qquad A \in \mathcal{L}(E_{;}F).$$

Then :
i) Problem (P) has a unique $x_\circ - (MNS)\ \hat{x}$.
ii) For every $n \in \mathbb{N}$, problem (P_n) is quadratically wellposed on the neighborhood $\vartheta = F$ of $\varphi(C)$, with the stability estimate $\varepsilon_n \parallel \hat{x}_n - \hat{x}'_n \parallel_E \leq \parallel z_n - z'_n \parallel_F$ for any $z_n, z'_n, \in \vartheta$.
iii) If moreover :
. $\hat{x}$ satisfies the regularity condition

$$(LR) \qquad x_\circ - \hat{x} \in rangeA^* + T(C,\hat{x})^-,$$

. ε_n *is chosen such that* $\varepsilon_n^2 = 0(\delta_n)$,
then :

$$(2) \qquad \| \varphi(\hat{x}_n) - z \|_F = 0(\varepsilon_n^2), \qquad \| \hat{x}_n - \hat{x} \|_E = 0(\varepsilon_n).$$

We refer for the proof of these classical results, for example to [8] in the unconstrained case, and to [12] or [13] or [5] for the constrained case.

3 Regularization of non-linear problems : the general situation

We examine here the situation in the nonlinear case when one makes hypothesis only on the regularity of φ, but not on the shape of $\varphi(C)$. In the remaining of this paper, we shall suppose that φ is at least uniformly $W^{2,\infty}$ on all segments of C, *ie.*

$$(H_1) \begin{cases} \text{there exists } \alpha_M > 0 \text{ and } \beta > 0 \text{ such that, for any } x_\circ, x_1 \in C \text{ and } t \in [0,1], \\ \text{the mapping (path) } P : t \to \varphi((1-t)x_\circ + tx_1) \text{ is in } W^{2,\infty}[0,1]; F) \text{ and} \\ \| V(t) \|_F \leq \alpha_M \| x_1 - x_\circ \|_E, \| A(t) \|_F \leq \beta \| x_1 - x_\circ \|_E^2 \\ \text{where } V(t) = P'(t) \text{ and } A(t) = P''4(t). \end{cases}$$

When φ is twice differentiable, then α_M is an uniform bound on the norm of $\varphi'(x)$ and β a uniform bound on the norm of $\varphi''(x)$.

Hypothesis $(H_\circ)(H_1)$ however are not enough to ensure even existence of solutions to (P) and/or (P_n).A first way to obtain existence is to use weak compactness arguments, which leads to ([7] [5]) :

THEOREM 3.1. *(Regularization of nonlinear problems : compactness)*
Let hypothesis $(H_\circ)(H_1)$ hold, and suppose that

$$(WSC) \qquad \varphi \text{ is weakly sequentially closed.}$$

Then :
i) Problem (P) has at least one solution, but an $x_\circ-$ MNS does not necessarily exist.
ii) For every $n \in \mathbb{N}$, problem (P_n) has at least one solution, and the solutions of (P_n) depend continuously on z_n in the following weak sense : if $z^k \to z_n$ and x^k is a solution of (P_n) with z_n replaced by z^k, then $\{x^k\}$ contains a convergent subsequence, and the limit of every convergent subsequence of $\{x^k\}$ is a minimizer of (P_n).
iii) If more over :
. *an $x_\circ-$MNS $\hat{x}$ to (P) exists,*
. *φ is Gateaux-derivable at $\hat{x}$,*
. *$\hat{x}$ satisfies the regularity condition*

$$(NLR) \qquad x_\circ - \hat{x} \in \textit{Range}\, \varphi'(\hat{x})^* + T(C,\hat{x})^-,$$

. *the a-priori guess $x_\circ$ is close enough to $\hat{x}$ so that*

$$(3) \qquad \begin{cases} \beta \| w \|_F < 1 \textit{ where } w \in F \textit{ and } \mu \in T(C,\hat{x})^- \\ \textit{satisfy } x_\circ - \hat{x} = \varphi'(\hat{x})^* w + \mu \end{cases}$$

. ε_n *is chosen such that* $\varepsilon_n^2 = 0(\delta_n)$,
then any global minimizer $\hat{x}_n$ *of* (P_n) *satisfies :*

$$(4) \qquad \| \varphi(\hat{x}_n) - z \|_F = 0(\varepsilon_n^2), \qquad \| \hat{x}_n - \hat{x} \|_E = 0(\varepsilon_n)$$

The situation described in the above theorem is quite far from the idyllic picture of the linear case : in particular, for a given n, the regularized problem can have parasitic local minima and many global minima, and hence are not quadratically wellposed.

One alternative way to ensure existence of solutions at least for some of the regularized problems (P_n) is to use a large enough regularization term, which will guarantee that the output set associated to the regularized problem (P_n) has a finite and small enough curvature, thus allowing to use the geometric theory of quasiconvex sets ([2]) to handle the projection onto it. This leads to ([1]) :

THEOREM 3.2. *(Regularization of nonlinear problems : geometric theory)*
Let hypothesis $(H_\circ - iii), (H_1)$ *hold, and suppose that :*

$$(CBD) \qquad diamC < +\infty.$$

Then :
i) Problem (P) *may have no solution.*
ii) There exists $\varepsilon_{min} > 0$ *such that* (P_n) *is quadratically wellposed as soon as* $\varepsilon_n > \varepsilon_{min}$*. When* $\varepsilon_{min} \geq \varepsilon_n > 0$, (P_n) *may have no solution.*
iii) Void !

The coefficient ε_{min} can be evaluated by simple formula from $\alpha_M, \beta, diamC$ and the position of the a-priori guess $x_\circ$ in the admissible set C.

We see from theorems 3.1 and 3.2 that the classical Tikhonov regularization will produce, when applied to quite general nonlinear least squares problems, quadratically well posed problems when the regularization coefficient ε_n is large enough.But in general, this property will not subsist when ε_n goes to zero. If we want to maintain it for all ε's, we have to be more specific and consider only regularized problems where the output set has a finite curvature (in order to mimick the linear case where the output set is a linear subspace) and to localize the problem in order to prevent the output set to fold over onto itself. This is investigated in the next three paragraphs.

4 Weakly and strongly nonlinear problems

We are interested in this paragraph in a special class of nonlinear least squares problems (P) which we define now [6].

DEFINITION 4.1. *(Weakly nonlinear problems)*
The problem (P) *is called weakly nonlinear if it satisfies* $(H_\circ - i), (H_1)$ *and*

$$(WNL) \qquad \begin{cases} \exists\, 1/R \geq 0 \text{ such that, } \forall x_\circ, x_1 \in C, \forall t \in [0,1] \\ \| A(t) \|_F \leq 1/R \| V(t) \|_F^2 \end{cases}$$

Naturally, when (WNL) holds, we will suppose without loss of generality, that β in (H_1) is given by

$$\beta = \alpha_M^2 / R. \tag{5}$$

The geometrical interpretation of weakly nonlinear problems is a follow : condition (WNL) ensures that the paths P of $\varphi(C)$ defined in (H_1) can be reparametrized as $W^{2,\infty}(O, L)$ functions of the arc length ν along the path P (where L is the length of the path) with

$$\| P''(\nu) \|_F \leq 1/R \tag{6}$$

But $\| P''(\nu) \|_F$ is nothing but the curvature of the path at abscissa ν, so that (WNL) means that all paths P of $\varphi(C)$ defined in (H_1) have a curvature smaller than $1/R$. This is the nonlinear counterpart of the fact that, in the linear case, all paths P are segments of $\varphi(C)$ and hence have a zero curvature !

However there is one property of segments that finite curvature paths fail to mimick : segments never pass twice through the same point, but finite curvature path can do that if they are long enough and hence can turn enough ! In order to quantify this we introduce, using (H_1) :

$$\triangle = \alpha_M \, diamC = \text{ upper bound to the length of paths } P \text{ of } \varphi(C) \tag{7}$$

and

$$\Theta = \frac{\triangle}{R} = \frac{\alpha_M \, diamC}{R} = \text{ upper bound to the deflection of paths } P \text{ of } \varphi(C) \tag{8}$$

where the deflection of a path is the maximum angle between two tangents to the path. Requiring $\Theta < 2\pi$ would prohibit paths of $\varphi(C)$ to pass twice through the same point, and hence $\varphi(C)$ to cross over itself.

But this is still not enough, as there is one other property of segments that the finite curvature paths P of $\varphi(C)$ still fail to reproduce : from any given point z of the space F, the "distance to z"; function is convex along any segment, and hence has an unique global minimum and no parasitic local minimum. In order to reproduce this property for each path P of $\varphi(C)$ for all z in a neighborhood of the path, one is led to require that

$$\Theta < \pi \tag{9}$$

(it is easy to convince oneself by a small drawing that, if P is an arc of circle with a deflection $\Theta \in [\pi, 2\pi[$, one can find points z as close as desired to P and from which the "distance to z" function has one parasitic local minimum).

DEFINITION 4.2. *(deflection size $\times$ curvature condition)*
The weakly nonlinear problem (P) is said to satisfy a deflection size $\times$ curvature condition if

$$\Theta = \triangle / R < \pi \iff \alpha_M diamC < \pi R. \tag{SCC}$$

When this size $\times$ curvature condition holds, the size of a neighborhood of each path P of $\varphi(C)$ on which the "distance to z" has a unique global minimum and no parasitic local minimum is given by

$$(10) \qquad 0 < R_G = \begin{cases} R & if\ 0 \leq \Theta \leq \pi/2, \\ R\sin\Theta & if\ \pi/2 \leq \Theta < \pi. \end{cases}$$

With these definitions and notations, we can now state the

THEOREM 4.1. *(properties of weakly nonlinear problems)*
Let (P) be a weakly nonlinear problem satisfying a deflection size $\times$ curvature condition, ie such that $(H_\circ - i), (H_1)$, (WNL), (SCC) hold.
Then :
i) The preimages of φ are (closed) convex :

$$(11) \qquad \forall y \in \varphi(C), \qquad \varphi^{-1}(y) = \{x \in C : \varphi(x) = y\} \ \ is\ convex,$$

which implies that :

$$(P) \qquad admits\ a\ unique\ x_\circ - MNS\ \hat{x}.$$

ii) If $y \in F$ is in $B_\circ(\varphi(C), R_G) = \{y \in F,\ such\ that\ dist(y, \varphi(C)) < R_G\}$ and admits a projection Y on $\varphi(C)$, this projection is unique, and the "distance to y"function admits no parasitic local minima over C.
iii) The "projection on $\varphi(C)$"is, whenever it exists, Lipschitz continuous on $B_\circ(\varphi(C), R_G)$: if $y_\circ, y_1 \in B_\circ(\varphi(C), R_G)$ admit projections $Y_\circ, Y_1$ on $\varphi(C)$ and are sufficiently close so that there exists $d \geq 0$ such that

$$(12) \qquad \| y_\circ - y_1 \|_F + Max_{j=0,1}\ dis(y_j, \varphi(C)) \leq d < R_G$$

then, for any path P going from $y_\circ$ to y_1 one has

$$(13) \qquad \| Y_\circ - Y_1 \|_F \leq L \leq (1 - \frac{d}{R})^{-1} \| y_\circ - y_1 \|_F$$

where L is the length of the path P.
iv) If moreover :

the Gateaux derivative $\varphi'(x)$ exists for every $x \in C$ and is injective,

then :

φ is injective over C.

Notice that, when specialized to the linear case where $\varphi(x) = Ax$, theorem 4.1 reproduces exactly the linear results. Hence weakly nonlinear problems satisfying a deflection size $\times$ curvature condition can really be seen as perturbations of linear problems. This will be confirmed in the next paragraph where we shall see that the results of theorem 2.1 on the regularization of linear problems carry over nicely.

We conclude this paragraph by giving a simple example of a weakly nonlinear problem. Let y be the solution of the elliptic equation

$$(14) \qquad \begin{cases} -\Delta y + K(y) = f & \text{in } \Omega \\ y = 0 & \text{on } \partial\Omega_D \\ \dfrac{\partial y}{\partial \nu} = g & \text{on } \partial\Omega_N \end{cases}$$

where Ω is a bounded domain of $\mathbb{R}^n, n \leq 3$, with $\mathcal{C}^\infty$ boundary $\partial\Omega = \partial\Omega_D \cup \partial\Omega_N$ with $\partial\Omega_D$ and $\partial\Omega_N$ both open and closed relatively to $\partial\Omega, \nu$ denotes the outward normal to $\partial\Omega$, and where $K \in W^{2,\infty}(\mathbb{R}), K(0) = 0, K'(\zeta) \geq 0 \; \forall \zeta \in \mathbb{R}$.
The nonlinear mapping to be inverted is there

$$(15) \qquad \begin{cases} \varphi : (f,g) \to y \text{ where} \\ (f,g) \in L^2(\Omega) \times L^2(\partial\Omega_N) = E \\ y \in L^2(\Omega) = F \end{cases}$$

One can then show, by writing the equations satisfied by V and A, that the nonlinear least squares problem (P) associated to the inversion of φ on any closed convex subset C of E is weakly nonlinear. Sufficient reduction of the size of C will then ensure that the size $\times$ curvature condition is satisfied.

5 Regularization of weakly nonlinear problems

We consider here [6] the collection of regularized problems (P_n) constructed on a weakly nonlinear problem (P) satisfying the size $\times$ curvature condition of definition 4.2 :

$$(SCC) \qquad \Theta = \Delta/R < \pi \Longleftrightarrow \alpha_M diamC < \pi R.$$

But (P_n) can be seen as a nonlinear least square problem associated to the inversion of $\varphi_n(x) = (\varphi(x), x)$ on C given a data $(z_n, x_\circ) \in F_n = F \times E$ equipped with the adequate norm. The geometrical attributes of $\varphi_n(C)$ are then :

$$(16) \qquad \begin{cases} \Delta_n = \Delta(1 + \dfrac{\varepsilon_n^2}{\alpha_M^2})^{\frac{1}{2}} & > \Delta \\ \Theta_n = \Theta(1 + \dfrac{\varepsilon_n^2}{\alpha_M^2})^{-\frac{1}{2}} & < \Theta \quad < \pi \\ R_n = R(1 + \dfrac{\varepsilon_n^2}{\alpha_M^2}) & > R \quad > 0 \end{cases}$$

so that (P_n) is clearly quadratically well posed in F_n for all n, on an open neighborhood of $\varphi_n(C)$ of size

$$(17) \qquad R_{G,n} = \begin{cases} R_n & \text{if } \; 0 \leq \Theta_n \leq \pi \\ R_n \sin\Theta_n & \text{if } \; \pi/2 \leq \Theta_n < \pi. \end{cases}$$

In order to deduce from this the quadratical wellposedness of (P_n) in F, one needs to introduce the radius of C seen from the a-priori guess $x_\circ$:

$$(18) \qquad Rad = sup_{x \in C} \parallel x - x_\circ \parallel_E,$$

satisfies

$$(19) \qquad Rad \geq \frac{1}{2} diamC$$

and

$$(20) \qquad Rad \leq diamC \text{ as soon as } x_\circ \in C.$$

With this notation, the size d_n of an open neighborhood of $\varphi(C)$ on which (P_n) will be quadratically wellposed is given by

$$d_n^2 = R_{G,n}^2 - \varepsilon_n^2 Rad^2. \tag{21}$$

LEMMA 5.1. *Equations (21) (18) (17) (16) define uniquely* $d_n > 0$ *for all* n *such that* $\varepsilon_n \in [0, \bar{\varepsilon}[$, *where* $\bar{\varepsilon}$ *is given by :*

$$\begin{array}{lll} \bar{\varepsilon} = +\infty & & \text{if } \alpha_M Rad < 2R_G \\ 1 + \dfrac{\bar{\varepsilon}^2}{\alpha_M^2} = & \dfrac{2}{1 + \sqrt{1 - (\frac{2R}{\alpha_M Rad})^2}} & \text{if } \alpha_M Rad \geq 2R_G \end{array}$$

When the a-priori guess $x_\circ$ is chosen in C (which seems reasonnable !), one checks easily using (19) (20) that $\bar{\varepsilon} = +\infty$ as soon as the deflection Θ of (P) is smaller than $\Theta_\circ \in]\pi/2, \pi[$ solution of $2\sin\theta = \theta$. When Θ is between $\Theta_\circ$ and Θ_1 solution of $4\sin\theta = \theta$, the situation will depend on the location of $x_\circ$ in C : the closest $x_\circ$ is to the "center"of C, the strongest the chances are that $\bar{\varepsilon} = +\infty$. Finally, when Θ is between Θ_1 and π, one always has $\bar{\varepsilon}$ finite independantly of the position of $x_\circ$ - but small Rad will give larger $\bar{\varepsilon}$! So we see that,in all cases, one always has advantage to choose the a-priori guess $x_\circ$ close to the center of C.

THEOREM 5.1. *(Regularization of weakly nonlinear problems)*
Let hypothesis $(H_\circ)(H_1)$ *hold, and suppose that* (P) *satisfies*

(WNL) *weak nonlinearity*

(SCC) *deflection size* $\times$ *curvature condition*

Then :
i) Problem (P) *has a unique* $x_\circ -$*MNS* $\hat{x}$.
ii) For every $n \in \mathbb{N}$ *such that* $\varepsilon_n \in [0, \bar{\varepsilon}[$ *where* $\bar{\varepsilon}$ *is defined in lemma 5.1, problem* (P_n) *is quadratically wellposed on an open neighborhood* ϑ_n *of* $\varphi(C)$ *of size* d_n*, with the stability estimate :*

$$\varepsilon_n \parallel \hat{x}_n - \hat{x}_n' \parallel_E \leq (1 - \frac{1}{R_n}(d^2 + \varepsilon_n^2 Rad^2)^{\frac{1}{2}})^{-1} \parallel z_n - z_n' \parallel_F$$

provided that

$$\parallel z_n - z_n' \parallel_F + max\{dist(z_n, \varphi(C)), dist(z_n', \varphi(C))\} \leq d < d_n.$$

iii) If moreover :
. φ *is Gateaux-derivable at* $\hat{x}$,
. $\hat{x}$ *satisfies the regularity condition*

(NLR) $$x_\circ - \hat{x} \in Range\varphi'(\hat{x})^* + T(C, \hat{x})^-,$$

. *the a-priori guess* $x_\circ$ *is close enough to* $\hat{x}$ *so that* $\beta \parallel w \parallel_F = \dfrac{\alpha_M^2}{R} \parallel w \parallel_F < 1$, *where* $w \in F$ *and* $\mu \in T(C, \hat{x})^-$ *satisfy* $x_\circ - \hat{x} = \varphi'(\hat{x})^* w + \mu$,

. *ε_n is chosen such that $\varepsilon_n^2 = 0(\delta_n)$*
then the unique solution $\hat{x}_n$ of (P_n) satisfies :

$$\| \varphi(\hat{x}_n) - z \|_F = 0(\varepsilon_n^2), \quad \| \hat{x}_n - \hat{x} \|_E = 0(\varepsilon_n)$$

When φ is linear and satisfies $(H_\circ)$, then hypothesis (H_1), (WNL)(SCC) are automatically satisfied with $\beta = 0, R = \infty$ and $\Theta = 0$, so that $\bar{\varepsilon} = +\infty$, and theorem 5.1 reduces exactly to theorem 2.1 ! So, weakly nonlinear problems which satisfy a size $\times$ curvature condition are perturbations of linear problems, which retain their main properties.

Comparing now theorem 5.1 to theorem 3.1 for the general non linear case, we see that the additional properties generated by the weak nonlinearity and size $\times$ curvature hypotheses concern the points i) (existence of a minimum norm solution) and ii) (quadratical wellposedness of the regularized problems), but not the point iii) as the convergence results are the same in the two theorems.

6 State space regularization

We suppose here that (P) satisfies the minimum set of hypothesis $(H_\circ - i)$, but that the mapping φ to be inverted satisfies

$$\varphi = A \circ \tilde{\varphi}$$

where A is an injective continuous linear operator, and where $\tilde{\varphi}$ is a nonlinear mapping which is continuously invertible on a neighborhood of its range $\tilde{\varphi}(C)$ [3] [4].

This covers situations where $\varphi(C)$ may have infinite curvature, so that problem (P) is strongly nonlinear, but where this sad situation results only from the loss of information incurred by the linear operator A, which may have no continuous inverse : if the observation were made in the intermediate Hilbert space $\tilde{F}$ such that $\tilde{\varphi} : C \to \tilde{F}$ and $A \in \mathcal{L}(\tilde{F}; F)$, the problem would be stable by construction. The space $\tilde{F}$ is called the state space, by opposition to the observation space F, and A represents an observation operator.

In order to ensure the continuous invertibility of $\tilde{\varphi}$, one uses the geometric techniques of paragraphs 4 and 5, and suppose that $\tilde{\varphi}$ satisfies $(H_\circ - i)$ (H_1) (WNL) and (SCC). Then theorem 4.1 together with the additional assumption

$$(H_2) \qquad \begin{cases} \exists \alpha_m > 0 \; st, \forall x_\circ, x_1 \in C, \forall t \in [0,1] \\ \alpha_m \| x_1 - x_\circ \|_E \leq \| V(t) \|_F \end{cases}$$

provides the desired stability result :

LEMMA 6.1.
Let $\tilde{\varphi}$ satisfy $(H_\circ - i)(H_1)(H_2)$ (WNL) and (SCC). Then the problem

$$(\tilde{P}) \qquad \textit{find } \hat{x} = \textit{Arg} \min_{x \in C} \tilde{J}(x) = \tfrac{1}{2} \| y - \tilde{\varphi}(x) \|_{\tilde{F}}^2$$

is quadratically wellposed on the open neighborhood of $\varphi(C)$ *of size* R_G*, with the stability property*

$$(22)\qquad \begin{cases} \alpha_m \parallel \hat{x}_1 - \hat{x}_\circ \parallel_E \leq (1 - \dfrac{d}{R})^{-1} \parallel y_\circ - y_1 \parallel_{\tilde{F}} \\ \text{as soon as} \\ \parallel y_\circ - y_1 \parallel_{\tilde{F}} + Max_{j=0,1} d_{\tilde{F}}(y_j, \tilde{\varphi}(C)) \leq d < R_G \end{cases}$$

where R, Θ, Δ, R_G *are the geometrical attributes of* $\tilde{\varphi}(C)$*, given by*

$$(23)\qquad \begin{cases} R \text{ defined in (WNL)} \\ \Delta = \alpha_M \, diam \, C, \alpha_M \text{ defined in } (H_1) \\ \Theta = \Delta / R < \pi \text{ because of (SCC)} \\ R_G = R \, r_G(\Theta) \end{cases}$$

where

$$(24)\qquad r_G(\theta) = \begin{cases} 1 & if \, 0 \leq \theta \leq \pi/2, \\ \sin\theta & if \, \pi/2 \leq \theta \leq \pi. \end{cases}$$

From Lemma 6.1, we see that solving the original problem (P) is equivalent to solve

$$(Q)\qquad find \; \hat{y} = Arg \; \min_{y \in D} \; K(y) = \tfrac{1}{2} \parallel z - Ay \parallel_F^2,$$

where

$$(25)\qquad D = \tilde{\varphi}(C) \subset \tilde{F}.$$

Problem (Q) has a unique solution, as we know from hypothesis $(H_\circ - i)$ that $z \in A\tilde{\varphi}(C) = A(D)$, and as A is supposed injective. It consists in the inversion of a linear operator A on a non convex set D, and is illposed as A is not supposed to have a continuous inverse.

By analogy with the linear case of paragraph 2, one may first try to approximate (Q) by the following sequence of regularized problems :

$$(\tilde{Q}_n)\qquad \text{find } \tilde{y}_n = Arg \; \min_{y \in D} \; \tilde{K}_n(y) = \tfrac{1}{2} \parallel z_n - Ay \parallel_F^2 + \frac{\varepsilon_n^2}{2} \parallel y - y_\circ \parallel_{\tilde{F}}^2$$

where

$$(H_\circ - ii \; bis)\qquad y_\circ \in \tilde{F}$$

is a given a-priori guess of the optimal state

$$(26)\qquad \hat{y} = A^{-1} z$$

and where $z_n, \varepsilon_n > 0$ are chosen according to $(H_\circ - iii)$.

Problem $(\tilde{Q}_n)$ is nothing but searching for the projection, in $F_n = F \times \tilde{F}$ of $(z_n, y_\circ)$ on $A_n(D)$, where $A_n(y) = (Ay, y)$ and where F_n is equipped with the norm $\| (z,y) \|^2 = \| z \|_F^2 + \varepsilon_n^2 \| y \|_{\tilde{F}}^2$.

The geometrical attributes of $A_n(D)$ are easily found to be

$$(27) \qquad \begin{cases} R_n = \dfrac{\varepsilon_n^2}{(\varepsilon_n^2 + \| A \|^2)^{\frac{1}{2}}} \\ \Theta_n = (1 + \dfrac{\| A \|^2}{\varepsilon_n^2})\Theta \\ \Delta_n = (\varepsilon_n^2 + \| A \|^2)^{\frac{1}{2}} \Delta \end{cases}$$

which shows that when $n \to \infty$, the curvature of $A_n(D)$ tends to infinity, whereas its size remains always larger than that of $A(D)$. Hence $(\tilde{Q}_n)$ has no chance of being quadratically wellposed for all $n \in N$.

So one has first to localize $(\tilde{Q}_n)$ by replacing D by a smaller subset D_n such that the size of $A_n(D_n)$ is small enough to ensure a deflection smaller than π, and hence quadratical wellposedness of the projection on $A_n(D_n)$. So one defines

$$(28) \qquad D_n = \begin{cases} D = \varphi(C) & \text{for } n = 0 \\ \{y \in D : (\| Ay - z \|_F^2 + \varepsilon_n^2 \| y - \hat{y} \|_{\tilde{F}}^2)^{\frac{1}{2}} \leq \chi_D R_n\} & \text{for } n = 1, 2 \end{cases}$$

where $\chi_D \in]0, 1[$ will be defined later.

But from (27) we see that the size $R_{G,n}$ of the neighborhood of D_n on which the projection is wellposed will be necessarily smaller than R_n, which tends to zero as ε_n^2. So if the regularized problems are to have a unique stable solution, necessarily the distance d_n, in F_n, of $(z_n, y_\circ)$ to $A_n(D)$ has to be smaller than $R_{G,n}$, and hence than R_n, which implies that necessarily

$$(29) \qquad d_n = 0(\varepsilon_n^2).$$

But one checks easy that

$$(30) \qquad d_n \leq (\delta_n^2 + \varepsilon_n^2 \| y_\circ - \hat{y} \|_{\tilde{F}}^2)^{\frac{1}{2}}.$$

The first term can easily be made $0(\varepsilon^2)$ by requiting that $\delta_n = 0(\varepsilon_n^2)$. But the second term is only of order $0(\varepsilon_n)$, unless one enhances the a-priori guess $y_\circ$ by replacing it by y_n obtained by solving the regularized problem associated to the inversion of A on the whole space $\tilde{F}$ with the a-priori guess $y_\circ$:

$$(Q_n^*) \qquad \text{find } y_n = Arg \ \min_{y \in \tilde{F}} \tfrac{1}{2} \| z_n - Ay \|_F^2 + \frac{\varepsilon_n^2}{2} \| y - y_\circ \|_{\tilde{F}}^2 .$$

Then theorem 2.1 implies that $\| y_n - \hat{y} \|_{\tilde{F}} = 0(\varepsilon_n)$ provided that $y_\circ - \hat{y} \in$.*Range* A^* and $\delta_n = 0(\varepsilon_n^2)$, which will ensure that $d_n = 0(\varepsilon_n^2)$, and hence let open the possibility that the regularized problems have a unique stable solution.

So we are led to replace $(\tilde{Q}_n)$ by

$$(Q_n) \qquad find\ \hat{y}_n = Arg \ \min_{y \in D_n} K_n(y) = \tfrac{1}{2} \| z_n - Ay \|_F^2 + \frac{\varepsilon_n^2}{2} \| y_n - y \|_{\tilde{F}}^2$$

The state-space regularization of problem (P) consists then in the resolution, for each $n \in \mathbb{N}$,of the auxiliary problem (Q_n^*) followed by that of (Q_n).

In practice, the resolution of (Q_n) is made by coming back to the x domain, and is equivalent to the solution of

$$(P_n^*) \qquad \textit{find } \hat{x}_n = Arg\ \min_{x \in C_n} J_n^*(x) = \frac{1}{2} \parallel z_n - A\tilde{\varphi}(x) \parallel_F^2 + \frac{\varepsilon_n^2}{2} \parallel y_n - \tilde{\varphi}(x) \parallel_{\tilde{F}}^2$$

where :

$$(31) \qquad C_n = \tilde{\varphi}^{-1}(D_n) = \{x \in C : \parallel A\tilde{\varphi}(x) - z \parallel_F^2 + \varepsilon_n^2 \parallel \tilde{\varphi}(x) - \hat{y} \parallel_{\tilde{F}}^2 \leq \chi_D R_n\}.$$

Of course, practical implementation of the constraint $x \in C_n$ in our optimization algorithm is pratically unfeasible, as C_n is defined by reference to $\hat{y}$, whose knowledge is equivalent, by Lemma 6.1, to that of the true solution $\hat{x}$ of (P) ! But we shall see in theorem 6.1 that the constraint $x \in C_n$ will not be active during the resolution of (P_n^*), so that one can ignore it in the numerical resolution.

THEOREM 6.1. *(state space regularization)*
Let $(H_\circ - i)$ $(H_\circ - iibis)$ $(H_\circ - iii)$ hold, and suppose that

$$(SS) \qquad \begin{cases} \varphi = A \circ \tilde{\varphi} \\ \tilde{F} = \textit{Hilbert space (state space)} \\ A \in \mathcal{L}(\tilde{F}; F), A \textit{ injective} \\ \tilde{\varphi} \textit{ satisfies } (H_1)(H_2)(WNL)(SCC) \textit{(so that lemma 6.1 holds)}. \end{cases}$$

Then :
i) (P) has a unique solution $\hat{x}$.
ii) If moreover :
.$\varepsilon_\circ$ is large enough so that

$$\Theta_\circ \triangleq (1 + \frac{\parallel A \parallel^2}{\varepsilon_\circ^2})\Theta < \pi,$$

. the ε_n are strictly decreasing, but not too quickly, so that

$$\exists \overline{\chi} \in]0,1[, \textit{ such that } \varepsilon_{n-1}^2 > \varepsilon_n^2 \geq \overline{\chi}\varepsilon_{n-1}^2,$$

. the localization constant χ_D in the definition of C_n is chosen such that

$$(32) \qquad \begin{cases} \chi_D = min\{\chi_D^*, r_G(\Theta_\circ)\} \\ \textit{where} \chi_D^* \textit{ is defined by the equation} \\ \chi_D^* = r_G(\frac{1}{\overline{\chi}} \frac{\chi_D^*}{(1 - \chi_D^*)^{\frac{1}{2}}}), \end{cases}$$

then, for all $n \in \mathbb{N}, C_n$ is closed and convex, and (P_n^*) is quadratically wellposed in $F_n = F \times \tilde{F}$ on a neighborhood of $A_n(\varphi(C))$ of size $\chi_D^* R_n$, where R_n is given by (27).
iii) If moreover
. $\hat{y}$ satisfies the regularity condition

$$y_\circ - \hat{y} \in \textit{ Range } A^*,$$

. *the a-priori guess* $y_\circ$ *is close enough to* $\hat{y}$ *so that*

$$(33)\qquad \left\{\begin{array}{l} \exists\eta \in]0,\frac{1}{2}[\textit{ such that} \\ \| w \|_F \leq (1-2\eta)\chi_z \dfrac{R}{(\varepsilon_\circ^2 + \| A \|^2)^{\frac{1}{2}}} \textit{ where} \\ \chi_z = \min(\chi_D, 1-\chi_D), \end{array}\right.$$

. *the error levels* δ_n *on* z_n *decrease quickly enough so that*

$$0 \leq \delta_n \leq \eta\chi_z \frac{\varepsilon_n^2}{(\varepsilon_n^2 + \| A \|^2)^{\frac{1}{2}}}$$

then

. $\forall_n \in \mathbb{N}$: (z_n, y_n) *is at a distance, in* $F_n = F \times \tilde{F}$, *of* $A_n(\varphi(C))$ *strictly smaller than* $\chi_D^* R_n$, *so that* (P_n^*) *has a unique solution* $\hat{x}_n$ *(stable with respect to* z_n, *with no local minima as shown in ii)), which satisfies :*

$$\begin{array}{l} \| \varphi(\hat{x}_n) - z \|_F = O(\varepsilon_n^2) \\ \| \tilde{\varphi}(\hat{x}_n) - \hat{y} \|_{\tilde{F}} = O(\varepsilon_n) \\ \| \hat{x}_n - \hat{x} \|_E = O(\varepsilon_n), \end{array}$$

. $\forall n = 1, 2 \ldots$ *one has*

$$J_n^* \textit{ is strictly convex over } C_n,$$

$$\begin{array}{l} \hat{x}_{n-1} \in C_n \\ \hat{x}_n \notin \partial C_n \end{array}$$

where ∂C_n *is the set of points where the constraint is saturated :*

$$\partial C_n = \{x \in C : \{\| \varphi(x) - z \|_F^2 + \varepsilon_n^2 \| \tilde{\varphi}(x) - \hat{y} \|_{\tilde{F}}^2\}^{\frac{1}{2}} = \chi_D R_n\}.$$

References

[1] G. Chavent, *A new sufficient condition for the wellposedness of non-linear least-square problems arising in identification and control,* in Analysis and Optimization of Systems, A. Bensoussan and J. Lions, eds., Springer, 1990, pp. 452–463. Lecture Notes in Control and Information Sciences 144 (Preprint : report INRIA n° 1121, nov. 89).

[2] ——, *New size × curvature conditions for strict quasiconvexity of sets,* SIAM Journal on Control and Optimization, 29 (1991), pp. 1348–1372. (Preprint rapport INRIA n°1120, nov. 89).

[3] G. Chavent and K. Kunisch, *Regularization in state space,* M2AN, 27 (1993), pp. 535–564. (Preprint : report INRIA *n*° 1730, jun. 92).

[4] ——, *State space regularization : geometric theory,* submitted to Journal of Applied Math. and Opt., (1993).

[5] ——, *Convergence of Tikhonov regularization for constrained ill-posed inverse problems,* Inverse Problems, 10 (1994), pp. 63–76.

[6] ——, *On weakly nonlinear inverse problems,* Submitted to Siam JL on Applied Mathematics, (1994). (Preprint : report Ceremade 9422, apr. 94).

[7] H. W. Engl, K. Kunisch, and A. Neubauer, *Convergence rates for Tikhonov regularization of non-linear ill-posed problems,* Inverse Problems, 5 (1989), pp. 523–540.

[8] C. Groetch, *The Theory of Tikhonov Regularization for Fredholm Equations of the First Kind,* Pitman Boston, (1984).

[9] K. Levenberg, *A method for the solution of certain nonlinear problems in least squares.*, Applied Mathematics, 11 (1944), pp. 164–168.

[10] D. W. Marquardt, *An algoritm for least squares estimation of nonlinear parameters.*, Journal of the Society of Industrial and Applied Mathematics, 11 (1963), pp. 431–441.

[11] V. A. Morozov, *Methods for Solving Incorrectly Posed Problems*, Springer New York, (1984).

[12] A. Neubauer, *Finite dimensional approximation of constrained Tikhonov-regularized solutions of illposed linear operator equations*, Math. Computing, 48 (1987), pp. 565–583.

[13] ——, *Tikhonov-regularization of illposed linear operator equations on closed convex sets*, J. Approximation Theory, 53 (1988), pp. 304–320.